RECENT ADVANCES IN
CHIRAL SEPARATIONS

CHROMATOGRAPHIC SOCIETY SYMPOSIUM SERIES

RECENT ADVANCES IN CHIRAL SEPARATIONS

Edited by

D. Stevenson

Robens Institute of Health and Safety
University of Surrey
Guildford, Surrey
United Kingdom

and

I. D. Wilson

ICI Pharmaceuticals
Macclesfield, Cheshire
United Kingdom

PLENUM PRESS • NEW YORK AND LONDON

Proceedings of a Chromatographic Society International Symposium on
Chiral Separations, held September 12–15, 1989, at the University of
Surrey, Guildford, Surrey, United Kingdom

ISBN 978-1-4684-8284-3   ISBN 978-1-4684-8282-9 (eBook)
DOI 10.1007/978-1-4684-8282-9

PREFACE

This volume represents the proceedings of the second international meeting on chiral separations held at the University of Surrey between the 12th and 15th of September 1989. Like the preceding meeting, it was jointly organised by the Chromatographic Society and the Robens Institute of the University of Surrey in response to the continued interest in this area of separation science.

Of particular interest to the organisers was the very clear change in the nature of the delegates attending this second symposium as compared with the first. At the previous meeting the majority of the delegates were composed of chromatographers with problems in the area of chiral separations who were keen to learn as much as possible about these techniques from the handful of recognised experts in this area. In this second symposium the divide between expert and novice was much less apparent, with the latter providing many interesting and useful contributions to the scientific programme in terms of both oral and poster presentations.

It was also readily apparent that, in the intervening period between the two symposia, enormous progress has been made in the separation of enantiomers. This has been the case for all areas, from the development of new chromatographic stationary phases to the understanding of the mechanisms underlying the separations and their application to "real" as well as to "academic" problems. It was also abundantly clear from this symposium that much remains to be done to improve the quality and reliability of methods for chiral separations, which still remains one of the most demanding areas of modern chromatography.

I. D. Wilson and D. Stevenson

CONTENTS

THE CHROMATOGRAPHIC SOCIETY

The Chromatographic Society is the only international organisation devoted to the promotion of, and the exchange of information on, all aspects of chromatography and related techniques.

With the introduction of gas chromatography in 1952, the Hydrocarbon Chemistry Panel of the Hydrocarbon Research Group of the Institute of Petroleum, recognising the potential of this new technique, set up a Committee under Dr. S. F. Birch to organise a symposium on "Vapor Phase Chromatography" which was held in London in June 1956. Almost 400 delegates attended this meeting and success exceeded all expectation. It was immediately apparent that there was a need for an organised forum to afford discussion of development and application of the method and, by the end of the year, the Gas Chromatography Discussion Group had been formed under the Chairmanship of Dr. A. T. James with D. H. Desty as Secretary. Membership of this Group was originally by invitation only but, in deference to popular demand, the Group was opened to all willing to pay the modest subscription of one guinea and in 1957 A. J. P. Martin, Nobel Laureate, was elected inaugural Chairman of the newly-expanded Discussion Group.

In 1958 a second Symposium was organised, this time in conjunction with the Dutch Chemical Society, and since that memorable meeting in Amsterdam the Group, now Society, has maintained close contact with kindred bodies in other countries, particularly France (Groupement pour l'Avancement des Methodes Spectroscopiques et Physico-chimiques d'Analyse) and Germany (Arbeitskreis Chromatographie der Gesellschaft Deutscher Chemiker) as well as interested parties in Eire, Italy, The Netherlands, Scandinavia, Spain and Switzerland. As a result chromatography symposia, in association with instrument exhibitions, have been held biennially in Amsterdam, Edinburgh, Hamburg, Brighton, Rome, Copenhagen, Dublin, Montreux, Barcelona, Birmingham, Baden-Baden, Cannes, London, Nurnburg, Paris and Vienna.

In 1958 "Gas Chromatography Abstracts" was introduced in journal format under the Editorship of C. E. H. Knapman; first published by Butterworths, then by the Institute of Petroleum, it now appears as "Gas and Liquid Chromatography Abstracts" produced by Elsevier Applied Science Publishers and is of international status — abstracts, covering all aspects of chromatography, are collected by Members from over 200 sources and collated by the Editor Mr. E. R. Adlard assisted by Dr. P. S. Sewell.

Links with the Institute of Petroleum were severed at the end of 1972 and the Group established a Secretariat at Trent Polytechnic in Nottingham, Professor Ralph Stock playing a prominent part in the establishment of the Group as an independent body. At the same time, in recognition of expanding horizons, the name of the organisation was changed to the Chromatography Discussion Group.

In 1978, the "Father" of Partition Chromatography, Professor A. J. P. Martin was both honoured and commemorated by the institution of the Martin Award which is designed as testimony of distinguished contribution to the advancement of chromatography. Recipients of the award include:

E. R. Adlard	Professor J. H. Knox
Professor E. Bayer	Dr. E. Kovats
Professor U. A. Th. Brinkman	Professor A. Liberti
Dr. L. S. Ettre	Dr. C. S. G. Phillips
Professor J. C. Giddings	Professor W. H. Pirkle
Professor G. Guiochon	Dr. G. Schomburg
Professor J. F. K. Huber	Dr. R. P. W. Scott
Professor E. Jellum	Dr. L. Snyder
Dr. C. E. R. Jones	Professor R. Stock
Dr. R. E. Kaiser	Dr. G. A. P. Tuey
C. E. H. Knapman	

The Group celebrated its Silver Jubilee in 1982 with the 14th International Symposium held, appropriately, in London. To commemorate that event the Jubilee Medal was struck as means of recognising significant contributions by younger workers in the field. Recipients of the Jubilee Medal include: Dr. J. Berridge, Dr. H. Colin, Dr. K. Grob Jr., Dr. J. Hermannson, Dr. E. D. Morgan, Dr. P. G. Simmonds, Dr. P. Schoenmachers, Dr. R. Tijssen, Dr. K. D. Bartle, and Dr. H. Lingeman.

In 1984 the name was once again changed, this time to The Chromagraphic Society, which title was believed to be more in keeping with the role of a learned society having an international membership of some 1000 scientists drawn from more than 40 countries. At that time, the Executive Committee instituted Conference and Travel Bursaries in order to assist Members wishing to contribute to, or attend, major meetings throughout the world.

The Society is run by an Executive Committee elected by its Members, in addition to the international symposia, seven or eight one-day meetings covering a wide range of subjects are organised annually. One of these meetings, the Spring Symposium, is coupled with the Society's Annual General Meeting when, in addition to electing the Society's Executive Committee, Members have the opportunity to express their views on the Society's activities and offer suggestions for future policy.

Regular training courses in all aspects of chromatography are run in conjunction with the Robens Institute of the University of Surrey and it is hoped that this particular service will eventually include advanced and highly specialised instruction.

Reports of the Society's activities, in addition to other items of interest to its members (including detailed summaries of all papers presented at its meetings), are given in the Chromatographic Society Bulletin which is produced quarterly under the editorship of I. W. Davies.

At the time of writing three grades of membership are offered: Membership with Abstracts, Membership, and Student Membership (includes Abstracts). Members receive the Bulletin free of charge, benefit from concessionary Registration Fees for all Meetings and Training Courses and are, of course, eligible to apply for Travel and/or Conference Bursaries.

For further information and details of subscription rates please write to:

Mrs. J. Challis
Executive Secretary
THE CHROMATOGRAPHIC SOCIETY
Trent Polytechnic
Burton Street
Nottingham NG1 4BU
United Kingdom

THE ROBENS INSTITUTE OF INDUSTRIAL AND ENVIRONMENTAL HEALTH AND SAFETY

The Robens Institute is Europe's largest University-based health and safety organisation, with an international reputation for its contribution to resolving problems resulting from our technological society. Since its foundation in 1978 the aims of the Robens Institute have been to advance health and safety worldwide by contributing in three ways:

1. By providing direct practical help to governments, industry, commerce and any others with a particular health and safety problem.

2. By conducting longer term research into fundamental health and safety issues.

3. By acting as a focus for discussion and training on health and safety matters.

In the battle for a safer world — a matter of increasing public concern — three fronts can be identified. There are problems in the Environment, in the Workplace, and with man-made Products, and the Institute's three main divisions correspond to these areas. It offers independent expert assessments, backed by high quality laboratory and field investigations. The Institute's clearly established independence, charitable status and wide experience in successfully dealing with emotive and confidential issues makes it ideally suited to meet the challenges of the future.

The Institute's hundred or more highly qualified staff include internationally recognised authorities in toxicology, occupational and environmental medicine, epidemiology, ergonomics, microbiology, human physiology, occupational hygiene, safety, analytical chemistry, radiation physics, industrial psychology, information science and statistics. Many of the staff are members of international or national committees concerned with environment, workplace, and product health and safety.

Being part of the University of Surrey, the Institute is also able to draw very readily on the University's extensive specialised resources, including its excellent library and computing facilities. The Institute has also established a network of consultants to supplement its own expertise. Thus it can provide a comprehensive service on health and safety issues. Current and future legislation, particularly arising from the European Community, will necessitate that expert advice on these health and safety issues assumes a high priority in all plans for the future. The Robens Institute provides a truly international service, supervising projects or collaborating with other organisations in all five continents.

Our effectiveness in being able to offer advice and practical solutions stems from extensive experience gained in fundamental and applied research in the field of health and safety. Among sponsors of this work are many government agencies, international organisations, large industrial concerns and charitable trusts.

The final key to continued progress in health and safety matters is Education and Training, and in recognition of this the Robens Institute offers very comprehensive programmes ranging from one-day seminars to modular MSc courses. These attract participants from all over the world.

In 1984 the Chromatographic Society set up a series of training courses (in GLC and HPLC) now run annually at the Robens Institute.

The Robens Institute is sited on the campus of the University of Surrey at Guildford. The main Institute building contains twenty well-equipped laboratories and supporting accommodation, with laboratories and offices in other parts of the campus and on the adjacent University of Surrey Research Park. It is planned to unite all these into one building in the near future. In the centre of the south east of England but with motorway and rail links to the rest of the country, the Institute is ideally placed to provide a rapid and effective response to the needs of industry and government. For the international traveller the London airports of Heathrow and Gatwick are less than an hour's drive away.

A *NOTE ON* A REGULATORY VIEW OF THE IMPORTANCE

OF CHIRAL DISCRIMINATION

J. W. Bridges

Robens Institute of Health and Safety
University of Surrey
Guildford, Surrey GU2 5XH, UK

INTRODUCTION

Many of the pesticides, human and veterinary drugs and industrial chemicals in current use incorporate one or more chiral centres. The vast majority of these compounds are marketed as racemates. In most cases no information is available regarding the biological properties of the individual isomers. It is timely to review whether or not this is acceptable for two reasons: (1) there are now many well substantiated cases in which one isomer differs markedly in biological potency from the other and (2) advances in analytical techniques now permit the separate analysis of individual isomers.

Differences in biological activity between isomers are not surprising since the beneficial and adverse effects of chemicals in most cases depends on their ability to bind selectively to a particular macromolecule(s) (e.g., "receptors") located either at the cell surface or within the cell. The majority of these "binding sites" require a three-dimensional fit and consequently are likely to discriminate to a greater or lesser extent between enantiomers [1].

DIFFERENCES IN PROPERTIES OF ENANTIOMERS

Before considering the requirements for new products and existing ones, a very brief review is appropriate of situations in which major differences between enantiomers may occur.

Enantiomers with Different Biological Properties

Differing biological properties between isomers may arise from differences in absorption, distribution, metabolism or excretion (ADME) and/or from discrimination at the "receptor" level [2,3]. Examples of ADME effects are given in Table 1 and "receptor" effects in Table 2. The relative paucity of examples is almost certainly a reflection of lack of research in the area rather than the rarity of the event as such. In some cases rapid interconversion of forms occurs and therefore any difference in biological effects of isomers is largely academic.

Most widely studied have been differences in rates of metabolism between isomers because most drug metabolising enzymes are stereoselective and therefore discrimination in the fate of individual isomers is very likely to occur. Among chemicals where there is a difference in biological activity between the different isomers there are a number of cases where one isomer is largely responsible for the desirable biological action while the other predominates in the toxicity/undesirable action.

Metabolism Leading to Enantiomers

Even if a chemical does not have a chiral centre, metabolism may give rise to enantiomers which differ in their biological properties [4]. Most drug metabolising

Recent Advances in Chiral Separations, Edited by D. Stevenson and
I. D. Wilson, Plenum Press, New York, 1991

Table 1. Differences in Distribution, Metabolism and Clearance of Enantiomers

DISTRIBUTION	COMPOUND
Binding to albumin	Verapamil, warfarin
Binding to α-1 acid glycoprotein	Propranalol

METABOLISM	
Configurational inversion	Ibuprofen
Rates	Prilocaine, hexobarbitone, bufuralol

CLEARANCE	Acenocoumarol, ephidrine

Table 2. Differences in Pharmacodynamic Action

ISOMERS PRODUCING DIFFERENT TYPES OR DEGREES OF TOXICITY

Chemical	Effect of (+)-isomer	Effect of (-)-isomer
Thalidomide	Teratogenic	Nonteratogenic?
Hyoscyamine	Peripheral effect least potent	Most potent (ten-fold)

ISOMERS PRODUCING DIFFERENT TYPES OF THERAPEUTIC EFFECT

Hydroxy-N-methyl morphinan	Analgesic	Antitussive
Bupivacaine	Local anaesthetic and vasoconstrictor	Local anaesthetic

ONE ISOMER RESPONSIBLE FOR THE "THERAPEUTIC ACTION", THE OTHER FOR THE TOXICITY

	"Therapeutic" effect	Toxic effect
Ketamine	Analgesic/hypnotic ((+)-form)	CNS stimulation ((-)-form)
Timolol	Reduction of intraocular pressure (RS-forms)	β-Adrenergic antagonism (S-form)
Dialkylphosphates	Pesticide in insects (no therapeutic benefit)	Inhibition of neurotoxic esterase leading to neuropathy ((-)-form)

enzymes are capable of producing stereoselective products. Oxidative and reduction reactions lead commonly to the insertion of a chiral centre. Prilocaine is an illustration of the introduction of a chiral centre resulting in differential toxicity between isomers. The amidases involved in the metabolism of prilocaine produce the (-)-isomer rapidly, the (+)-isomer slowly. Consequently it is the (-)-isomer which is primarily responsible for toxicity. From a practical standpoint, the main concern regarding stereoselectivity in isomer formation is that, because the drug metabolising enzymes vary substantially between species and even between tissues, it could be an important contributor to species differences in both therapeutic and toxic action. This can thereby cause difficulties in extrapolating animal data to man.

There are a number of possible situations in which one isomer may modify the therapeutic or toxicological properties of the other, and some examples are given below.

1. A complex between the (+) and (-) forms may arise. This has been claimed to occur for thalidomide with consequent changes in acute toxicity.
2. One enantiomer may reduce the metabolism of the other, e.g., Levomethorphan metabolism is inhibited by Dextromethorphan.
3. One isomer may change either the therapeutic or the toxic properties of the other through a pharmacodynamic effect.

AREAS OF PARTICULAR PRACTICAL CONCERN

There are a number of situations in particular in which justification should be sought from the producers of racemic products before marketing, or for the continued marketing of such products, namely situations in which:

1. The therapeutic/desired biological activity is due entirely, or primarily, to one isomer. In this situation the non-active isomer must be regarded as an impurity. Why should we accept up to 50% impurity in a drug or pesticide?
2. The toxicological effects are entirely or primarily due to one isomer.
3. One isomer has a major influence on the therapeutic or toxicological activity of the other.

For most chemicals in present use there is no information available that would enable the identification of whether any of these scenarios is likely. It is my view that this situation is unacceptable and that a progressive change must occur. This should begin by requiring relevant information on racemates which are proposed as new products with priority being given to those which fall into one of the following categories:

1. Chemicals with a low therapeutic index.
2. Chemicals which are particularly potent. (NB. High potency implies a precise structural fit with a "receptor".)
3. Agents of natural origin or very closely related to compounds of natural origin. (NB. Naturally occurring compounds tend to be stereoselective in their biological effects).
4. Chemicals which are closely related structurally to compounds which show marked differences in the biological properties of their enantiomers.

The least acceptable situation is one in which either the toxicity resides in one isomer, while the therapeutic or other benefits is due to the other isomer or one isomer exacerbates the toxicity of the other.

For some compounds it may be acceptable to market them as racemates even if there are considerable differences in the biological properties between isomers if it can be demonstrated that there is a rapid interconversion between isomeric forms or there is insufficient benefit in manufacturing the requisite isomer in relation to the cost involved.

Consideration of specific categories of proposed products will undoubtedly broaden to embrace ultimately all new products and eventually the reassessment of existing ones. The priority in reviewing those products already on the market should be given to those falling within categories 1-4 above. These requirements will prove an exciting and demanding challenge both to the analytical and the synthetic organic chemist.

REFERENCES

1. M. Simonyi, I. Fitos and J. Vizg, *Trends in Pharmacol. Sci.*, 7:112-116 (1986).
2. F. P. A. Lehmann, *Trends in Pharmacol. Sci.*, 7:281-285 (1986).
3. T. Walle and K. W. Walle, *Trends in Pharmacol. Sci.*, 7:155-158 (1986).
4. B. Testa, *Trends in Pharmacol. Sci.*, 7:60-64 (1986).

IS THERE A RATIONAL BASIS FOR THE SELECTION AND/OR DESIGN OF CHIRAL

STATIONARY PHASES FOR HIGH-PERFORMANCE LIQUID CHROMATOGRAPHY?

David R. Taylor

Chemistry Department
UMIST, PO Box 88
Manchester M60 1QD, UK

SUMMARY

One of the major problems facing the analyst in chiral high-performance liquid chromatography (HPLC) is that of selecting an appropriate chiral stationary phase (CSP) for a given analysis. Here an attempt is made to address the problem by proposing a simple algorithm for use in CSP selection, based upon Wainer's classification of the five main types of CSP.

The prospect of using computer-assisted molecular modelling (CAMM) to assist in this area has aroused much interest, and several case histories of such investigations are described. These case histories highlight not only the power of CAMM in modelling diastereoisomeric interaction complexes, but also the associated problems which can arise.

In addition, suggestions for other approaches to the design of new CSPs are made, including work based upon both reciprocity of interaction and the development of new families of CSPs.

INTRODUCTION

The selection of an appropriate CSP for the HPLC analysis of a particular racemic mixture has become a serious problem for the busy analyst, since the number of commercially available CSPs is now over thirty and continues to rise. It is therefore relevant to discuss whether a rational basis can be found for such a selection process. The closely related problem of rational design procedures, such as the use of CAMM, for developing improved CSPs will also be considered.

SELECTION OF CHIRAL STATIONARY PHASES IN HPLC

The presently available CSPs [1] were developed in response to an urgent need for enantioselective HPLC analysis, which arose as the implications of the unwanted teratogenic properties of one enantiomer of the racemic ingredient of Thalidomide were appreciated by the scientific community and by drug regulatory authorities [2]. It cannot be sound judgement to permit the registration of a 50:50 mixture of two biologically active substances (i.e., a racemate) unless both have separately established desirable therapeutic effects.

To generate enantioselective packings for the HPLC methods essential for bringing about changes in production control and pharmacological studies of pure enantiomers, CSP designers turned first to the so-called chiral pool, the natural and synthetic optically-active substances available for immediate purchase. They first adopted as chiral selectors quite simple asymmetric substance such as (i) chiral acids and amines, already of proven use in indirect analytical procedures for synthesis of diastereoisomeric derivatives, and (ii) chiral metal complexes containing amino acid ligands. Next used were natural asymmetric

Recent Advances in Chiral Separations, Edited by D. Stevenson and
I. D. Wilson, Plenum Press, New York, 1991

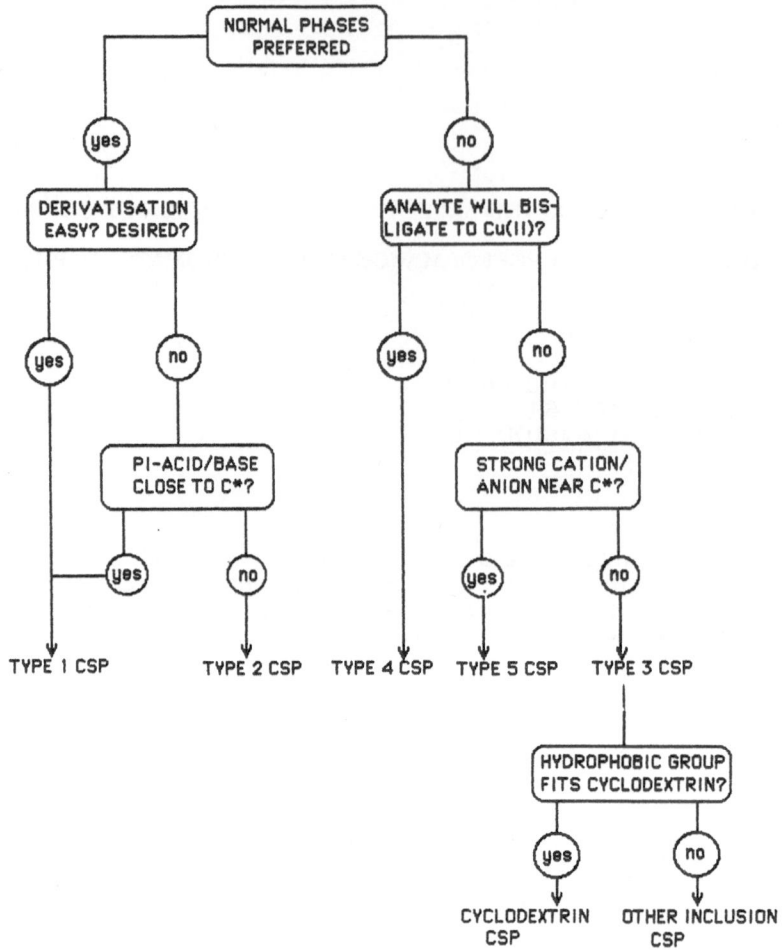

Fig. 1. A possible selection procedure for CSPs.

macromolecules including (iii) proteins, such as bovine serum albumin and human α-glycoprotein, and (iv) carbohydrate derivatives, such as cyclodextrins and cellulose triesters. Finally, in imitation of these natural materials, (v) chiral polymers were developed from chiral monomers or by polymerisation over chiral catalysts.

Now that such groups of asymmetric selectors have proliferated, there is a need for a rational classification scheme (Table 1) such as that proposed by Wainer [3], which is adopted here with little modification. This enables the problem of choice to be restricted initially to one of selecting from five broad classes of CSP. As a simple aid to this selection process, the attached algorithm (Figure 1) can be used: it proceeds by a stepwise question-and-responses sequence, as follows.

1. *Is a Non-aqueous Mobile Phase Preferred/Possible?*

A "Yes" response indicates a CSP within Classes 1 or 2, since only rarely can separations on CSPs within these two classes be undertaken with reversed-phase solvents.

2. *For Non-aqueous Analyses, Can the Analyte be Derivatised?*

Derivatisation by an achiral reagent can be used to increase analyte-CSP interactions and hence improve resolution of enantiomers. It can also bring other benefits, such as increased detectability by UV or fluorescence detectors, or enhanced solubility in normal phase eluents. A "Yes" response indicates the use of a Class 1 CSP (donor-acceptor), a "No" response favouring a Class 2 CSP (silica-supported cellulose triesters, etc), unless a π-acid or a π-base is already located close to a chiral centre in the analyte (as for example in an arylsulphoxide), in which case a Class 1 CSP should be selected.

6

Table 1. Classification of Existing Chiral Stationary Phases

Class	Description	Specific Examples	Usual Mode
1	"Pirkle-type" (donor-acceptors)	DNB-Glycine DNB-Leucine Naphthylalanine	Normal phase (polar modifier)
2	Silica-supported cellulose tri- esters, -carbamates	Chiralcel-OA, -OB -OF, -OJ, etc	Normal phase (polar modifier)
3	Inclusion CSPs e.g., Cyclodextrins Polyacrylates Polyacrylamides Crown ethers	Cyclobond 1-3 Chiralpak-OP, OT Merck grafted Chiralcel-CR	Reversed-phase e.g., aqueous MeCN aqueous MeOH
4	Ligand exchangers	Proline HO-proline	Reversed-phase buffered aqueous
5	Immobilised proteins	Albumin Glycoprotein	Reversed-phase buffered aqueous

Within analytes identified as candidates for separation with a Class 1 CSP, two subdivisions can be defined, those with π-donor groups and those with π-acceptor groups. Hence the next selection criterion.

3. *In a Class 1 Analyte or its Derivative, Is There a π-Donor Group (e.g., Phenyl or Naphthyl) Close to the Asymmetric Centre?*

A "Yes" response would lead to the initial selection of a π-acceptor Class 1 CSP, such as the dinitrobenzoyl (D)-phenylglycine phase. On the other hand, a "No" response would only indicate a π-donor Class 1 CSP (e.g., naphthylvaline) if a π-acceptor group was similarly located in the analyte or its derivative. These requirements are not overriding, however, since Class 1 CSPs also operate by dipole-dipole, H-bonding, and steric interactions.

4. *For Reversed-phase (RP) Analytes, Can Bidentate Ligation to a Metal Ion Such as Cu(II) be Foreseen?*

If a "Yes" response is possible, Class 4 CSPs (ligand exchange) are indicated. In the absence of such potential ligation, the selection process must devolve towards CSPs from Classes 3 and 5.

5. *For Non-ligating RP Analytes, Is a Strongly Cationic or Anionic Site Located Close to the Asymmetric Centre?*

A "Yes" response indicates that the analyte may be resolvable on a protein-based CSP (Class 5), a "No" suggesting that trials should first be undertaken with inclusion CSPs (Class 3).

Within potential candidate analytes for separation on a Class 3 CSP, two broad subdivisions exist, leading to the next selection criterion.

6. *Does the Analyte Contain a Non-polar Group of Such Dimensions That it Will Insert Snugly into a Cyclodextrin Cavity?*

Such a non-polar includable group (ideally 5-9 Å across), which should not lie too far from an asymmetric centre, indicates the selection of one of the Cyclobond range of cyclodextrin CSPs. Its absence would indicate selection of other types of inclusion CSPs such as the chiral acrylate polymers currently marketed by Merck and Diacel.

The operation of this simple structure-screening routine could form the basis for a miniature expert system if the structural features of the analyte were "perceived" by a computer, in much the same way that they are perceived by currently available chemical expert systems such as LHASA. This approach to CSP selection can be exemplified by consideration of many reported separations, e.g., those of Ibuprofen [4-6] and Propranolol [7, 8] (structures 1 and 2, respectively).

Structure 1. Ibuprofen.   Structure 2. Propranolol.

RATIONAL DESIGN OF NEW CHIRAL STATIONARY PHASES FOR HPLC

Following on from the rational selection of CSPs comes the problem of their rational design. Much interest has been aroused in the use of CAMM protocols to explain how enantioselective retention occurs on CSPs. We have, for example, applied this approach to several novel CSPs, especially N-formyl (S)-aminoacids bound covalently to silica via an amide linkage [9, 10], and similarly bound anilides of (R,R)-tartaric acid [11]. Other groups have also reported similar investigations recently (Table 2). Here we wish to sound a note of caution about this approach to designing CSPs by reference to several studies of Class 1 CSPs, which one might expect to be the easiest to model and rationalise.

CAMM proceeds at a high resolution VDU-based workstation and commences with structure building, preferably using X-ray crystallographic data to obtain a valid starting point for separate molecular models of the CSP and the analyte. The study then proceeds with an often lengthy conformational analysis of both model structures, in order to identify those low-energy conformers which are likely to be well-populated, taking account of charge distribution, intramolecular hydrogen bonds and van der Waals' repulsions. None of the studies of this type so far reported makes any allowance for the presence of solvent molecules.

Table 2.   Known Case Studies of Computer-aided Molecular Modelling of Recognition Mechanisms on Chiral Stationary Phases

No.	Compounds Modelled	Reference
1	Cyclodextrin and Propranolol	Armstrong et al. [8]
2	N-Formylphenylalanine and N-acylaminocid esters	Bridger and Taylor, unpublished (1987)
3	DNB-Phenylglycine and trifluoranthrylethanol	Lipkowitz et al. [12]
4	Various CSPs and analytes	Norinder and Sundholm [15]
5	Tartaric anilide and aryloxypropionates	Bridger, Garner and Taylor, unpublished (1988)
6	Naphthylalanine and DNB-leucine	Topiol et al. [16]
7	Pivaloylnaphthylethyl-amide and DNB-amides	Daeppen et al. [20]

Selected conformers of the CSP and each analyte enantiomer are then brought into close proximity, to determine whether molecular interactions arise, such as π-π charge transfer or intermolecular H-bonding, as well as to investigate the operation of steric repulsions. The particular objective at this stage is to effect a comparison between, in the case of an (R)-configuration CSP for example, the (R,R) analyte-CSP association complex and the diastereoisomeric (S,R) analyte-CSP complex. Thus, absolute interaction energies are much less important than the relative energies for these two complexes. Nevertheless, several very serious difficulties arise which may make even such a simple approach to a rational analysis of relative retention very uncertain of success.

The first problem is that of limitations in the software currently available: only van der Waals' repulsions are efficiently computed, and accurate assessment of the importance of dipole-dipole and π-π interactions is often not easily achievable. Our ChemX software, for example, does not itself identify potential hydrogen bonds although it will compute their effect if instructed to do so. The second problem, well-illustrated in two of the case histories which follow, is that when more than one conformation of a partner in the complex is appreciably populated, the number of likely association complexes rapidly escalates and meaningful comparisons become both difficult and time-consuming to achieve. Thirdly, even if energy differences are substantial between possible (R,R) and (S,R) association complexes, it is not necessarily at all obvious what the effect of structural variation in either partner in the association complex will be. Hence, the use of CAMM for improving the selectivity of a given CSP, or of identifying other analytes which it might resolve, is potentially unreliable. Clearly, one should not overvalue the role of CAMM in this area of scientific study, but rather we should treat it as just one of the currently available resources to probe the interesting phenomenon of enantioselective HPLC retention.

CASE HISTORIES

One of the phases which we have studied in depth has been that derived from (R,R)-tartaric acid (see structure **3**) by the sequence of reactions shown in Scheme 1. When finally unmasked by de-acetylation of the silica-bound selector, it contains a terminal amide residue which is advantageously, though not necessarily, a π-acceptor such as a 3,5-dinitroanilide. Our CAMM studies of this structure were conducted using the n-propylamide, molecular parameters for which were first obtained by an in-house X-ray crystallographic study (Figure 2). We were particularly interested in explaining how this type of CSP provided enantioselectivity for aryloxypropionates such as structure **4** (the separation of which is shown in Figure 3), although this CSP has much wider potential and should shortly be commercially available. Initially using a variety of CAMM facilities, we computed the energies of around 4000 analyte conformers using standard molecular mechanics procedures, identifying four low-lying structures (Figure 4). These were found to offer a common association mechanism with the optimised CSP model, essentially arrived at on the basis of π-π complex formation between the relevant π-acidic and π-basic groups at a distance of about 3 Å.

(3) Ar = 3,5-(NO$_2$)$_2$ C$_6$H$_3$

(4)

Structure **3**. Tartaric acid CSP. Structure **4**. Aryloxypropionate.

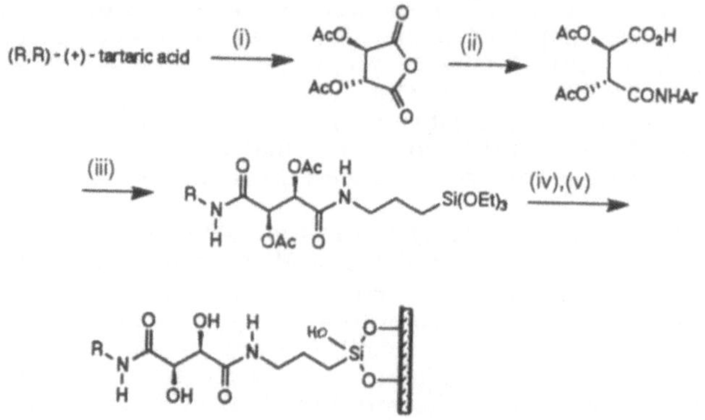

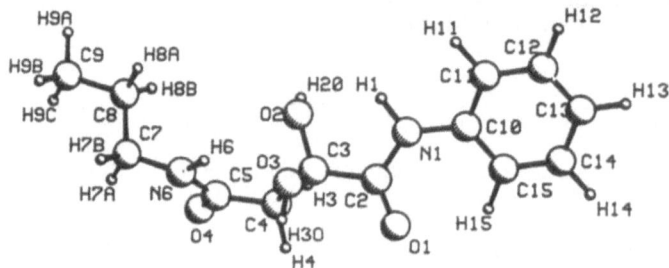

Scheme 1. Synthesis of tartaric-based CSPs. Reagents: (i) Ac_2O, H_2SO_4; (ii) RNH_2 (eg, R = Ph, 3-$NO_2C_6H_4$, 3,5-$(NO_2)_2C_6H_3$); (iii) $(EtO)_3Si(CH_2)_3NH_2$, peptide coupling; (iv) 5 μm silica gel; (v) NH_3, ammonia.

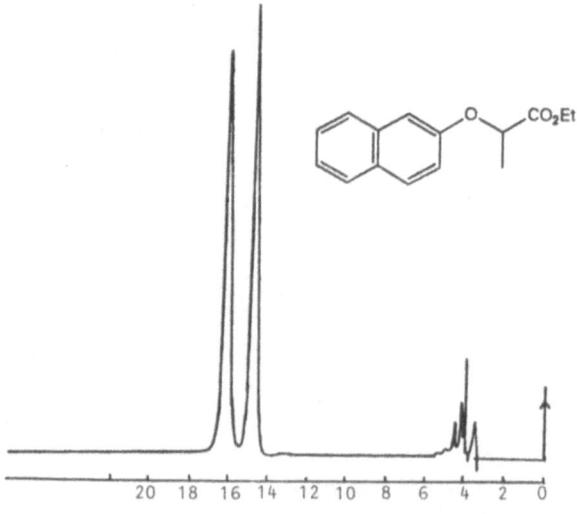

Fig. 2. X-Ray analysis of 3,5-dinitroanilide of (R,R)-tartaric acid n-propylamide.

Fig. 3.   HPLC analysis of racemic aryloxypropionate on CSP **3**. Conditions: 3% ethylacetate in n-hexane, flow rate 1 ml min^{-1}.

10

A

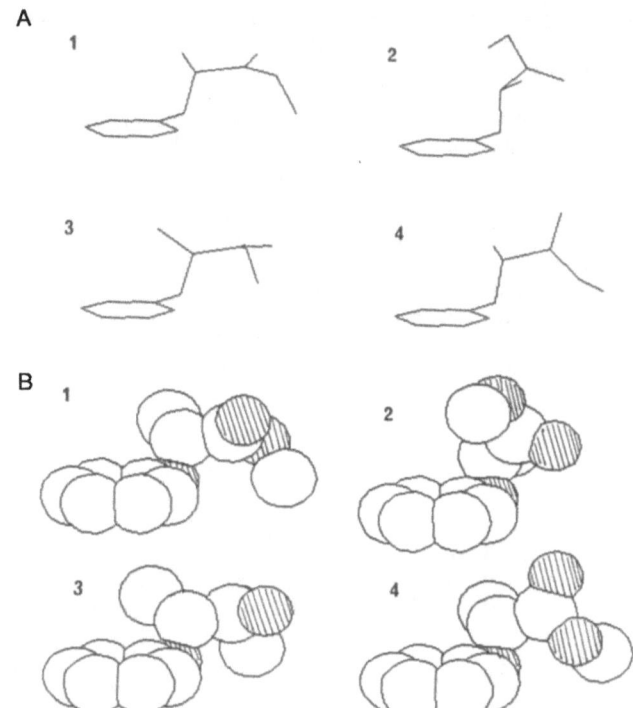

B

Fig. 4. (A) Framework display of ChemX-optimised methyl phenoxypropionate (more retained enantiomer): (1) lowest energy conformer as docked with (R,R)-tartramide CSP; (2)-(4) three other low-energy conformers. (B) Space-filling display (H atoms not shown, oxygens shaded) of ChemX-optimised methyl phenoxypropionate (more retained enantiomer): (1) lowest energy conformer as docked with (R,R)-tartramide CSP; (2)-(4) three other low-energy conformers.

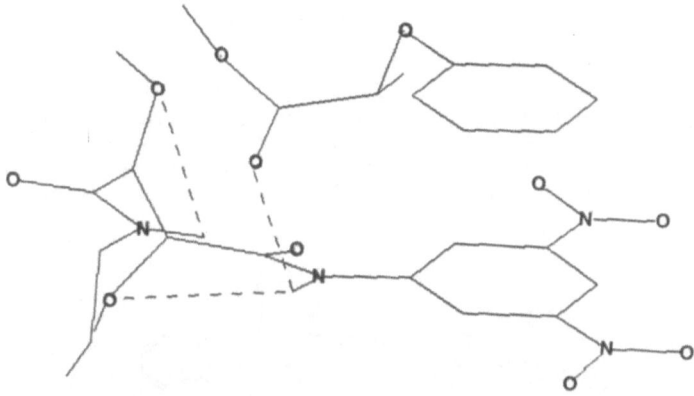

Fig. 5. Interaction complex (ChemX, framework with added heteroatoms, H atoms shown if connected to heteroatoms) of lowest-energy conformer of methyl (S)-phenoxypropionate (more retained, above) with ChemX-optimised (R,R)-tartaric dinitroanilide (below).

This association we believe is accompanied by hydrogen bond formation, since enantioselectivity is lost if H-bonding solvents are present. In addition to intramolecular H-bonding within the CSP structure, which appears to hold the CSP in a rather rigid conformation in which the two hydroxyl groups are located in approximately orthogonal planes, intermolecular hydrogen bonding is believed to occur between the ester carbonyl oxygen of the analyte and the anilide NH of the CSP (Figure 5). Inversion of the analyte greatly increases steric repulsion in this situation, and association from the other side of the dinitroanilide is discouraged by repulsion of the ester group with the curving (R,R)-CSP's chain. Hence CAMM has given us a very useful working hypothesis of the interactions arising for one type of analyte which predicts correctly the effect of substantial chain extension at one analyte terminus, for example.

In contrast to the quite evident effect of analyte inversion in the above case, two other published studies reveal very clearly the problems which can arise when using CAMM in this area. The first, reported by Lipkowitz et al. in a series of papers [12-14], and studied also by Norinder and Sundholm [15] and by us in rather less detail (Garner and Taylor, unpublished observations), involves the interaction complexes arising between the widely-used Pirkle phase based on dinitrobenzoyl-(R)-phenylglycine (structure **5**) and its test analyte trifluoroanthrylethanol (structure **6**). Lipkowitz's group first of all predicted the wrong elution order on the basis of their very detailed and sophisticated computational procedure, in which the association complexes were generated by the computer on a systematically varied basis. Only much more detailed computations revealed that two different conformations were being preferred in the (R,S) and (R,R) association complexes, with a very slight but probably crucial preference for the complex of the more retained analyte enantiomer.

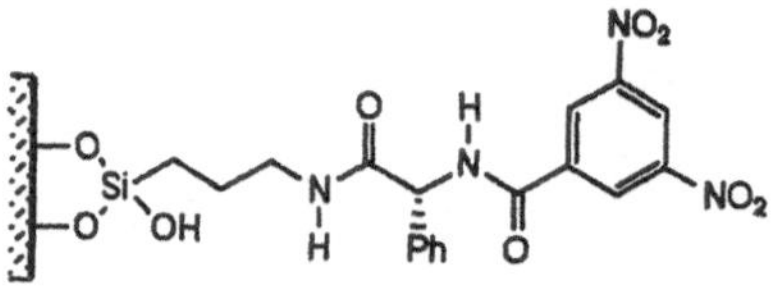

(5)

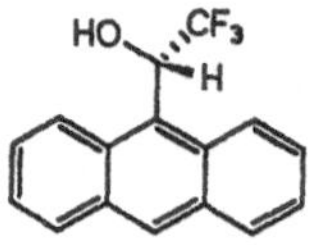

(6)

Structure **5.** Dinitrobenzoyl-R-phenylglycine (Pirkle) CSP.
Structure **6.** Trifluoroanthrylethanol.

In a very recent study of another Class 1 CSP [16, 17], models were computed for the association complexes between N-naphthyl-(S)-valine, a model for the third-generation Pirkle phase (structure **7**), and a very well-resolved analyte (R/S) N-(3,5-dinitrobenzoyl)leucine n-propylamide (structure **8**). This pair of interacting structures has been studied experimentally by visible and NMR spectroscopy by Pirkle's group [18] and a

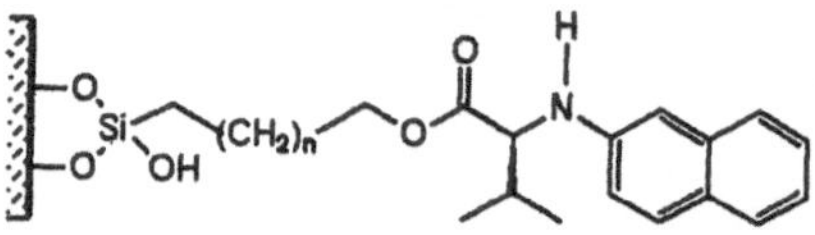

(7)

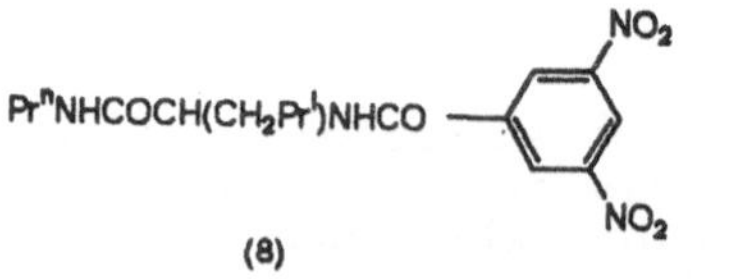

(8)

Structure **7.** N-Naphthylvaline CSP.
Structure **8.** N-(3,5-Dinitrobenzoyl)leucine π-propylamide.

well-established interaction mechanism was thus tested by Topiol and Sabio's CAMM analysis. The outcome seems to be that both (R)- and (S)-enantiomers can associate in almost exactly the same manner by a combination of π-π interaction and H-bonding. The authors therefore came down very strongly against the widely accepted three-point interaction process, which they suggested degenerates to a pseudo-two-point interaction in this example, since the three contact points lie along only two of the bonds to the chiral centre. Their study also pinpointed the difficulty of computing the effect of π-π interactions, which this type of software frequently indicates are repulsive rather than attractive.

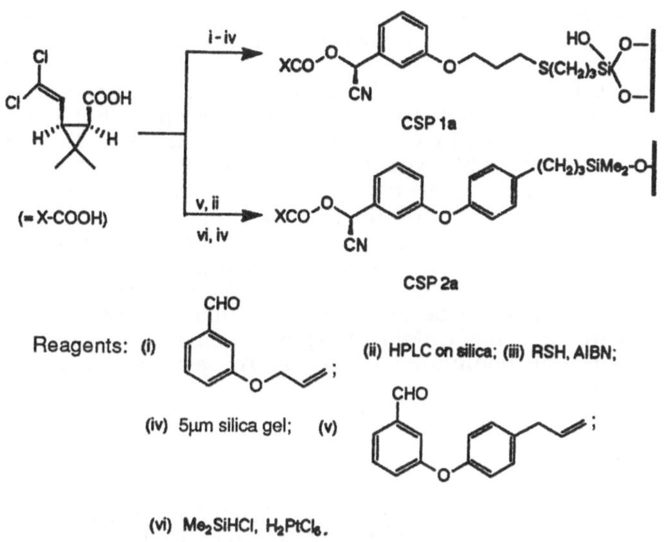

Scheme 2. Synthesis of Cypermethrin-based CSPs.

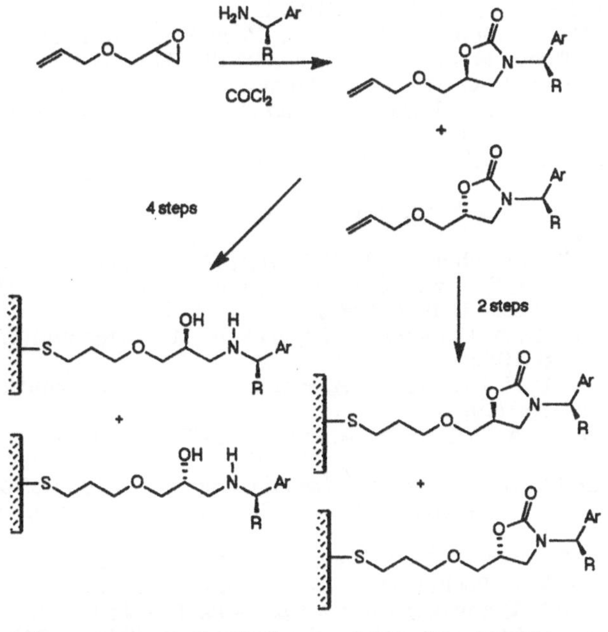

Scheme 3. Synthesis of a family of CSPs from epoxides.

13

OTHER STRATEGIES FOR DESIGN OF NEW CSPs

If CAMM does not offer an easily implemented and logical route to the design of new CSPs for HPLC, to what alternatives can we turn? Two rational alternatives have been used by ourselves and others, and may be termed the reciprocity route and the CSP-family route.

The well-tried principle of reciprocity of interaction [19] states that, if a CSP based upon a compound (R)-A resolves the racemic mixture of (S)- and (R)-B, then binding compound (R)-B to silica should afford a CSP capable of resolving the racemic mixture of (R)- and (S)-A. The accompanying proviso is that the mode of attachment does not disrupt the essential interaction mechanism(s). As discussed elsewhere in this volume (Taylor and Clift, pp 39-41), we are using this approach to find CSPs which might be fairly generally useful for the analysis of racemic pyrethroids. Substantial progress in this programme has been made by attaching a modified Cypermethrin constituent to silica (Scheme 2).

The family-route to new CSPs is also exemplified elsewhere in this volume (Ferraz Lourenço and Taylor, pp 77-83) where we describe the development of a number of structural related CSPs, the precursors for which were prepared by reaction of racemic epoxides with optically-active amines (Scheme 3). This powerful technique is essentially empirical in its approach, but instead of affording only one type of CSP it can lead to a range of novel CSPs which are interrelated and possess in-built scope for structural variety. In the particular case reported here (Scheme 3), the CSPs may be obtained from diastereoisomeric aminoalcohols, as well as from their derived oxazolidones and trifluoroacetates. Clearly, not all such phases will or should reach the marketplace; however, our understanding of their mode of action will aid the design of further phases which may be commercially exploited. Although we shall use CAMM to explore their mode of operation, the rationale behind their design is one dominated by the stereoselective synthesis, rather than a mechanistically-led one.

ACKNOWLEDGEMENTS

We are indebted to ICI Agrochemicals plc, Shell Chemicals plc, and Glaxo Group Research for financial support for this work, which was carried out by Dr. David Garner, Dr. Gary Bridger, Mr. Wagner Lourenço and Miss Elizabeth Clift.

REFERENCES

1. I. W. Wainer, "Practical Guide to Selection and Use of HPLC Chiral Stationary Phases", J. T. Baker Inc., Phillipsburg, NJ (1988).
2. G. Blaschke, H.-P. Kraft and H. Markgraf, Chem. Ber., 113:2318-2322 (1980).
3. I. W. Wainer, Trends in Analytical Chem., 6:125-134 (1987).
4. D. M. McDaniel and B. G. Snider, J. Chromatogr., 404:123-132 (1987).
5. J. Hermansson and M. Eriksson, J. Liquid Chromatogr., 9:621-639 (1986).
6. I. W. Wainer and T. D. Doyle, J. Chromatogr., 284:117-124 (1984).
7. I. W. Wainer, T. D. Doyle, K. H. Donn and J. R. Powell, J. Chromatogr., 306:405-411 (1984).
8. D. W. Armstrong, T. J. Ward, R. D. Armstrong and T. E. Beesley, Science, 232:1132-1135 (1986).
9. J. N. Akanya, S. M. Hitchen and D. R. Taylor, Chromatographia, 16:224-227 (1982).
10. J. N. Akanya and D. R. Taylor, J. Liquid Chromtogr., 10:805-817 (1987).
11. G. Bridger, PhD Thesis, UMIST (1987).
12. K. B. Lipkowitz, D. A. Demeter, R. Zegarra, R. M. Larter and T. Darden, J. Comput. Chem., 9:63-78 (1988).
13. K. B. Lipkowitz, D. A. Demeter, C. A. Parish, J. M. Landwer and T. Darden, J. Comput. Chem., 8:753-760 (1987).
14. K. B. Lipkowitz, D. A. Demeter, C. A. Parish and T. Darden, Anal. Chem., 59:1731-1733 (1987).
15. U. Norinder and E. G. Sundholm, J. Liquid Chromatogr., 10:2825-2844 (1987).
16. S. Topiol, M. Sabio, J. Moroz and W. B. Caldwell, J. Am. Chem. Soc., 110:8367-8376 (1988).
17. S. Topiol and M. Sabio, J. Chromatogr., 461:129-148 (1989).
18. W. H. Pirkle and T. C. Pochapsky, J. Am. Chem. Soc., 109:5975-5982 (1987).
19. W. H. Pirkle and R. Dappen, J. Chromatogr., 404:107-115 (1987).
20. R. Daeppen, V. R. Meyer and H. Arm, J. Chromatogr., 464:39-47 (1989).

14

CHROMATOGRAPHIC CHIRAL SEPARATIONS ON IMMOBILISED PROTEINS:

DOES HIGH-PERFORMANCE LIQUID CHROMATOGRAPHY MIRROR NATURE?

I. W. Wainer

St. Jude Children's Research Hospital
Memphis, TN 38105, USA

SUMMARY

A commercially available high-performance liquid chromatography (HPLC) chiral stationary phase (CSP) based upon immobilised bovine serum abumin, the BSA-CSP, has been used to investigate the protein binding of enantiomeric solutes. The changes in the chromatographic retention which resulted from the addition of mobile phase modifiers were compared with parallel *in vitro* protein binding studies. The results of these studies indicate that the commercially available BSA-CSP can be used to investigate the binding interactions between a series of solutes and the protein. The use of mobile phase modifiers has demonstrated a direct relationship between *in vitro* protein binding and chromatographic retention. In addition, the use of enantiomeric solutes has shown that the BSA-CSP can be used to probe the number and relative affinities of various binding sites on the protein.

INTRODUCTION

HPLC is an effective analytical tool which is employed in chemical, pharmacological and biological laboratories. While the common applications of HPLC technology involve the identification, quantification and preparation of compounds, this technique can also be used to probe the biological properties of molecules. An example of this type of application is the use of an HPLC stationary phase composed of octadecylsilane bonded to silica (a C_{18} column) in the determination of molecular hydrophobicity.

The hydrophobicity of a drug substance plays an important role in the compound's biological activity. This physicochemical property is the "driving force" for liquid-liquid distribution processes, micelle formation and passive membrane transport [1]. Because of its biological importance, there have been a number of attempts to quantify molecular hydrophobicity. The initial correlation between hydrophobicity and biological activity was reported in 1983 [1]. This work and a large body of subsequent activity involved the measurement of the differential solubility of a compound between an aqueous and an organic phase, usually n-octanol. The octanol-water partition coefficients (log P) have become an accepted model for lipophilicity in quantitative structure-activity relationships [2]. The original measurements of log P were accomplished by physically mixing the n-octanol and water layers. This "shake-flask" method is time consuming, wasteful of sample, and subject to errors from impurities, poor detectability, dissociation, decomposition and stable emulsion formation [2,3]. In addition, compounds of very high or low partition cannot be measured by this method [2,3].

Another approach to the determination of log P values is the use of reversed-phase HPLC using a C_{18} stationary phase and an aqueous mobile phase [1-5]. In this method, the stationary phase replaces the n-octanol and the partition between the aqueous and organic phases is measured by the chromatographic retention of the solute on the HPLC column (k'). A number of investigations have demonstrated that there is a linear correlation between log

Recent Advances in Chiral Separations, Edited by D. Stevenson and
I. D. Wilson, Plenum Press, New York, 1991

k' and log P and the HPLC method is an accepted method for the determination of the hydrophobicity of an number of classes of compounds [1-5].

The same approach has been used by Ganansia et al. [6] to investigate the correlation between retention on a reversed-phase HPLC system and plasma protein binding. In this study, the percent protein binding to serum albumin or α_1-acid glycoprotein was used instead of the n-octanol/water partition coefficient (log P) and the HPLC stationary phase was cyanopropyl-bonded silica instead of octadecyl silica. The correlation between the chromatographic retention (k') of betaxolol and its O-alkyl analogues, and the percent plasma protein binding of these compounds was investigated by plotting log k' versus percent unbound compound (free fraction). The relationship between the free fraction and log k' was sigmoidal instead of linear. This result illustrates the inherent problem in determining the extent of protein binding with an approach which only measures hydrophobic interactions since electrostatic, steric and stereochemical factors also play a role in this process. One way to consider all the possible protein/substrate interactions is to use the protein in the chromatographic process either as a mobile phase modifier or as part of the stationary phase.

Marle et al. [7] have investigated the relationship between chromatographic retention and protein binding in an HPLC system using human serum albumin (HSA) as a mobile phase additive. This approach was used to determine the binding affinities of the enantiomers of omeprazole and tryptophan (Table 1). The calculated binding affinities were in good agreement with those obtained by other methods. While the results from this study are encouraging, this approach has some limitations. Three solid phases were used in the experiments, LiChrosorb RP-8, Phenyl Hypersil and LiChrosorb Diol, and each gave different results due to adsorption of the HSA. Thus, the choice of the solid phase is important and will vary from solute to solute. For example, the studies with omeprazole and tryptophan reported in Table 1 were accomplished using different solid supports, the diol and phenyl columns, respectively.

The UV absorbance of HSA is also a problem and this limits the concentrations which can be added to the mobile phase. In addition, the technique seems to be restricted to compounds with affinity constants in the range 10^3-10^5.

The other approach, the use of immobilised proteins, has been made feasible by the development of two HPLC chiral stationary phases (HPLC-CSPs) - the bovine serum albumin CSP (BSA-CSP) developed by Allenmark [8] and the α_1-acid glycoprotein CSP (AGP-CSP) developed by Hermansson [9]. Both of these phases resolve enantiomeric compounds by utilising the stereoselective binding properties of the respective proteins including electrostatic, steric and stereochemical factors as well as hydrophobic interactions. Thus, the chromatographic retention of a compound on the BSA-CSP or AGP-CSP should correspond to its *in vitro* protein binding and some studies of this relationship have been reported.

Jewell et al. [10] studied 15 compounds and found no significant correlation between chromatographic retention on the AGP-CSP and the extent of the binding of the compounds to AGP (expressed as the free fraction after equilibrium dialysis). However, when the BSA-CSP was studied by Lammers et al. [11] there was a significant correlation between percent protein binding to BSA and the reciprocal of the chromatographic retention ($1/k'$). This work was carried out using a series of analogous piperazines and the use of this technique with different families of compounds was not investigated.

Table 1. Determination of the Binding Affinity (n_1K) of Omeprazole and Tryptophan via HPLC with Human Serum Albumin in the Mobile Phase (data from Ref. [7])

Compound	$n_1K \pm$ SD	
	HPLC Method	Literature
Omeprazole		
Enantiomer 1	6.6 ± 0.4 x 10^4	2 x 10^4
Enantiomer 2	4.6 ± 0.2 x 10^4	
L-Tryptophan	1.29 ± 0.02 x 10^4	1.1 x 10^4
D-Tryptophan	4.4 ± 0.1 x 10^4	3 x 10^4

The study of the correlation of protein binding with chromatographic retention on the BSA-CSP has also been undertaken in our laboratory and the initial results of this study are discussed below.

STEREOSELECTIVE BINDING TO SERUM ALBUMINS

Serum albumins, including bovine serum albumin (BSA), are globular, hydrophobic proteins which have been shown to stereoselectively bind small enantiomeric molecules [12]. The various mammalian forms of serum albumin are structurally different, but these differences are usually small and not very impressive [12]. However, there are some dramatic differences between species in the stereoselective binding of small molecules. Most of the stereospecific protein binding studies have been carried out using human serum albumin (HSA). There are two major drug-binding sites on HSA: the warfarin-azapropazone binding area and the indole and benzodiazepine binding site which both display some stereoselectivity [13-15]. An additional one (digitoxin [16]) to three sites (digitoxin, bilirubin and fatty acid [17]) have also been proposed.

The stereoselectivity of the warfarin-azapropazone binding area has been investigated by a number of laboratories. Sellers and Koch-Weser [18] have demonstrated that S-warfarin is more highly bound to HSA than is R-warfarin although there is only a slight difference between the isomers (S, 99.47% bound; R, 99.15% bound [19]). A much higher stereoselectivity has been found for phenprocoumon with an almost two-fold higher affinity of S-phenprocoumon over R-phenprocoumon [20].

The indole and benzodiazepine binding site has also been the subject of a number of investigations and appears to be more stereospecific than the warfarin-azapropazone binding area. For example, the ratio of the affinity constants of (+)-oxazepam hemisuccinate/(-)-oxazepam hemisuccinate is 49.5 [21] while the ratios of the affinity constants of the S-enantiomers of N-desmethyl-3-methyldiazepam and N-desmethyl-3-methylmedazepam to the corresponding R-isomers are 135 and 28, respectively [22,23]. A less dramatic example is the difference between the affinity constants of the enantiomers of ketoprofen where the affinity of the (+)-enantiomer is three times greater than that of the (-)-isomer [24].

The studies of the different binding sites on HSA have utilised the ability of competing molecules to displace a marker compound from the protein. For example, a variety of substances have been shown in *in vivo* and *in vitro* studies to displace warfarin from HSA [12,16-18]. These results suggest that compounds which displace warfarin from HSA *in vitro* should have the same effect when the warfarin is bound to the immobilised BSA in the BSA-CSP. The result of the displacement would be a reduction of the chromatographic retention (k') of warfarin on the BSA-CSP. The extent of this reduction should be a reflection of the differential binding affinities of warfarin and the displacing substance and the sites at which these compounds bind. In this manner the BSA-CSP can be used to investigate the protein binding of various compounds.

EFFECT OF MODIFIERS ON RETENTION AND STEREOSELECTIVITY ON THE BSA-CSP

Wainer and Chu [25] have studied the effect of a variety of compounds on the k' and stereoselectivity (α) of warfarin (structure, see Figure 1) on the BSA-CSP. Trichloroacetic acid (TCA), cyclamic acid (CYCL) and lauric acid (LA) were chosens modifiers based on previous work by Sellers and Koch-Weser [18] in which these compounds were shown to displace warfarin from HSA binding. The effect of lorazepam (structure, see Figure 1) and D- and L-tryptophan on k' and α were also examined. However, it was assumed that these compounds would not affect warfarin binding to BSA since they are known to bind to another site on HSA [12,16,17]. The effect of TCA and CYCL on k' and α of other solutes was also investigated.

In Table 2 the maximum observed effect of the mobile phase modifiers used in this study on the retention and stereoselectivity of warfarin are summarised. The observed effects can be the result of one or both of the following mechanisms: (1) direct competition between warfarin and the mobile phase modifier for binding at nonspecific sites on the BSA molecule and at the warfarin binding site; and (2) alteration of the affinity and stereoselectivity of BSA for warfarin through the binding of the mobile phase modifier to other sites on the protein (allosteric interactions).

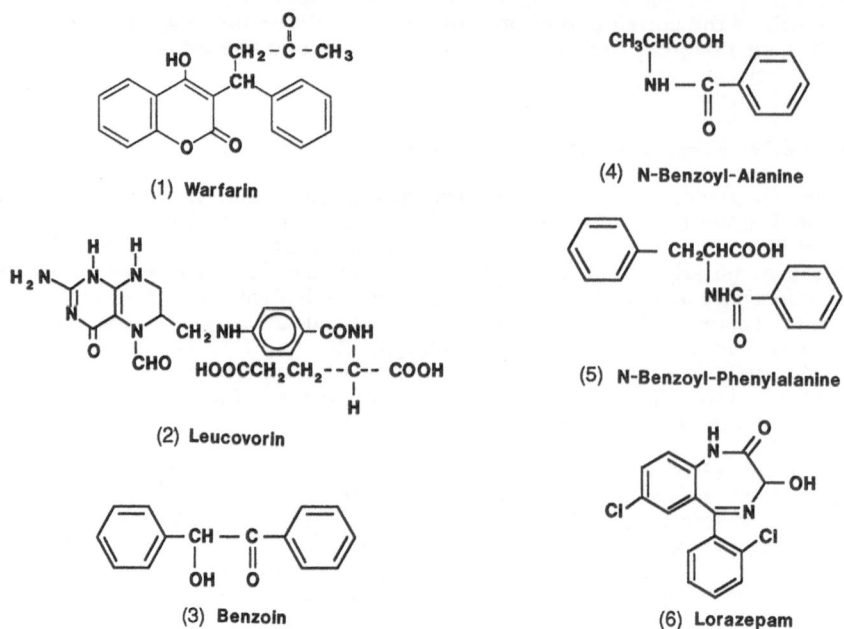

Fig. 1. Structures of the compounds used in these studies.

Table 2. Effect of Mobile Phase Modifiers on the Retention and Stereochemical Resolution of Warfarin (data from Ref. [25])

Modifier	Concentration [mM]	k_1'*	α†
Trichloroacetc Acid	0.0	18.1	1.20
	5.0	9.4	1.16
Cyclamic Acid	0.0	23.9	1.20
	7.0	18.3	1.16
Lauric Acid	0.0	20.7	1.20
	0.5	6.1	1.0
Lorazepam	0.0	23.6	1.20
	0.4	19.4	1.04
L-Tryptophan	0.0	19.8	1.31
	3.0	18.3	1.31
D-Tryptophan	0.0	18.2	1.26
	3.0	17.3	1.29

* Capacity factor first eluted enantiomer (S-warfarin).
† Stereoselectivity.

When TCA was consecutively added to the mobile phase up to a concentration of 5.0 mM, there was a significant effect on the retention of S- and R-warfarin but the effect on the stereoselectivity was not significant. The k's of S- and R-warfarin were reduced by about 50%, while α was diminished by only 3% from 1.20 to 1.16 (Table 2). Similar results were obtained through the addition of CYCL to the mobile phase although the magnitude of the effect on k' was not as large. A 7.0 mM concentration of CYCL reduced k' for both enantiomers by an average of 25% while α was reduced by only 3% (Table 2).

These results suggest that the primary interaction mechanisms for TCA and CYL involves competitive displacement of warfarin from nonspecific and specific binding sites on the immobilised BSA. This assumption is based upon the fact that R- and S-warfarin are both highly bound to serum albumin (R, 99.15% bound; S, 99.47% bound) [12]. Therefore, direct competition at the binding site should affect both enantiomer to the same extent leading to a reduction in k' but not α.

18

The existence of direct competition between a solute and a mobile phase additive for binding sites on an immobilised protein can be determined using a method developed by Muller and Carr [26,27]. In this work, immobilised concanavalin A was the stationary phase; p-nitrophenyl-α-D-mannopyranoside, p-nitrophenyl-α-D-glucopyranoside and p-nitrophenyl-α-D-glacatopyranoside were the solutes; and, α-methyl-D-mannoside and α-methyl-D-glucoside were the mobile phase modifiers. As with the warfarin/TCA experiments, an increase in concentration of the mobile phase modifier resulted in a decrease in the retention of the solutes. A plot of $1/k'$ versus modifier concentration yielded straight lines and the authors were able to demonstrate that this result indicated that the mobile phase modifiers and the solutes were competing for the same binding site on the protein.

A plot of $1/k'$ warfarin versus the concentration of TCA yielded straight lines for both (R)- and (S)-warfarin as illustrated in Figure 2. This indicates that the same binding site is involved for both enantiomers and that TCA also binds at this site [28]. Similar results were obtained with CYCL [28].

A series of *in vitro* protein binding studies were performed to determine whether the decline in k' produced by the TCA reflected a decrease in the *in vitro* BSA protein binding of warfarin and, therefore, whether k' and protein binding are related [28]. The binding studies were carried out using 1.5×10^{-5} M BSA, 1.3×10^{-5} M warfarin in phosphate buffer (0.2 M, pH 7.5)-propan-1-ol (97:3 v/v) at a temperature of 30°C. The total amount of protein was equivalent to the quantity of "active" protein (i.e., binding sites) on the BSA-CSP as determined by warfarin concentration versus k' studies and the warfarin concentration was the on-column concentration used during the k'/modifier studies. The aqueous phase and the temperature were identical to the mobile phase and temperature used during the chromatographic studies.

Trichloroacetic acid (1-7 mM) was added to the protein binding solution and the amount of each of the warfarin enantiomers bound to the BSA was determined using ultracentrifugation followed by chromatography on the BSA-CSP. The results of this study are presented in Table 3. When log k' was plotted against percent of the drug bound to the BSA, straight lines were observed for both (S)- and (R)-warfarin with correlation coefficients (R^2) of 0.991 and 0.999, respectively (see Figure 3). The slope of the line from (S)-warfarin was significantly greater than that of the line from (R)-warfarin. This may reflect the fact that the affinity constant for the (S)-enantiomer is more than twice that of the (R)-enantiomer, 570×10^{-3} and 250×10^{-3}, respectively [12].

These results are dramatically different from the sigmoidal curves obtained with the cyanopropyl-bonded silica [6] and are consistent with the earlier studies involving the BSA-CSP [11]. It is clear that, for this solute, k' and protein binding are directly related and that the BSA-CSP can be used to evaluate *in vitro* protein binding.

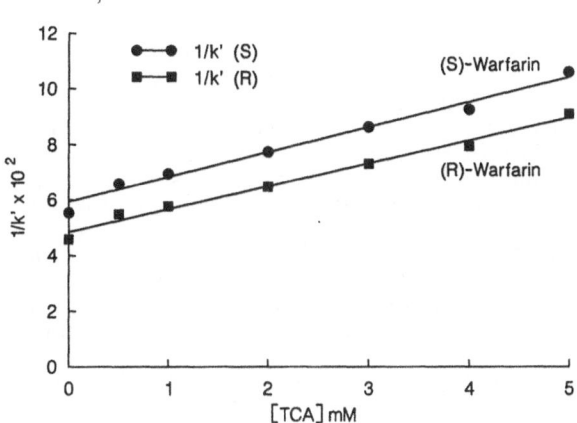

Fig. 2.   Determination of the number of warfarin binding sites on bovine serum albumin using the method of Muller and Carr [26,27], where ■ = (R)-warfarin, ● = (S)-warfarin. For chromatographic conditions see Ref. [25].

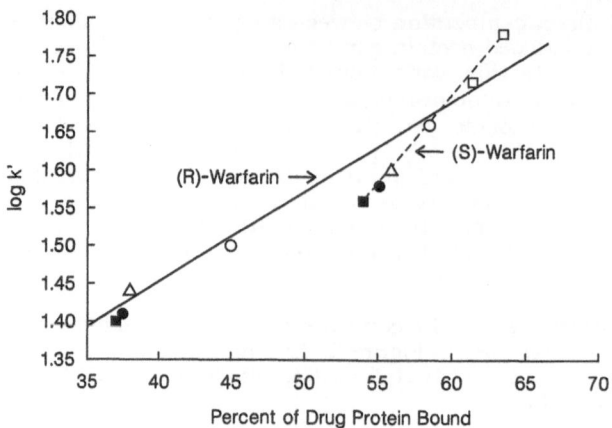

Fig. 3.   Evaluation of the correlation between the effect of tricholoroacetic acid (TCA) on the chromatographic retention (k') of (R)- and (S)-warfarin on the BSA-CSP and its effect on the *in vitro* protein binding of warfarin enantiomers to free BSA, where ☐ = 0 mM TCA, ○ = 1 mM TCA, Δ = 3 mM TCA, ● = 5 mM TCA and ■ = 7 mM TCA. For chromatographic conditions see Ref. [25] and for protein binding conditions see text.

Table 3.   Effect of Tricholoroacetic Acid (TCA) on the Binding of Warfarin (WAR) Enantiomers to Human Serum Albumin (HSA) (see text for experimental conditions)

Concentration TCA [mM]	Percent of WAR Bound to HSA	
	(S)-WAR	(R)-WAR
0.0	60.5	63.5
1.0	45.0	58.5
3.0	38.0	55.9
5.0	37.5	55.1
7.0	37.0	54.0

The use of modifiers other than TCA and CYCL produced different results which suggested alternative interaction mechanisms between the modifier and the protein. For example, when lauric acid (LA) was the mobile phase modifier, the addition of 0.5 mM LA to the mobile phase reduced the k' of both enantiomers of warfarin by over 70% with an accompanying loss of stereoselectivity (Table 2). The loss of both retention and stereoselectivity suggests a change in the affinity of the binding site rather than displacement by direct competition. LA is a C_{11}-carboxylic acid which is assumed to bind at the fatty acid binding site on serum albumin which differs from the warfarin binding area. In HSA, binding at the fatty acid site has been shown to affect the binding affinity at other ligand binding sites via allosteric interactions [17]. These interactions can result in both enhanced and reduced affinities at the affected site (in this case the warfarin binding area) [17]. The loss of stereoselectivity at the warfarin site is consistent with this mechanism.

When a 0.4 mM concentration of lorazepam was added to the mobile phase the effect was similar to the result obtained with LA. There was an 18% reduction in k' for S-warfarin, a 29% reduction in k' for R-warfarin and stereoselectivity fell from $\alpha = 1.20$ to 1.04. Larger concentrations of the modifier were not studied due to the poor solubility of lorazepam in the mobile phase. These results were unexpected since lorazepam is a benzodiazepine derivative and these molecules bind to a different site on HSA, the benzodiazepine binding site [12,16,17]. In addition, in *in vitro* studies, oxazepam, which is structurally similar to lorazepam, did not displace warfarin from HSA [16]. The lowering of k' and α suggests the existence of an allosteric interaction between the warfarin and benzodiazepine binding sites on the BSA-CSP.

D- and L-Tryptophan also bind to serum albumin at the benzodiazepine binding site [12] but their effect on the retention and stereoselectivity of warfarin on the BSA-CSP was not significant (Table 2). A 3.0 mM concentration of L-tryptophan in the mobile phase produced an 8% reduction in k' for both warfarin enantiomers and no change in α. With the same concentration of D-tryptophan, the k' of S-warfarin was reduced by 5%, the k' of the R-isomer was increased by 2% and there was no significant change in α (an increase from 1.26 to 1.29). The effect of D- and L-tryptophan on the k' and α of warfarin is consistent with the binding of solute and modifier to separate sites on the BSA molecule and the lack of secondary interactions between these sites. The difference between the effect of lorazepam and D- and L-tryptophan may arise from differences in the protein binding affinities of the compounds.

In order to assess the effect of TCA and CYCL on the chromatographic retention of other compounds, and thereby their effect on protein binding, five other solutes were investigated. The solutes were leucovorin (2), benzoin (3), N-benzoyl-alanine (4), N-benzoyl-phenylalanine (5) and lorazepam (6) (structures in Figure 1). The results of the studies with solutes 2 to 5 are summarised in Figure 4. For solutes 2 to 5 the addition of TCA and CYCL to the mobile phase had a greater effect on the k's of the most retained enantiomers producing a significant decrease in the observed stereoselectivities. This differed from the results obtained with warfarin where the reduction of k' for both enantiomorphs was approximately the same and the decrease in α was minimal. An examination of the change in k' corresponding to each modifier concentration indicated that unlike warfarin, solutes 2 to 5 were bound to more than one site on the BSA molecule and that there were differences in the binding affinities of the enantiomers at these sites. This is illustrated by the data for (S)- and (R)-N-benzoyl-phenylalanine (Table 4).

The addition of 0.5 mM TCA to the mobile phase resulted in a decrease in the chromatographic retention (k') of both enantiomers, 45% for the (S)-isomer and 35% for the (R)-isomer, relative to the k' determined without the TCA. However, with 5 mM TCA in the mobile phase, the magnitude of the reduction was greater for the (R)-enantiomer, 72% (R) versus 66% (S). These results can be explained by assuming: (1) the existence of two protein binding sites, PBS1 and PBS2; (2) a greater affinity for (R)-N-benzoyl-phenylalanine at PBS1 relative to the (S)-enantiomer; (3) a greater or equal affinity for (S)-N-benzoyl-

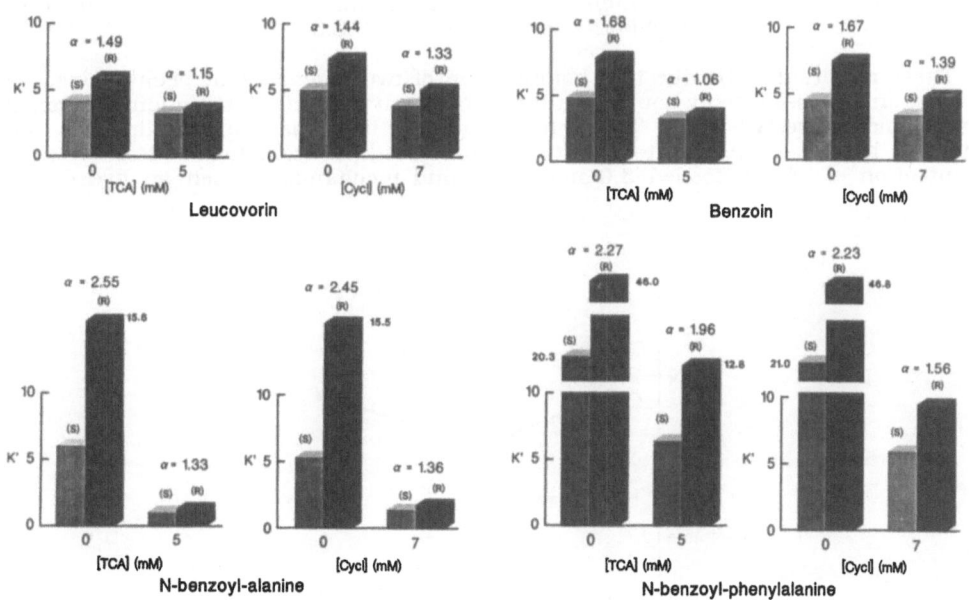

Fig. 4. The effect of trichloroacetic acid (TCA) and cyclamic acid (CYCL) on chromatographic retention (k') and stereoselectivity (α) of leucovorin, benzoin, N-benzoyl-alanine and N-benzoyl-phenylalanine on the BSA-CSP. For chromatographic conditions see Ref. [25].

Table 4. Effect of Trichloroacetic Acid (TCA) on the Chromatographic Retentions of (S)- and (R)-N-Benzoyl-Phenylalanine (NBP) (see Ref. [25] for experimental conditions)

Concentration TCA [mM]	Capacity Factor (k')	
	(S)-NBP	(R)-NBP
0.0	20.25	46.00
0.5	11.13	29.25
1.0	9.25	23.25
3.0	7.13	15.00
5.0	6.50	12.75

phenylalanine at PBS2 relative to the (R)-enantiomer; and (4) a higher affinity of TCA for PBS1 than PBS2 [27]. The existence of at least two binding sites for (R)- and (S)-N-benzoyl-phenylalanine was supported by the results from the plot of $1/k'$ versus TCA concentration. In this instance, the resulting graph was curvilinear rather than the linear relationship found with (R)- and (S)-warfarin (Wainer et al., unpublished results).

In addition to solutes 2 to 5, the effect of TCA on the retention and stereoselectivity of lorazepam (compound 6) was investigated. Lorazepam was chosen because it is known to bind at a site other than the warfarin binding area [12]. The effect of the TCA mobile phase concentration on the retention of the two enantiomers of lorazepam is shown in Figure 5. Since the enantiomers of lorazepam rapidly interconvert [28], the injection of the individual isomers is not possible and the elution order is unknown. The enantiomers are referred to as "first" and "second" eluted.

The addition of up to a 3.0 mM concentration of TCA had very little effect on the k' of the first eluted enantiomer. However, the k' of the second eluted enantiomer passed through a maximum between 0.5 and 1.0 mM TCA and then returned to approximately the initial value at 3.0 mM TCA. As a result, the stereoselectivity also passed through a maximum rising from $\alpha = 1.48$ (0 mM TCA) to 1.95 (1.0 mM TCA) to 1.62 (3.0 mM TCA). When the concentration of TCA in the mobile phase was raised from 3.0 mM to 5.0 mM, there was a 70% increase in the k' of the first eluted enantiomer ($k' = 9.3$ to $k' = 15.8$) and the stereoselectivity was lost. An additional rise in TCA concentration from 5.0 mM to 7.0 mM resulted in an increase in the k' of both enantiomers to $k' = 18.5$.

These results may be due to a combination of two independent mechanisms. The initial rise in stereoselectivity could be a reflection of an allosteric interaction between the warfarin binding area where the TCA is bound and the benzodiazepine binding site where solute 6 is bound — the counterpart of the warfarin-lorazepam interaction. When the concentration of TCA exceeded 3.0 mM, a second mechanism based on hydrophobic

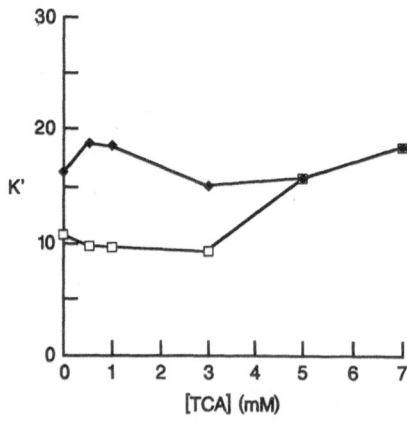

Fig. 5. The effect of trichloroacetic acid (TCA) on the chromatographic retention (k') of lorazepam enantiomers on the BSA-CSP. For chromatographic conditions see Ref. [25].

interactions may have predominated. An elevation in the ionic strength increased the polar environment surrounding the immobilised BSA which, in turn, potentiated the hydrophobic interactions between the solute and the protein. This should lead to the observed increase in k' and the loss of stereoselectivity. A similar mechanism has been suggested by Allenmark et al. [29] who observed increases in the k's of N-benzoyl-D,L-amino acids on the BSA-CSP when the phosphate buffer concentration of the mobile phase was raised above 0.2 mM.

CONCLUSIONS

The results of these studies indicate that the commercially available BSA-CSP can be used to investigate the binding interactions between a series of solutes and the protein. The use of mobile phase modifiers has demonstrated a direct relationship between *in vitro* protein binding and chromatographic retention. In addition, the use of enantiomeric solutes has shown that the BSA-CSP can be used to probe the number and relative affinities of various binding sites on the protein. The full extent of these properties and their uses are currently under investigation.

REFERENCES

1. D. A. Brent, J. J. Sabatka, D. J. Minick and D. W. Henry, *J. Med. Chem.*, 26:1014-1020 (1983).
2. N. El Tayar, H. Van de Waterbeemd and B. Testa, *J. Chromatogr.*, 320:305-312 (1985).
3. T. Braumann, *J. Chromatogr.*, 373:191-225 (1986).
4. T. L. Hafkensheid and E. Tomlinson, *J. Chromatogr.*, 292:305-317 (1984).
5. W. E. Hammers, G. J. Meurs and C. L. de Ligny, *J. Chromatogr.*, 247:1-13 (1982).
6. J. Ganansa, L. Bianchetti and J. P. Thenot, *J. Chromatogr.*, 421:83-90 (1987).
7. I. Marle, C. Pettersson and T. Arvidsson, *J. Chromatogr.*, 456:323-326 (1988).
8. S. Allenmark, B. Bomgren and H. Boren, *J. Chromatogr.*, 264:63-68 (1983).
9. J. Hermansson, *J. Chromatogr.*, 269:71-80 (1983).
10. R. C. Jewell, K. L. R. Brouwer and P. J. McNamara, *J. Chromatogr.*, 487:257-264 (1989).
11. N. Lammers,H. de Bree, C. P. Groen, H. M. Ruijten and B. J. de Jong, *in:* "Bioanalysis of Drugs and Metabolites, Especially Anti-inflammatory and Cardiovascular", E. Reid, J. D. Robinson and I. D. Wilson, eds., Plenum Press, New York, pp 301-303 (1988).
12. W. E. Muller, *in:* "Drug Stereochemistry: Analytical Methods and Pharmacology", I. W. Wainer and D. E. Drayer, eds., Marcel Dekker, New York, pp 227-244 (1988).
13. W. E. Muller and U. Wollert, *Pharmacology*, 19:59-67 (1979).
14. J. P. Tillement, G. Houin, R. Zini, S. Urien, E. Albengres, J. Barre, J. Lacomte, P. D'Athis and B. Sebille, *Adv. Drug Res.*, 13:59-94 (1984).
15. I. Sjoholm, *in:* "Drug Protein Binding", M. M. Ridenberg and S. Erill, eds., Prager Publishers, Philadelphia, pp 158-178 (1988).
16. U. Kragh-Hansen, *Pharmacol. Rev.*, 33:17-53 (1981).
17. K. J. Fehske, W. E. Muller and U. Wollert, *Biochem. Pharmacol.*, 30:687-692 (1981).
18. E. M. Sellers and J. Koch-Weser, *Ann. N. Y. Acad. Sci.*, 179:213-225 (1971).
19. N. A. Brown, E. Jahnchen, W. E. Muller and U. Wollert, *Mol. Pharmacol.*, 13:70-79 (1977).
20. M. Otagiri, J. S. Fleitman and J. H. Perrin, *J. Pharm. Pharmacol.*, 32:478 (1980).
21. W. E. Muller and U. Wollert, *Mol. Pharmacol.*, 11:52-60 (1975).
22. T. Alebic-Kolbah, F. Kajfez, S. Rendic, V. Sunic, A. Konowal and G. Snatzke, *Biochem. Pharmacol.*, 28:2457-2464 (1979).
23. G. Gratton, E. Decorte, F. Moimas, C. Angeli and V. Sunjic, *Il Farmaco Ed. Sci.*, 40:209-215 (1985).
24. A. Konowal, G. Snatzke, T. Alebic-Kolbah, F. Kajfez, S. Rendic and V. Sunjic, *Biochem. Pharmacol.*, 28:3109-3113 (1979).
25. I. W. Wainer and Y.-Q. Chu, *J. Chromatogr.*, 455:316-322 (1988).
26. A. J. Muller and P. W. Carr, *J. Chromatogr.*, 284:33-51 (1984).
27. A. J. Muller and P. W. Carr, *J. Chromatogr.*, 357:11-32 (1986).
28. G. Blaschke, *Angew. Chem. Int. Ed. Eng.*, 19:13 (1980).
29. S. Allenmark, B. Bomgren and H. Boren, *J. Chromatogr.*, 316:617-624 (1984).

A NOTE ON SOME EXAMPLES OF CHIRAL HIGH-PERFORMANCE LIQUID CHROMATOGRAPHIC RESOLUTION IN THE PHARMACEUTICAL INDUSTRY

R. J. Collicott

Glaxo Group Research Limited
Greenford, Middlesex, UK

INTRODUCTION

Examples of the methods by which enantiomers may be resolved chromatographically using either preliminary derivatisation with a chiral reagent, direct separation using a chiral mobile phase, or by direct separation using a chiral stationary phase (CSP) are outlined. These methods are illustrated by reference to chiral resolutions of compounds of interest to Glaxo Group Research Limited.

The Use of Chiral Derivatisation for Enantiomer Resolution

A racemic carboxylic acid (see inset in Figure 1) was derivatised with (R)-(+)-1-phenylethylamine using 1-ethoxycarbonyl-2-ethoxy-1,2-dihydroquinoline (EEDQ, Aldrich) (see inset in Figure 1) to promote the coupling. The resulting diastereoisomers were resolved by normal phase chromatography (Figure 1) on silica gel.

As little as 0.2% of the impurity enantiomer could be determined using this approach. The potential problems associated with this technique include: (1) enantiomeric impurity of reagent; (2) stereoselective reaction; (3) inversion of any chiral centre during reaction; and (4) difference in UV absorbance between diastereoisomers. However, in this case these problems were not encountered.

A second example of chiral derivatisation was the use of Marfey's reagent [1] (Pierce, Oud-Beijerland, The Netherlands) to derivatise a complex mixture of amino acids (e.g., a peptide hydrolysate). The general reaction is shown in Scheme 1. The 22 derivatives studied were mostly resolved in a single reversed-phase chromatographic run on a C_{18}-bonded column (Figure 2). Even though the complete separation of all 22 derivatives was not achieved it is worth noting that, for every amino acid (Cys, Ser, Arg, Asp, Gln, Ala, Tyr, Met, Ileu, Phe and Leu), the derivative of each enantiomer was well resolved from that of its enantiomeric counterpart. A useful feature of this reagent is that it allowed very sensitive and selective detection at 340 nm. Again problems of the type alluded to above were not encountered.

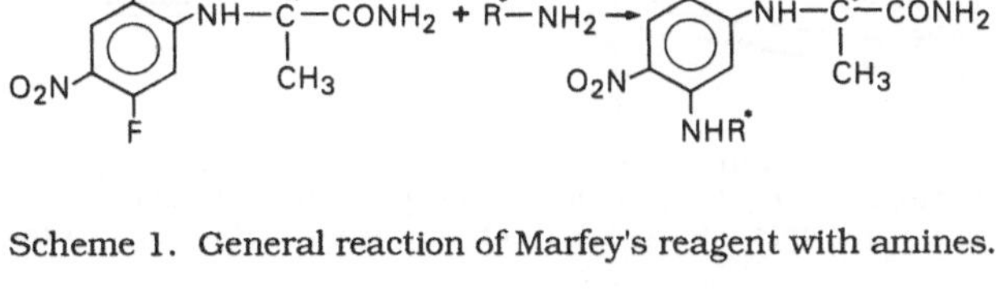

Scheme 1. General reaction of Marfey's reagent with amines.

Recent Advances in Chiral Separations, Edited by D. Stevenson and I. D. Wilson, Plenum Press, New York, 1991

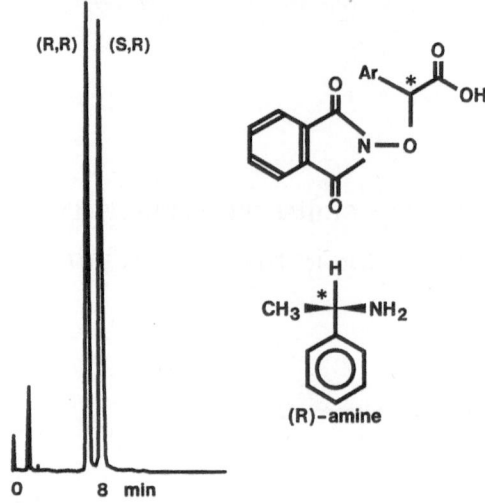

Fig. 1. Resolution of a racemic acid as its (R)-(±)-1-phenylethylamine derivative on a Hypersil 5 μm silica column. The mobile phase was 5% propan-2-ol-hexane.

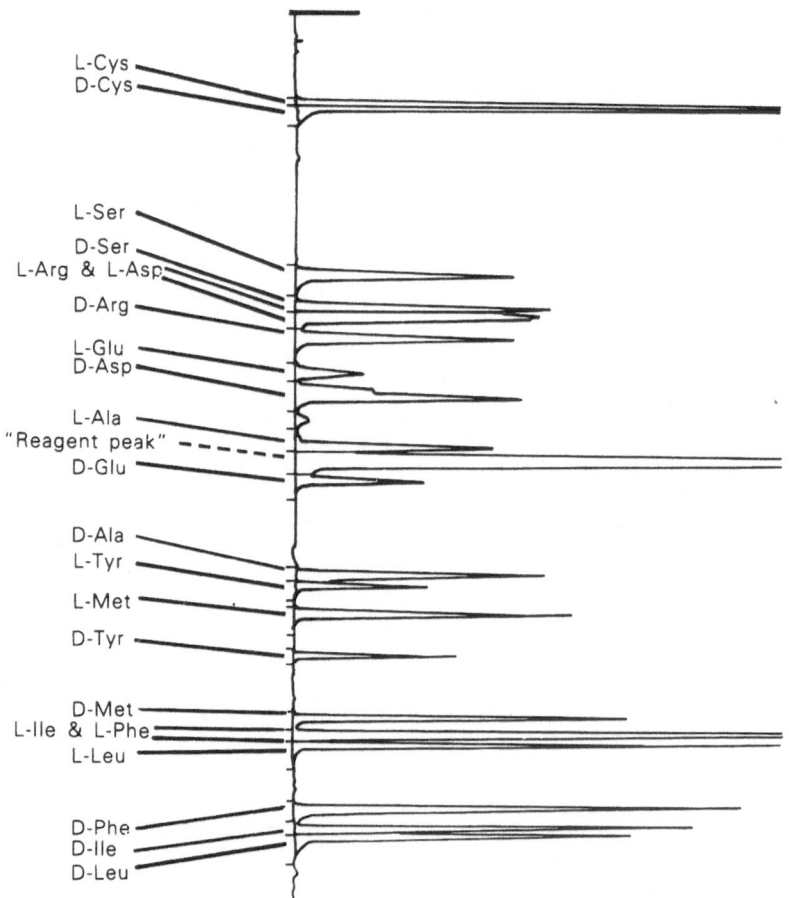

Fig. 2. Chromatographic resolution of 11 racemic amino acids derivatised with Marfey's reagent. Column: Spherisorb ODS-2. Mobile phase: propan-1-ol-acetonitrile-0.05 M ammonium dihydrogen phosphate (pH 3.0), gradient elution from 1.25:10:88.75 to 4:32:64 in 50 min at 40°C.

The antifungal agent (structure 1) was resolved using reversed-phase chromatography by enantioselective inclusion complexation with β-cyclodextrin in the mobile phase at a concentration of 40 mM. Chromatography was performed on a C_{18}-bonded column.

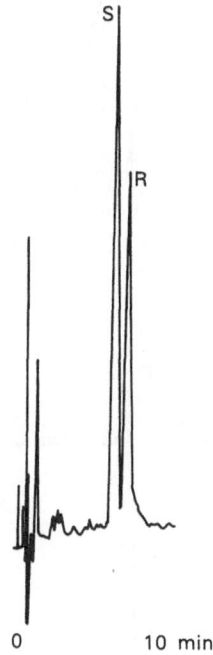

Structure 1

The addition of formamide to the mobile phase appeared to increase the solubility of β-cyclodextrin which has the benefit that a higher concentration of β-cyclodextrin is expected to increase enantioselectivity [2]. The enantioselectivity was also improved by low temperature operation. By using a β-cyclodextrin concentration of 40 mM and by running the column at -5°C the chromatography was improved from around 30% separation to nearly baseline resolution of racemic I (Figure 3). A Violet T-55 heater/cooler (Flowgen Instruments, Maidstone, UK) was used to control the column temperature.

Structure 2

Fig. 3.   Chromatographic resolution of an antifungal agent (structure 1) using a chiral mobile phase (40 mM β-cyclodextrin in acetonitrile-formamide-water, 10:5:85) on Spherisorb ODS-2 at -5°C.

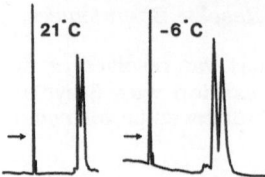

Fig. 4. The effect of temperature on the chromatographic resolution of an antibiotic intermediate (structure 2). A Pirkle-type phenylglycine column was used with a mobile phase of propan-2-ol-dichloromethane-n-hexane (2:15:83).

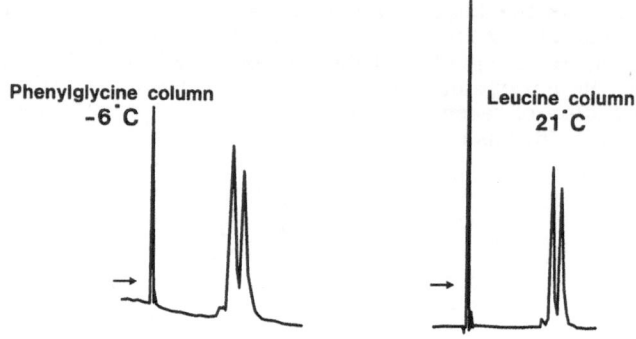

Fig. 5. Comparison of two Pirkle-type columns for the chromatographic resolution of the antibiotic intermediate 2. Mobile phase composition as in Figure 4.

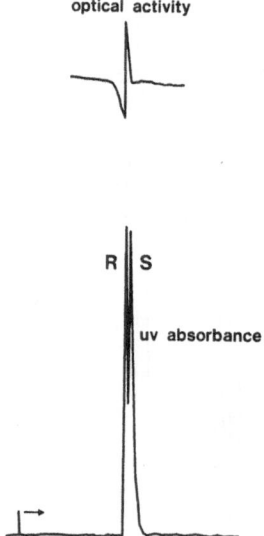

Fig. 6. Use of the ChiraMonitor optical activity detector in the chromatographic resolution of the antibiotic intermediate 2. Conditions as in Figure 5 (leucine column), 200 μg of compound 2 was injected onto the column.

28

The antibiotic intermediate (structure 2) contained the molecular prerequisites for chiral recognition on a Pirkle phase [3,4]. The phthalimide, ester and amide groups would allow hydrogen bonding, dipole stacking and π–π overlap with the stationary phase.

Chromatography on both phenylglycine and leucine Pirkle phases was examined at various temperatures. The phenylglycine Pirkle phase gave better separation at -6°C than at 21°C (Figure 4). The leucine phase offered a further improvement in enantiomer resolution compared to the phenylglycine phase (Figure 5). With the leucine phase resolution was not increased at lower temperature. A ChiraMonitor optical activity detector (ACS Ltd., Macclesfield, UK) showed that the laevo enantiomer was eluted first (Figure 6).

The antiherpetic intermediate (structure 3) contained an amine group adjacent to a bulky ring system, a molecular feature which often allows chiral resolution on an α_1-acid glycoprotein column [5,6]. When the racemate was chromatographed on such a column it was well resolved (Figure 7) with an enantioselectivity factor $\alpha = 2.7$.

Structure 3

A final example is the resolution of a compound related to the prostaglandin PGE_1. The prostaglandin was obtained as a mixture of four stereoisomers (structure 4). Initially, the mixture was separated into two racemates under achiral conditions. Inspired by the resolution of a number of chiral acids on the 3,4-dimethylphenylcarbamate derivatised cellulose (Daicel-OD) by Okamoto et al. [7], this chiral stationary phase was selected to attempt the prostaglandin racemates separation. Baseline resolution of both racemates was possible (Figure 8) with optical activity once again measured using the ChiraMonitor.

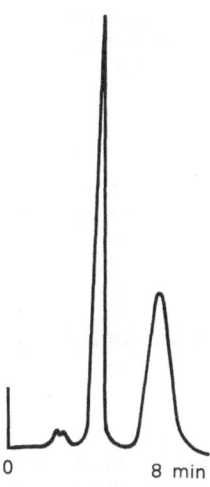

0   8 min

Fig. 7.   Chromatographic resolution of the antiherpetic intermediate compound on EnantioPac (LKB, Bromma, Sweden) with a mobile phase of 0.1 M ammonium dihydrogen phosphate adjusted to pH 7.5 with ammonia.

29

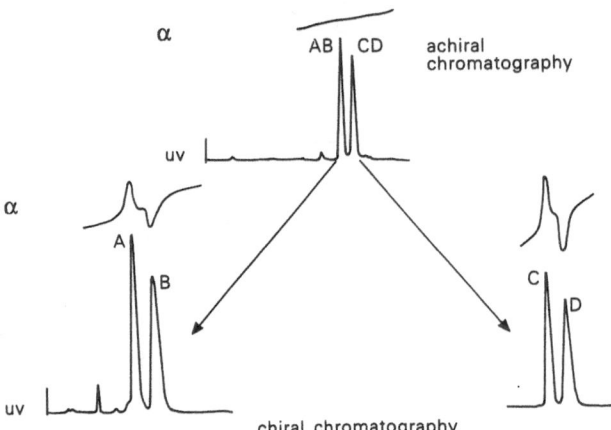

Structure 4. Four stereoisomers of PGE₁.

Fig. 8.   Achiral and chiral chromatographic resolution of the four stereoisomers of a prostaglandin compound. The achiral chromatography was performed on a Supelcosil DIOL column with cyclohexane-dichloromethane-methanol-glacial acetic acid (800:100:10:1) at -6°C. The chiral chromatography used a Chiralcel OD column (Daicel Chemical Industries, Tokyo, Japan) with n-hexane-propan-2-ol-trifluoroacetic acid (900:100:1) at 22°C. Optical activity (α) measured by ChiraMonitor detector.

Micropreparative chiral HPLC was then used to effect the resolution of milligram quantities of each of the four stereoisomers for biological evaluation.

ACKNOWLEDGEMENT

The author wishes to thank Amita Karia for her contribution to this work.

REFERENCES

1.   P. Marfey, *Carlsberg Res. Commun.*, 49:591 (1984).
2.   T. Takeuchi, H. Asai and D. Ishii, *J. Chromatogr.*, 357:409-415 (1986).
3.   W. H. Pirkle and J. M. Finn, *J. Org. Chem.*, 46:2935 -2938(1981).
4.   W. H. Pirkle and C. J. Welch, *J. Org. Chem.*, 49:138-140 (1984).
5.   G. Schill, I. W. Wainer and S. A. Barkan, *J. Liq. Chromatogr.*, 9:641-666 (1986).
6.   G. Schill, I. W. Wainer and S. A. Barkan, *J. Chromatogr.*, 365:73-88 (1986).
7.   Y. Okamoto, R. Aburatani, Y. Kaida and K. Hatada, *Chem. Lett.*, 1125-1128 (1988).

THE DESIGN OF CHIRAL SEPARATIONS FOR β-BLOCKER DRUGS ON PIRKLE
HIGH-PERFORMANCE LIQUID CHROMATOGRAPHY PHASES USING
ACHIRAL DERIVATISATION

A. M. Dyas, M. L. Robinson and A. F. Fell*

International Development Laboratory
Bristol-Myers Squibb
Moreton, Merseyside L46 1QW, UK, and
*Department of Pharmaceutical Chemistry
School of Pharmacy, University of Bradford
Bradford BD7 1DP, UK

SUMMARY

The chromatographic performance of a series of isocyanate and isothiocyanate
derivatives of (+/-)-propranolol on three Pirkle phases was investigated. For each phase the
choice of urea or thiourea derivative had a consistent effect on retention although not on
resolution. The relationship between structure and retention on π-acidic amino acid phases
is explicable although this cannot be extended to separations observed. Inversion of elution
order of the (+)- and (-)-enantiomers was observed for some derivatives on the covalent
1-leucine phase which further hinders the attempt to link structure to resolution for
predictive method development purposes. The α-naphthyl and phenyl derivatives generally
gave good resolution of enantiomers.

INTRODUCTION

The chromatographic separation of enantiomers frequently requires prior
derivatisation of the analyte with either chiral or achiral reagents. Derivatisation with
chiral reagents results in the formation of diastereomeric pairs which are separable on
conventional, achiral chromatographic systems, whilst achiral derivatisation is employed
to mask polar functions which would otherwise affect separation on certain chiral
stationary phases. Pirkle phases, in particular, often require achiral "protective"
derivatisation which can be used to both mask interfering functions and add beneficial
groups to enhance the chiral recognition processes.

In 1988 Yang et al. [1] reported the separation of a range of β-blockers (on a Pirkle
phase) after conversion to their urea derivatives with α-naphthylisocyanate according to
Scheme 1. The reaction proceeds at room temperature in a few minutes and thus offers the
opportunity to systematically evaluate the effect of analyte structure on chiral resolution,
as there is a wide range of commercially available isocyanates. Additionally, the use of
isothiocyanate reagent will yield the corresponding thiourea derivative whose chromatographic performance may be compared to that of the urea analogue. The carbon sulphur
(C=S) double bond has a larger dipole moment than that of the carbonyl bond (C=O) and a
comparison of the chromatographic performances of corresponding urea and thiourea
derivatives will furnish information on the importance of this part of the molecule in the
chiral recognition process.

The chromatographic performances of the various derivatives on different Pirkle
chiral stationary phases (CSPs) can shed light on the separation mechanisms involved and

---

* Author to whom correspondence should be addressed.

Recent Advances in Chiral Separations, Edited by D. Stevenson and
I. D. Wilson, Plenum Press, New York, 1991

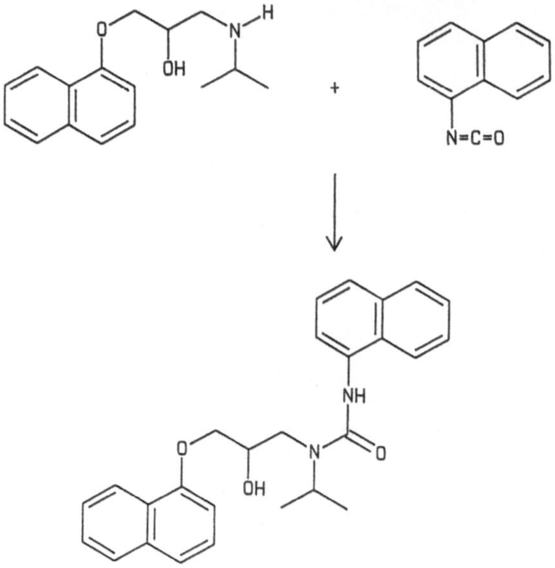

Scheme 1. Derivatisation scheme for propranolol with naphthylisocyanate.

lead to the development of rules for use in the predictive stages of method development. Thus, in this study, the chromatographic performances of a range of urea and thiourea derivatives of the β-blocker drug propranolol were compared on three different Pirkle-type CSPs. The resulting separation data were evaluated in order to try and discern any resulting structure-resolution relationships.

EXPERIMENTAL

Reagents and Materials

Chromatographic and reaction solvents were purchased from BDH (Poole, Dorset, UK) and were high-performance liquid chromatography (HPLC) or analytical grade as appropriate. Racemic (+/-)-propranolol hydrocholoride was obtained from ICI Pharmaceuticals (Macclesfield, Cheshire, UK). (R)-(+)-Propranolol hydrochloride (98%) and (S)-(-)-propranolol hydrochloride (98%) were obtained from Aldrich Chemical Company Ltd. (Poole, Dorset, UK) and (+/-)-, (+)- and (-)-propranolol bases were obtained from the appropriate salt by extraction from an alkaline solution with n-hexane. The following isocyanate and isothiocyanate reagents were obtained from Aldrich Chemical Company Ltd. at better than 97% purity: ethyl-, cyclohexyl-, phenyl-, 4-methoxyphenyl-, 4-nitrophenyl-, 4-fluorophenyl-, naphth-1-yl-.

Equipment

A Hewlett Packard HP1090M liquid chromatograph was used for the separation of the various derivatives. The instrument was fitted with a linear diode-array detector which allowed scans to be taken of the eluting peaks over the range 210-400 nm, assisting in peak identification. The three chiral columns used were as follows:

Column I: 3,5-Dinitrobenzoyl-1-leucine covalently bound to aminopropyl silica 5 μm, 250 mm x 4.6 mm id.

Column II: 3,5-Dinitrobenzoyl-1-phenylglycine ionically bound to aminopropyl silica 5 μm, 100 mm x 4.6 mm id.

Column III: (R)-N-Phenethyl-N-triethoxysilylpropylurea covalently bound to silica 5 μm, 250 mm x 4.6 mm id.

The structures of the phases are given in Figure 1.

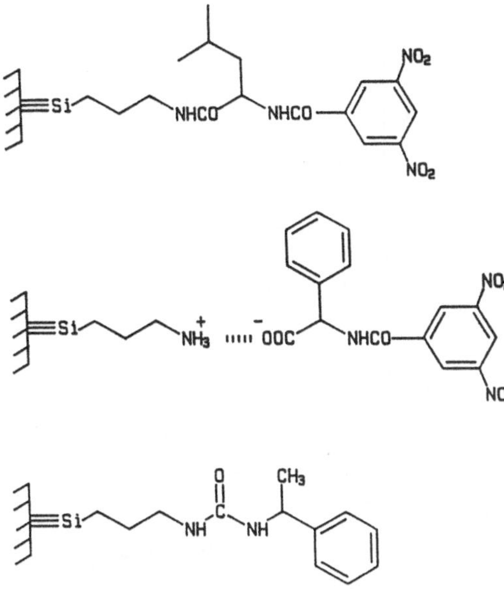

Fig. 1. Structures of Pirkle phases investigated.

Columns I and III were purchased commercially from Hichrom Ltd. (Reading, Berkshire, UK) and Phase Separations Ltd. (Deeside, Clwyd, UK), respectively, whilst Column II was prepared according to the method of Pirkle and Finn [2] and packed in our laboratory. The eluent used in all cases was a mixture of n-hexane-propan-2-ol-acetonitrile pumped at 1 ml min^{-1}. For Columns I and II a 45:5:1 v/v/v mixture was used whilst Column III required a 180:5:1 v/v/v mixture. Samples (20 µl) were injected onto each column and detection was by UV absorbance, the wavelength being optimised for each analyte.

Chromatographic Evaluation

The chromatographic performance of the 14 derivatives was assessed in terms of k', the mean capacity factor observed for the peak(s). Where separation was attained the selectivity factor, α, was calculated along with the peak separation factor Rp which is based on the Kaiser separation factor [3]. Rp is essentially an expression of the depth of the valley between the two peaks as a percentage of the mean peak height (Figure 2).

Experimental Design

Solutions of the 14 derivatives of (+/-)propranolol in n-hexane-propan-2-ol-acetonitrile (45:5:1 v/v/v) were prepared at 8 mg ml^{-1} concentration with separate solutions of the individual (+)- and (-)-enantiomers also being prepared at 10 mg ml^{-1}. All 42 samples were injected onto each of the three columns and the 126 resulting chromatograms evaluated as detailed above.

RESULTS

Chromatographic Performance: Retention

The mean capacity factors for the 14 derivatives on each column are given in Table 1 where it can be seen that there was a difference between the urea and thiourea derivatives. The urea derivatives were more strongly retained than their thiourea counterparts on Column I and Column III whilst the reverse was true on Column II.

Chromatographic Performance: Separation

The selectivity factors observed are given in Table 2 and the corresponding peak separation factors in Table 3. It is clear that selectivity is a relatively insensitive

33

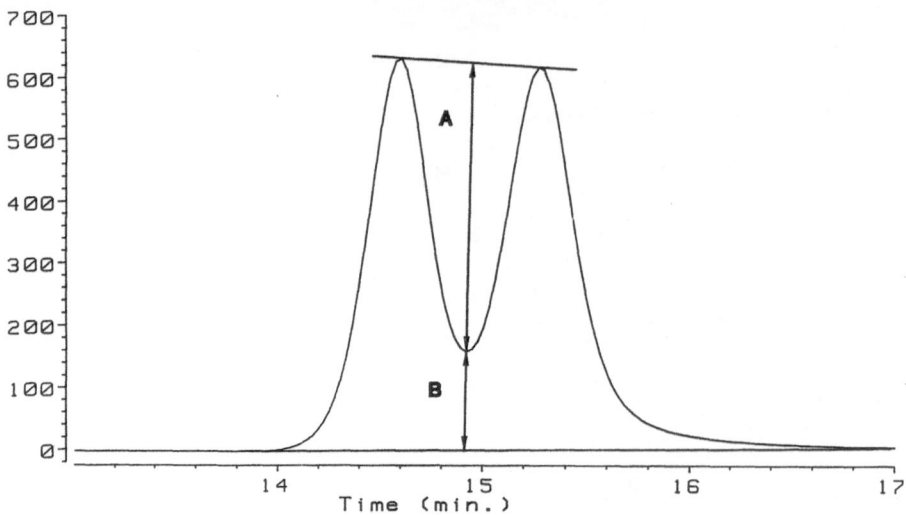

Fig. 2. Calculation of peak separation factor $Rp = 100 \times A/(A + B)\%$.

Table 1. Mean Capacity Factors for Propranolol Derivatives

| Derivative | Chiral Stationary Phase | | | | | |
| | 1-Leucine | | 1-Phenylglycine | | Phenethylpropyl Urea | |
	Oxo	Thio	Oxo	Thio	Oxo	Thio
Ethyl-	3.71*	2.41*	4.26*	7.39	12.49	5.52
Cyclohexyl-	2.23*	1.76*	2.78*	7.14*	7.54	3.84
Phenyl-	7.13	3.13	7.10	9.97	13.34	7.51
4-Methoxyphenyl-	7.46	4.98	12.28	16.50	14.10	10.80
4-Nitrophenyl-	7.56*	6.70	22.85	30.39	16.86	12.85
4-Fluorophenyl-	5.89	2.89	6.99	9.36	10.66	5.11
Naphth-1-yl-	10.43	5.98	15.77	21.42*	14.24	11.96

Oxo = Isocyanate derivative.
Thio = Isothiocyanate derivative.
* No resolution observed.

Table 2. Selectivity Factors for Propranolol Derivatives

| Derivative | Chiral Stationary Phase | | | | | |
| | 1-Leucine | | 1-Phenylglycine | | Phenethylpropyl Urea | |
	Oxo	Thio	Oxo	Thio	Oxo	Thio
Ethyl-	1.00	1.00	1.00	1.08	1.07	1.07
Cyclohexyl-	1.00	1.00	1.00	1.00	1.09	1.03
Phenyl-	1.05	1.06	1.03	1.09	1.14	1.07
4-Methoxyphenyl-	1.02	1.09	1.05	1.09	1.08	1.05
4-Nitrophenyl-	1.00	1.04	1.06	1.12	1.04	1.04
4-Fluorophenyl-	1.04	1.05	1.04	1.09	1.09	1.05
Naphth-1-yl-	1.06	1.05	1.08	1.00	1.13	1.07

Oxo = Isocyanate derivative.
Thio = Isothiocyanate derivative.

34

Table 3. Peak Separation Factors for Propranolol Derivatives

| Derivative | Chiral Stationary Phase | | | | | |
| | 1-Leucine | | 1-Phenylglycine | | Phenethylpropyl Urea | |
	Oxo	Thio	Oxo	Thio	Oxo	Thio
Ethyl-	0	0	0	46	75	19
Cyclohexyl-	0	0	0	0	70	82
Phenyl-	70	69	15	49	100	50
4-Methoxyphenyl-	9	90	15	63	98	8
4-Nitrophenyl-	0	50	31	87	19	30
4-Fluorophenyl-	56	62	7	79	87	31
Naphth-1-yl-	76	69	48	100	100	79

Oxo = Isocyanate derivative.
Thio = Isothiocyanate derivative.

parameter as a wide range in the separation factor (7-100%) related to a small range in the selectivity factor (1.02-1.13).

Enantiomeric Elution Order

The (-)-enantiomer of propranolol preceded the (+)-enantiomer for all derivatives on Columns II and III whilst the opposite was true for the thiourea derivatives on Column I. The urea derivatives showed varied response on Column I with (+)- preceding (-)- only for the methoxyphenylurea and naphthylurea derivatives.

DISCUSSION

The capacity factors given in Table 1 indicate the importance of the choice of derivatives on chromatographic retention. The seven structures eluted in essentially the same order for both urea and thiourea derivatives on both amino acid columns implying a similarity of mechanism. Some attempt at explanation of the elution order may be based on π-electron considerations. The ethyl derivatives lack the delocalised π-electron system found in the aromatic analogues, which is considered an important prerequisite for chiral interaction [4] and hence retention on Pirkle phases, and are thus less retained than their aromatic counterparts. The cyclohexyl derivatives also lack this function and, in addition, may suffer from steric hindrance, reducing their ability to interact with the phase, further reducing their retention. The methoxyphenyl and naphthyl analogues are more π-basic than phenyl and would thus be expected to interact more strongly with the π-acidic phases, increasing their retention, whilst the converse is true for fluorophenyl derivatives which display reduced retention. The nitrophenyl derivatives appear to contradict this model, being strongly retained, π-acidic moieties. However, it may be that the strong dipole of the nitro function enhanced the interaction sufficiently to compensate for any π-electron repulsion which may have occurred.

The retention data for the π-basic phenethylpropylurea phase is not simple to interpret. The derivatives eluted in essentially the same order, with the fluorophenyl structures suffering a relative reduction in retention, and eluting before the ethyl derivatives. The acidic analytes might have been expected to be more strongly retained on this basic phase with the basic analytes being less strongly bound, but this was not the case. This demonstrates the complexity of the chromatographic process which cannot simply be classified under a single type of interaction.

The similarity between CSP I and CSP III extended to the relative retentions of the urea and thiourea derivatives, with the former being more strongly retained on these phases. In contrast, the thiourea derivatives were more strongly retained on CSP II. This significant difference may be attributable to the mode of linkage of phase to support, with CSPs I and III being covalently bound whilst CSP II is an ionic phase.

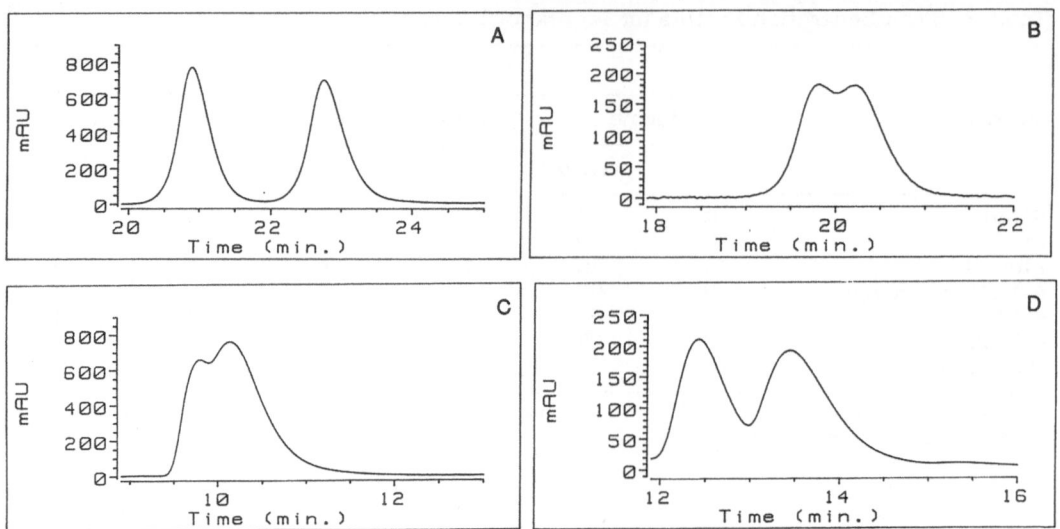

Fig. 3. Comparison of separation differences between urea and thiourea derivatives of propranolol: (A) methoxyphenylurea derivative on phenethylpropylurea phase; (B) methoxy-phenylthiourea derivative on phenethylpropylurea phase; (C) fluorophenylurea derivative on phenylglycine phase; and (D) fluorophenylthiourea derivative on phenylglycine phase.

Whilst the relationship between structure and retention may be postulated, the relationship between structure and resolution is not clear, this being evident from the data in Tables 2 and 3. The poor separation of the alkyl derivatives on amino acid phases is in contrast to that observed with the phenethylpropylurea phase, indicating a possible difference in the chiral recognition mechanisms between these two types of phase. On all three phases there appeared to be little correlation between retention and resolution with instances of short retention/high resolution (e.g., cyclohexylurea on Column III) and long retention/low resolution (e..g, methoxyphenylurea on Column III). The ability to predict retention from structure is of little value if it cannot be extended to separation.

A partial explanation of the poor correlation may lie in an inversion in the elution order of the enantiomers on CSP I for some derivatives. On Columns II and III the (-)-enantiomer always preceded the (+)-enantiomer whilst the reverse order was observed for thiourea derivatives on Column I. There was a gradual change in order for the urea derivatives on Column I, with (-) eluting first for the fluorophenyl and phenyl derivatives, coelution of the nitrophenyl, (+) just preceding (-) for methoxyphenyl and (+) clearly preceding (-) for the naphthyl derivatives. The inability to explain or predict this inversion is a major problem in the development of separation "rules".

The effect of selection of isocyanate or isothiocyanate as derivatising reagent on retention has been clearly shown but the effect is less uniform (and therefore less predictable) on separation. On CSP II, the phase where it resulted in increased retention, it always lead to improved resolution and this may be significant. On the two other phases it improved separation in five out of 14 instances. Thus, apart from CSP II, it is not possible to rationally select which type of reagent to use; the result of a "wrong" selection can clearly be seen in Figure 3.

CONCLUSION

The effect of structure on retention of the various derivatives studied here may be partially explained for the amino acid phases but the model proposed does not fit the π-basic phenethylpropylurea phase. Additionally the poor relationship between retention and peak separation for the data as a whole renders it difficult to predict resolution based on structure, which is the goal of method development. The inversion of elution order of the enantiomers on certain phases may be observed and, in general, the presence of naphthyl or phenyl moieties in the derivative confers a structure which enhances the chiral recognition process. The attempt to discern mechanistic information from structure/resolution data suffers from the drawback that such data is the result of complex, composite interactions and thus interpretation is fraught with difficulty. The procedure does have value however as an empirical, method development exercise which is easily carried out.

REFERENCES

1. Q. Yang, Z. Sun and D. Ling, *J. Chromatogr.*, 447:208-211 (1988).
2. W. H. Pirkle and J. M. Finn, *J. Org. Chem.*, 46:2935-2938 (1981).
3. R. E. Kaiser, "Gas Chromatographie", Geest and Portig, Leipzig (1960).
4. W. H. Pirkle, *in:* "Chromatography and Separation Chemistry: Advances and Developments", S. Ahujer, ed., American Chemical Society (1986).

A NOTE ON THE USE OF RECIPROCAL INTERACTION IN THE DESIGN OF PYRETHROID SPECIFIC CHIRAL STATIONARY PHASES

D. R. Taylor and E. Clift

Chemistry Department
UMIST, PO Box 88
Manchester M60 1QD, UK

INTRODUCTION

We have attempted to design chiral stationary phases (CSPs) to resolve all the stereoisomers of the insecticide Cypermethrin (Structure **1**) using the principle of reciprocal interaction. In order to do this Cypermethrin has been structurally modified to enable it to be bonded to silica, and the resulting CSP will be used to identify and optimise the interactions of racemic structures from which a single enantiomer will be bound to silica to afford a Cypermethrin-specific CSP.

The principle of reciprocal interaction [1] is as follows. If a CSP based on a pure enantiomer (+)-A resolves a racemate (+/-)-B, then a single pure enantiomer (+)- or (-)-B should resolve (+/-)-A. This is true if no major point of interaction is changed when attaching B to silica.

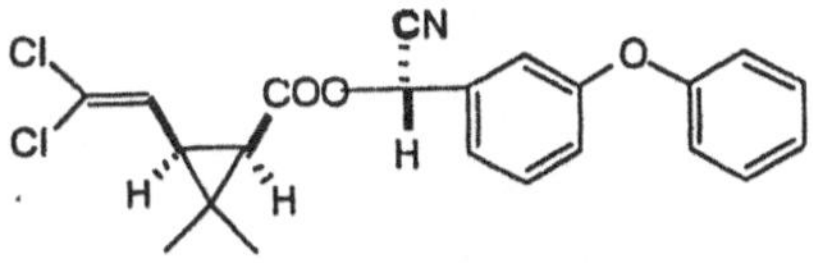

Structure **1**

MATERIALS AND METHODS

Cypermethrin is a highly active pyrethroid insecticide, the most biologically active isomer being the 1R *cis* S isomer. To enable its attachment to silica the structure of Cypermethrin had to be modified. We decided to incorporate an allylic group which would enable us to achieve either a free radical addition or a hydrosilylation across the double bond.

The aldehydes **2** and **3** were synthesised as shown in Reaction Schemes 1(a) and 1(b).

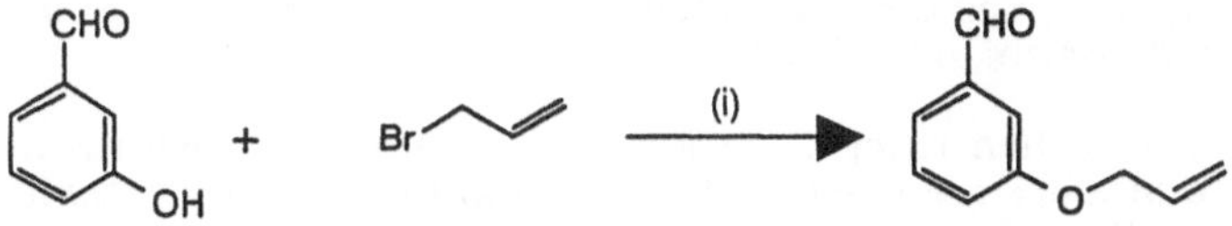

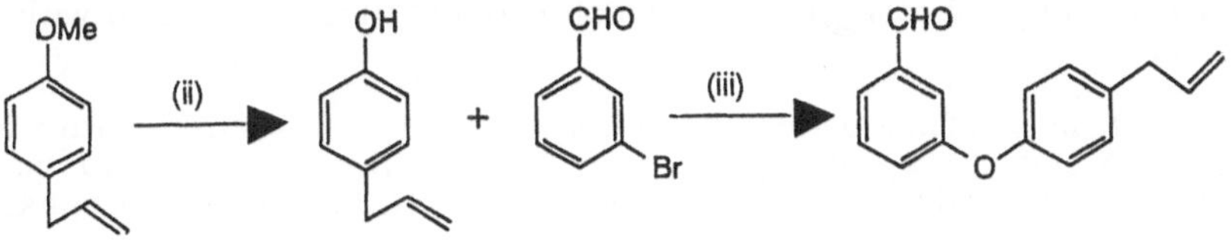

(i) K_2CO_3, MeOH, reflux.
(ii) BBr_3, (iii) $KOBu^t$, CuCl, pyridine, toluene, reflux.

Recent Advances in Chiral Separations, Edited by D. Stevenson and
I. D. Wilson, Plenum Press, New York, 1991

Aldehyde **2** was also modified compared to Cypermethrin to the extent that an aromatic ring was omitted.

Aldehydes **2** and **3** were then reacted, as shown in Reaction Scheme 2, with the chrysanthemic acid analogue in the presence of cyanide to yield the modified pyrethroids **5** and **6** as the 1R *cis* R/S mixtures. The diastereoisomers were then separated by preparative high-performance liquid chromatography (HPLC) using 5% tetrahydrofuran in n-hexane on silica gel.

Reaction Scheme 2

RESULTS AND DISCUSSION

It was important, for mechanistic studies, to know the absolute configuration of the isolated diastereoisomers at the benzylic carbon. A combination of [^{1}H]-NMR chemical shifts and the sign of the circular dichroism (CD) was used to obtain this information where the known stereoisomers of Cypermethrin were compared to the isolated isomers of **5** and **6**. Thus, comparing the signals of the benzylic proton, the chemical shifts of the second isomer eluted in the analyses of Cypermethrin and the modified pyrethroids were slightly downfield in those of the first eluted isomers (0.02 ppm). The first eluted isomers of Cypermethrin and the modified pyrethroids had negative CD curves over the range 210 to 250 nm, whilst the second eluted isomers had positive CD curves over the same range. Knowing that the first eluted isomer of Cypermethrin is 1R *cis* R and the second eluted isomer is 1R *cis* S, we assigned the first eluted isomers of compounds **5** and **6** to be 1R *cis* R and the second eluted isomers to be 1R *cis* S.

Compound **5** was then reacted via a free radical addition with 3-mercaptopropyl-trimethoxysilane to produce a long-chain trifunctional unit for reaction with silica to form a CSP. Compound **6** was converted via a hydrosilylation reaction with dimethyl-chlorosilane to yield a short-chain monofunctional unit for reaction with silica to provide a further CSP.

The reactive silane derivatives of **5** and **6** were then refluxed with silica in toluene to yield CSP **1** and CSP **2**, respectively (Figure 1). We then proceeded to use these phases to investigate reciprocal interaction as a means of designing a stationary phase.

As Cypermethrin can be resolved on a 3,5-DNB phenylglycine CSP, we attempted to demonstrate reciprocal interaction of our Cypermethrin CSPs with (+/-)-3,5-DNB phenyl-glycinepropylamide. This was successful and an example of the separation achieved on CSP **2** is shown in Figure 2 with another racemic analyte.

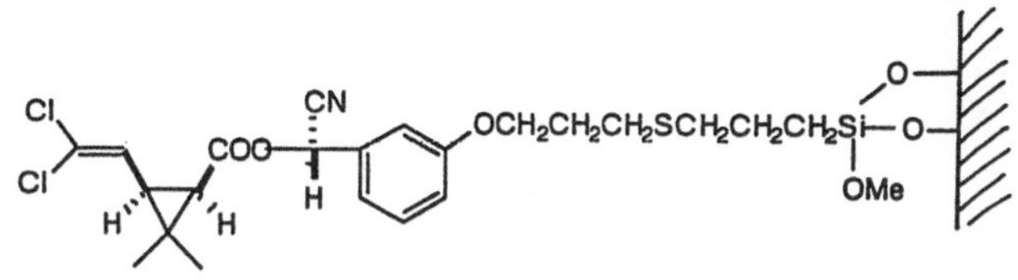

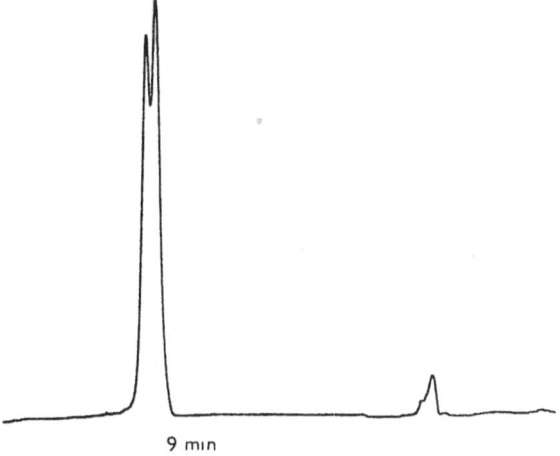

CSP 1

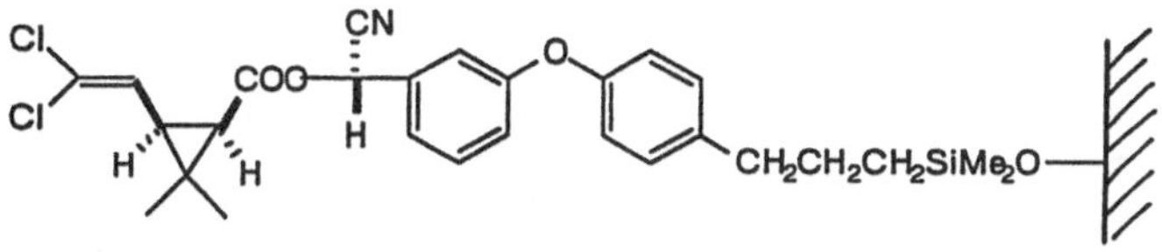

CSP 2

Fig. 1. Structures of CSP **1** and CSP **2.**

9 min

Fig. 2. CSP **2**, 18% tetrahydrofuran in hexane, flow rate 1.0 ml min^{-1}.

Future studies with these phases are planned and the mechanism of chiral recognition will be further studied with the aid of molecular graphics.

ACKNOWLEDGEMENTS

We would like to thank Shell Research Ltd. for financial support, and also Dr. Alexis Roberts-McIntosh and Mr. Edwin Cole for their support and encouragement throughout this project.

REFERENCE

1.   W. H. Pirkle, D. W. House and J. M. Finn, *J. Chromatogr.*, 192:143-158 (1980).

STEREOCHEMICAL ANALYSIS OF CHIRAL ALCOHOL, EPOXIDE AND SULPHOXIDE METABOLITES BY CHIRAL STATIONARY PHASE HIGH-PERFORMANCE LIQUID CHROMATOGRAPHY

D. R. Boyd* and R. A. S. McMordie

School of Chemistry
The Queen's University of Belfast
Belfast BT9 5AG, UK

SUMMARY

Several commercially available chiral stationary phase high-performance liquid chromatography (CSP-HPLC) columns have been used in the analysis of enantiomers resulting from enzyme-catalysed oxidations at carbon and sulphur atoms in cyclic systems.

Monohydroxylation reactions have been studied at both benzylic and allylic methylene groups. The metabolism of arenes, dihydroarenes and tetrahydroarenes in fungi, bacteria and animal liver systems have been investigated by the CSP-HPLC method. This procedure was particularly useful in the stereochemical analysis of arene hydrates, a new type of unstable metabolite, formed from dihydroarenes by allylic or benzylic hydroxylation.

A Chiralcel OB (Daicel Industries) column has been found to be successful in the resolution of non-K-region arene oxides and thus in the estimation of enzymatic stereopreference for one prochiral face during the metabolism of polycyclic aromatic hydrocarbons (PAHs) by hepatic monooxygenase enzymes.

The determination of optical purity of *cis*-dihydrodiols produced by bacterial metabolism of PAHs proved difficult by direct CSP-HPLC analysis. An indirect method involving hydrogenation, cyclic carbonate formation followed by CSP-HPLC analysis was more successful.

The ability of chiral sulphoxidising agents (chemical and enzymatic) to stereo-differentiate between prochiral lone pairs on a sulphur atom and prochiral sulphur atoms on a carbon atom has been investigated using a single CSP-HPLC system to separate both *cis*- and *trans*-sulphoxide isomers and individual sulphoxide enantiomers.

INTRODUCTION

The metabolic reactions of living cells are diverse and many of these steps may introduce new chiral centers. Here we describe the applications of CSP-HPLC systems in the study of selected oxidative metabolic steps where chirality has been introduced at one or two centers.

Enzyme-catalysed oxygen-atom-transfer reactions are commonly found in xenobiotic metabolism by both animals, plants and microorganisms. Introduction of a single oxygen atom (e.g., aliphatic and aromatic hydroxylation, heteroatom oxidation, epoxidation) or two oxygen atoms (e.g., *cis*- and *trans*-dihydroxylation) can thus occur during detoxification

---

*Author to whom correspondence should be addressed.

Recent Advances in Chiral Separations, Edited by D. Stevenson and
I. D. Wilson, Plenum Press, New York, 1991

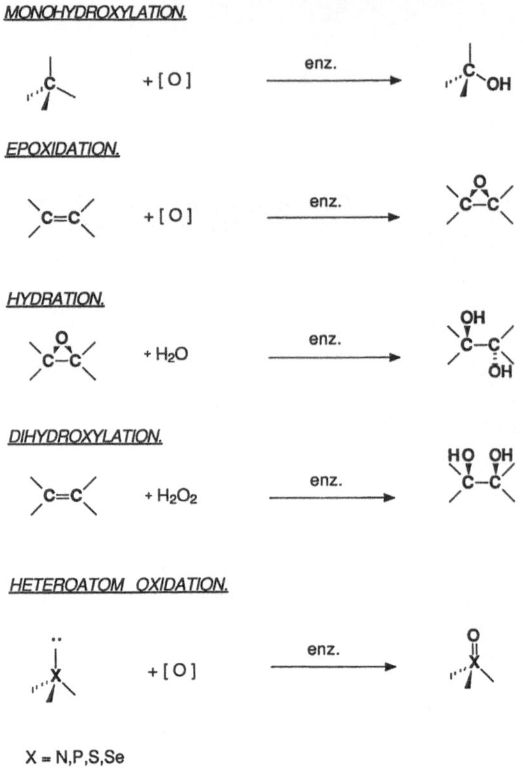

MONOHYDROXYLATION.

+ [O]   →ᵉⁿᶻ·   (product)

EPOXIDATION.

>C=C<   + [O]   →ᵉⁿᶻ·   (epoxide)

HYDRATION.

(epoxide)   + H_2O →ᵉⁿᶻ·   (diol)

DIHYDROXYLATION.

>C=C<   + H_2O_2 →ᵉⁿᶻ·   (diol)

HETEROATOM OXIDATION.

X   + [O]   →ᵉⁿᶻ·   (oxide)

X = N,P,S,Se

Fig. 1. Enzyme-catalysed oxygen-atom-transfer reactions in living systems.

by an organism to produce more water-soluble metabolic products. The stereoselectivity of alcohol, epoxide, cis-diol and sulphoxide metabolite formation (Figure 1) by CSP-HPLC analysis is discussed with examples from our current studies.

Monohydroxylation of Methylene Groups in Cyclic Molecules

Monohydroxylation at the benzylic position of tetralin has been observed in both fungi (e.g., *Mortierella isabellina*) and bacteria (e.g., *Pseudomonas putida*) (Scheme 1). These α-tetralol enantiomers can be separated by HPLC on a Chiralcel OB column (10% isopropyl alcohol in hexane), with an α value of 2.78 as shown in Figure 2. Optically pure [1R]-α-tetralol was obtained using *P. putida* [1] while *M. isabellina* has been reported to give a lower optical yield (33%) [2] (see Table 1).

β-Tetralol was not isolated as a metabolite in either of the microbial systems studied. This was fortunate since no satisfactory separation of β-tetralol enantiomers could be observed with the range of CSP-HPLC columns used in this study. β-Tetralol enantiomers have recently been resolved by CSP-GLC analysis of the trifluoroacetate derivatives using a 10 m dipentyl-β-cyclodextrin column (D. W. Armstrong, personal communication).

Scheme 1

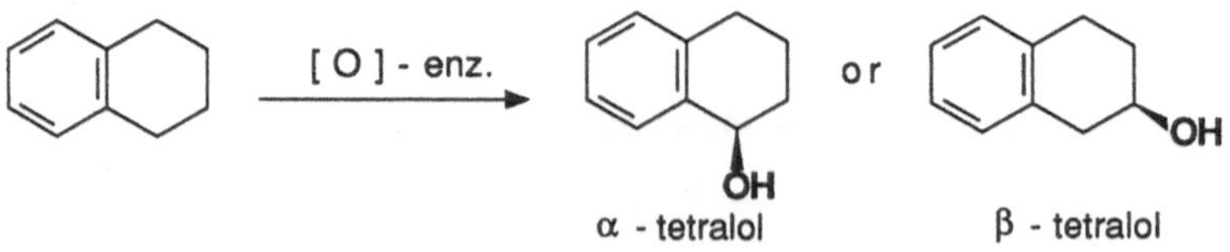

(tetralin) $\xrightarrow{\text{[O] - enz.}}$ (α-tetralol)   or   (β-tetralol)

α - tetralol   β - tetralol

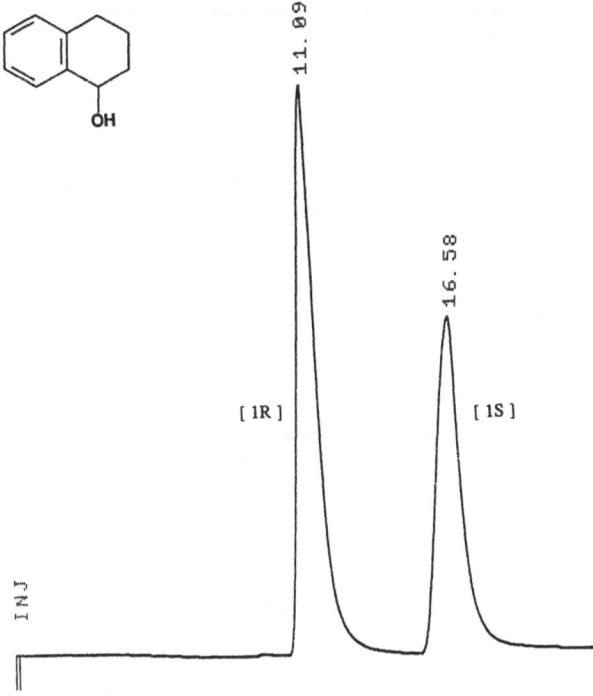

Fig. 2. Separation of [1R] and [1S] enantiomers of α-tetralol on a Chiralcel OB column. Chromatographic conditions were 10% isopropyl alcohol in hexane as eluent, and a flow rate of 0.5 ml min⁻¹. An α value of 2.78 was achieved.

Table 1. Relative Yields and Optical Yields of α- and β-Tetralol Obtained by Microbial Oxidation

Microbial Species	α-Tetralol Relative Yield (% ee; configurations)	β-Tetralol Relative Yield (% ee; configurations)
Fungi (*Mortierella isabellina*)	100 (33: R)	0
Bacteria (*Pseudomonas putida*)	90 (100:R)*	0

* 10% of α-tetralone also formed.

Scheme 2

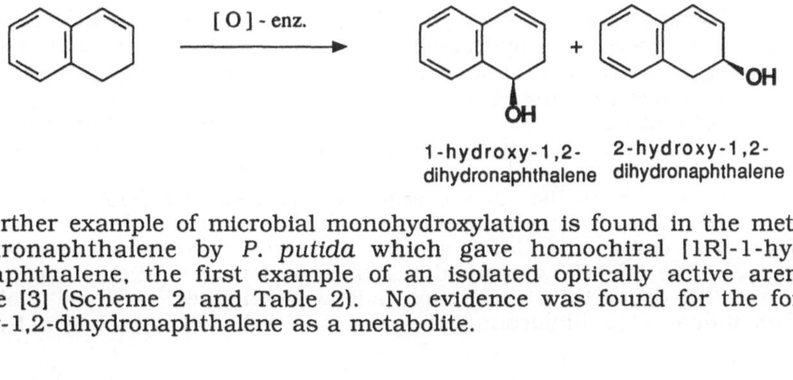

1-hydroxy-1,2-dihydronaphthalene   2-hydroxy-1,2-dihydronaphthalene

A further example of microbial monohydroxylation is found in the metabolism of 1,2-dihydronaphthalene by *P. putida* which gave homochiral [1R]-1-hydroxy-1,2-dihydronaphthalene, the first example of an isolated optically active arene hydrate metabolite [3] (Scheme 2 and Table 2). No evidence was found for the formation of 2-hydroxy-1,2-dihydronaphthalene as a metabolite.

45

Table 2. Relative Yields, Optical Yields and Absolute Configurations of 1- and 2-Hydroxy-1,2- dihydronaphthalene

Enzyme System	1-Hydroxy-1,2-dihydronaphthalene Relative Yield (% ee; configuration)	2-Hydroxy-1,2-dihydronaphthalene Relative Yield (% ee; configuration)
Bacteria (*Pseudomonas putida*)	100 (> 98: 1R)	0
Liver microsomes (3-methylcholanthrene induced)	28 (70: 1S)	72 (27:2R)

Table 3. Relative Yields, Optical Yields and Absolute Configurations of 2-Hydroxy-1,2-dihydronaphthalene and 1-Hydroxy-1,4-dihydronaphthalene

Enzyme System	2-Hydroxy-1,2-dihydronaphthalene Relative Yield (% ee; configuration)	1-Hydroxy-1,4-dihydronaphthalene Relative Yield (% ee; configuration)
Bacteria (*Pseudomonas putida*)	8 (2: 2S)	92 (> 95: 1R)
Liver microsomes (3-methylcholanthrene induced)	21 (*)	79 (*)

* Not yet determined.

Scheme 3

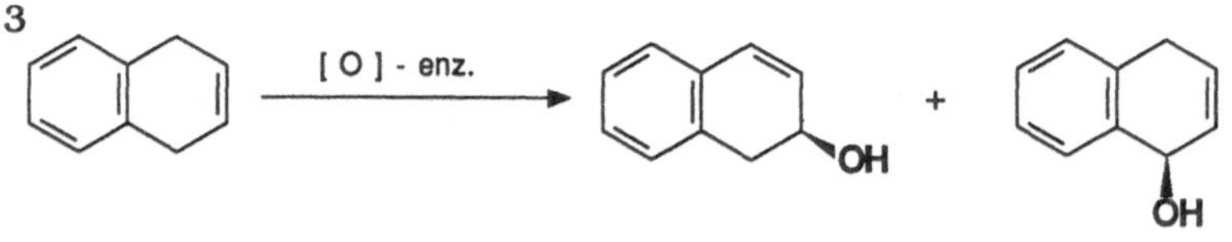

2-hydroxy-1,2-dihydronaphthalene   1-hydroxy-1,4-dihydronaphthalene

The detection and CSP-HPLC analysis of 1- and 2-hydroxy-1,2-dihydronaphthalene has been made possible by the development of new synthetic routes to both racemic and homochiral samples of these arene hydrates [1]. Using a Chiralcel OB column and 10% isopropyl alcohol in hexane as eluent, a base line separation of 1-hydroxy-1,2-dihydronaphthalene enantiomers ($\alpha = 1.32$) was observed (Figure 3). Metabolism of 1,4-dihydronaphthalene by *P. putida* gave 2-hydroxy-1,2-dihydronaphthalene (8%) and 1-hydroxy-1,4-dihydronaphthalene (92%) (Scheme 3). While 2-hydroxy-1,2-dihydronaphthalene was almost racemic (2% ee, 2S) the 1-hydroxy-1,4-dihydronaphthalene was essentially homochiral (> 95% ee, 1R) (Table 3).

Chiral HPLC analysis (Chiralcel OB, 10% isopropyl alcohol in hexane) clearly resolved the enantiomers of 1-hydroxy-1,4-dihydronaphthalene ($\alpha = 3.3$) whereas 2-hydroxy-1,2-dihydronaphthalene showed a partial resolution into enantiomers ($\alpha = 1.1$, Figures 3 and 4). When the latter CSP-HPLC analysis was recently repeated with a new Chiralcel OB column, a better separation of 2-hydroxy-1,2-dihydronaphthalene enantiomers was observed which gave a more accurate estimate of the percent ee value.

Liver microsomal metabolism of 1,2- and 1,4-dihydronaphthalene (Schemes 2 and 3) gave two types of arene hydrate from each dihydroarene substrate [4]. The quantities obtained from liver enzyme oxidation were much lower than from bacterial oxidation (Tables 2 and 3). The CSP-HPLC method was used to determine the optical purity of the 1-hydroxy- and 2-hydroxy-1,2-dihydronaphthalene metabolites. The metabolism of the cyclic olefins indene, 1,2-dihydronaphthalene, and 1,2-benzocyclohepta-1,3-diene by

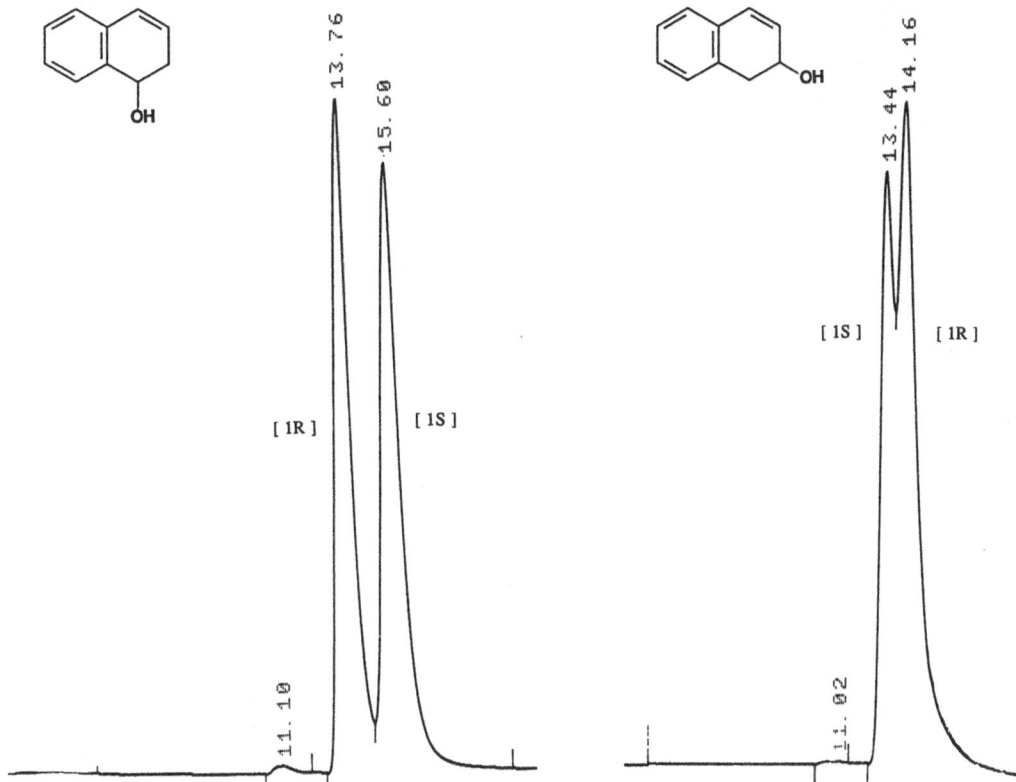

Fig. 3. Separation of [1R] and [1S] enantiomers of 1-hydroxy-1,2-dihydronaphthalene and 2-hydroxy-1,2-dihydronaphthalene on a Chiralcel OB column. Chromatographic conditions were 10% isopropyl alcohol in hexane as eluent at a flow rate of 0.5 ml min^{-1}. Under these conditions α values of 1.32 and 1.11, respectively, were obtained.

Scheme 4

> 98% ee   > 98% ee   > 98% ee

P. putida gave the corresponding benzylic alcohol products exclusively as the [R] enantiomers [3,5] (Scheme 4). Increasing the cyclic olefin ring size from five to seven carbons led to a corresponding decrease in separation factor, α (3.00 →1.32 →1.19) using the Chiralcel OB-CSP column (Figure 5).

Arene Oxide and cis-*Dihydrodiol Metabolite Formation from Arenes and Azaarenes*

The metabolism of polycyclic aromatic hydrocarbons by animal liver systems is known to proceed via an initially formed arene oxide to yield *trans*-dihydrodiols, and glutathione conjugates (Scheme 5) [6]. Non-K-region arene oxide metabolites cannot normally be detected by direct HPLC analysis due to low concentrations resulting from instability and competing metabolic processes and due to decomposition occurring on the column. Recently, it has proved possible to detect the 5,6-arene oxide metabolite of quinoline by having a epoxide-hydrolase inhibitor (3,3,3-trichloropropene-1,2-oxide) present [7] (Figure 6). Making use of the chiral HPLC method (Chiralcel OB, 50% isopropyl alcohol in hexane) it was found that the arene oxide metabolites of naphthalene

47

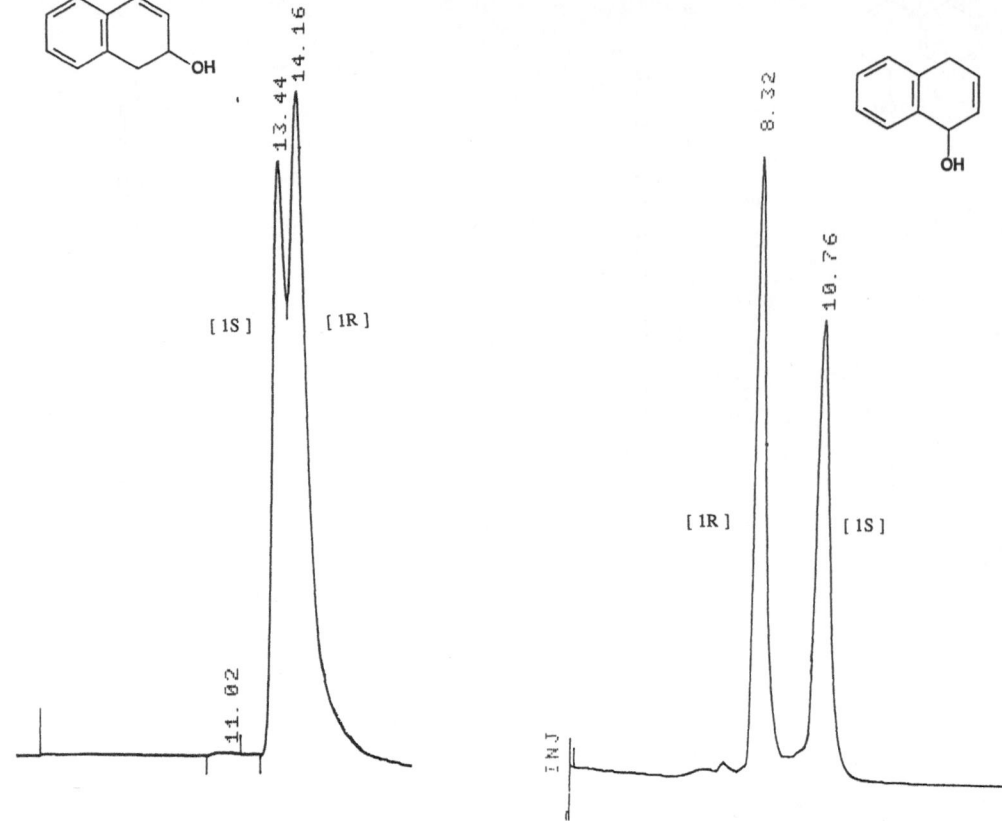

Fig. 4.   Separation of [1R] and [1S] enantiomers of 2-hydroxy-1,2-dihydronaphthalene and 1-hydroxy-1,4-dihydronaphthalene on a Chiralcel OB column. Chromatographic conditions were 10% isopropyl alcohol in hexane as eluent at a flow rate of 0.5 ml min^{-1}. Under these conditions α values of 1.11 and 3.3, respectively, were obtained.

(naphthalene 1,2-oxide, α = 2.0) and of quinoline (quinoline 5,6-oxide, α = 2.2) were both sufficiently stable under the chromatographic conditions employed to be separated into enantiomers. Thus, by making use of selective enzyme inhibition and HPLC on this chiral stationary phase, direct stereochemical analysis of quinoline epoxidation by liver enzymes has been successfully carried out [8].

Scheme 5

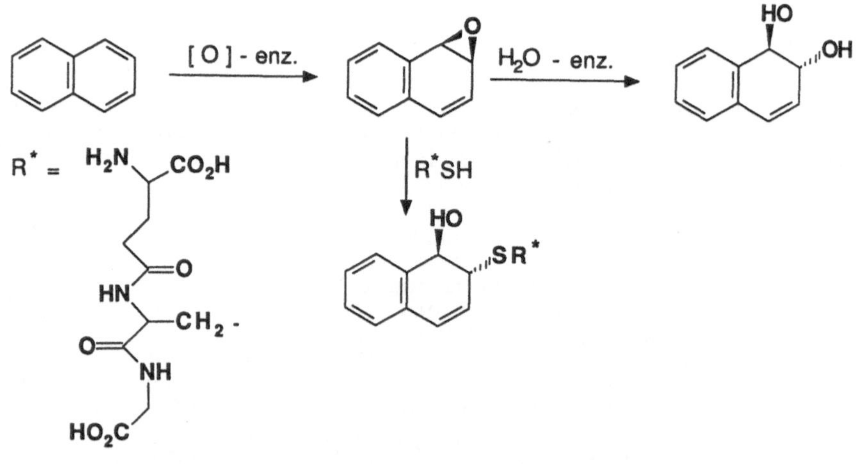

$R^* = $

(glutathione)

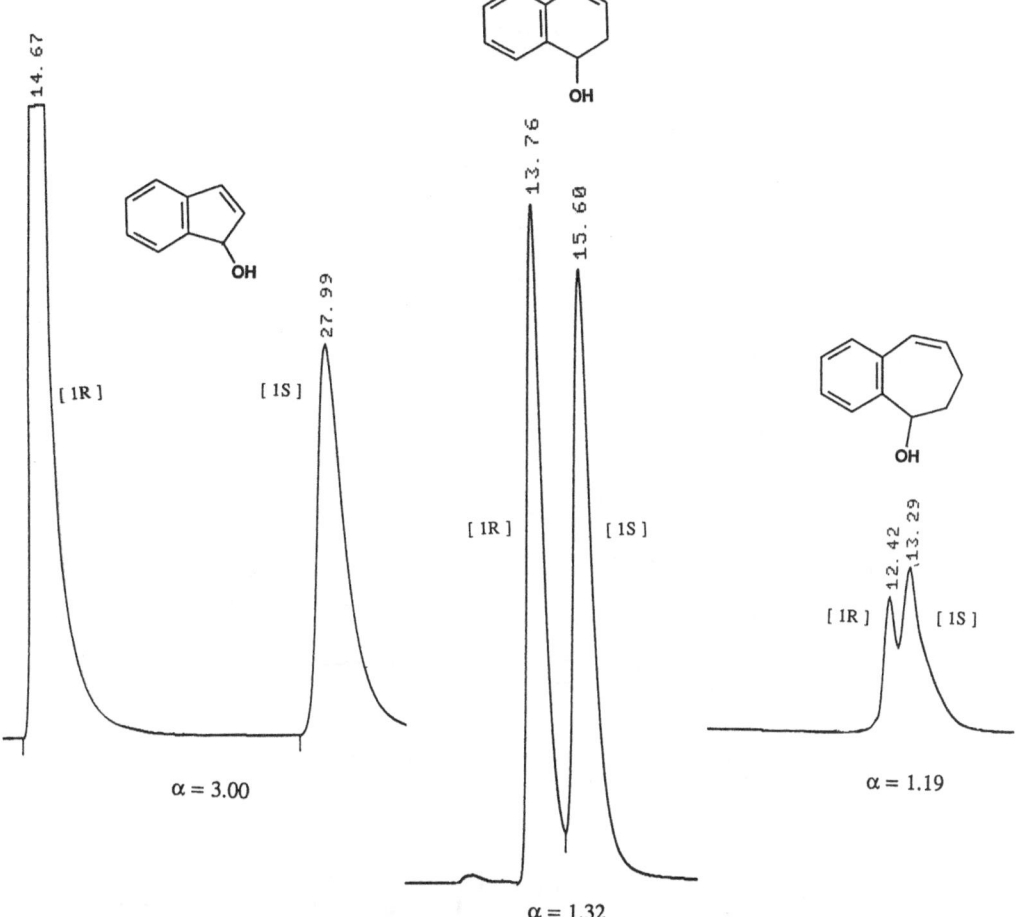

Fig. 5.   Separation of [1R] and [1S] enantiomers of 1-hydroxyindene, 1-hydroxy-1,2-dihydronphthalene and 3-hydroxy-1,2-benzocycloheptene on a Chiralcel OB column. Chromatographic conditions were 10% isopropyl alcohol in hexane as eluent at a flow rate of 0.5 ml min⁻¹. Under these conditions α values of 3.00, 1.32 and 1.19, respectively, were achieved.

Arene oxides are the initially formed metabolites of PAHs in eucaryotic organisms (e.g., animals and fungi). By contrast the metabolism of PAHs in procaryotic organisms (e.g., bacteria) has been found to yield cis-dihydrodiols as initial metabolites [9]. A wide range of cis-dihydrodiol metabolites from mono-, di-, tri-, tetra- and pentacyclic arenes has been isolated using mutant strains of bacteria (e.g., P. putida and a Beijerinckia species). However, the optical purity has to date only been obtained in a small number of examples.

From the total range of available Daicel columns only the Chiralcel OC (4% isopropylalcohol in hexane) column was found to give any separation of cis-1,2-dihydroxy-1,2-dihydronaphthalene enantiomers (Figure 7). The optimal separation of this column was however unsatisfactory (α = 1.11) for accurate ee analysis. Earlier CSP-HPLC separations (Figure 3) showed that benzylic alcohol enantiomers separated much more readily than allylic alcohols. The cis-dihydrodiol enantiomers (and the corresponding trans-dihydrodiol enantiomers) are thus closer in chromatographic characteristics to the allylic alcohols by CSP-HPLC analysis.

Catalytic hydrogenation of the cis-dihydrodiol bacterial metabolites of naphthalene (cis-1,2-dihydroxy-1,2-dihydronaphthalene) and quinoline (cis-5,6-dihydroxy-5,6-dihydroquinoline) to yield the corresponding tetrahydrodiols followed by reaction with 1,1'-carbonyldiimidazole gave the cyclic carbonate derivatives. The latter compounds could

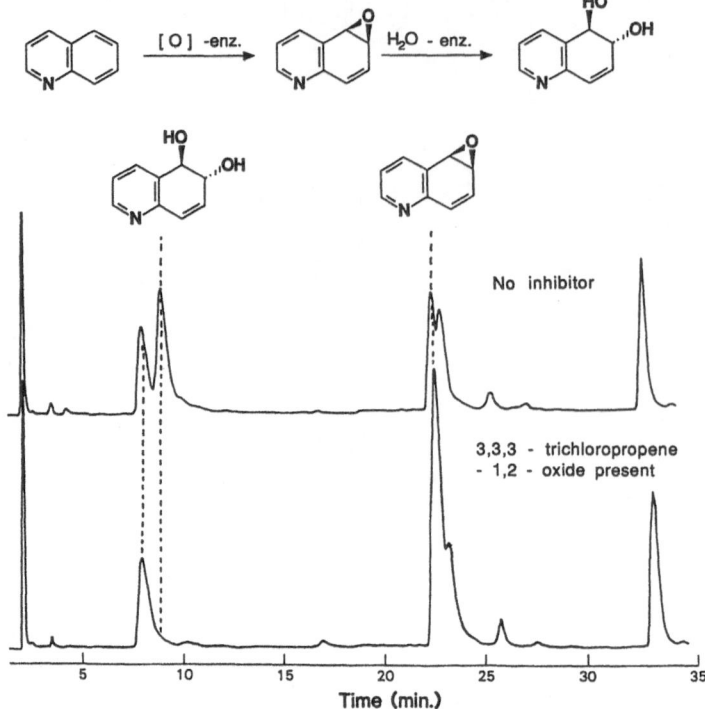

Fig. 6.   Achiral HPLC separation of *trans*-5,6-dihydroxy-5,6-dihydrodiol and quinoline-5,6-oxide metabolites of quinoline using liver microsomes on a Vydac ODS column.

Scheme 6

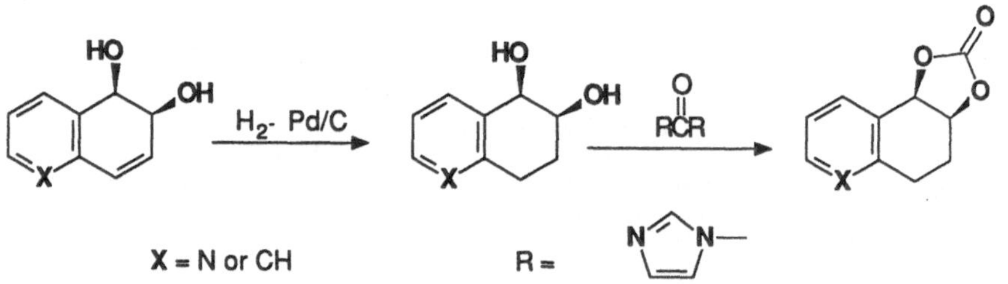

X = N or CH   R =

be readily separated using a Chiralcel OB column (50% isopropyl alcohol in hexane) with α values of 1.6 and 1.3 for the naphthalene and quinoline derivatives, respectively (Scheme 6 and Figure 8, respectively).

Monosulphoxidation of 1,3-Benzodithioles

The final aspect of the oxidative metabolism study concerns the oxidation of pro-R and pro-S lone pairs on a sulphur atom to yield chiral sulphoxides and the oxidation of pro-R and pro-S sulphur atoms on a carbon atom to again yield chiral sulphoxides (Scheme 7) [1,10,11].

The 1,3-benzodithiole system has been used to study this concept since both the sulphoxide enantiomers and diastereoisomers can be separated by HPLC on a CSP column. 1,3-Benzodithiole-1-oxide showed a clear separation into enantiomers using a Chiralcel OB column (50% isopropyl alcohol in hexane, α = 1.38), (Scheme 8, Figure 9) both on analytical (4.6 mm x 25 cm) and semipreparative (9.2 mm x 25 cm) columns.

The production of homochiral samples of the 1,3-benzodithiole-1-oxides either by chiral chemical oxidants or by microbial biotransformation methods had not previously been achieved. Thus, using 1,3-benzodithiole as substrate, the fungus *Aspergillus foetidus* and the bacterium *P. putida* gave 1,3-benzodithiole-1-oxide in 0% ee and 30% ee,

Scheme 7 PROCHIRAL LONE PAIRS

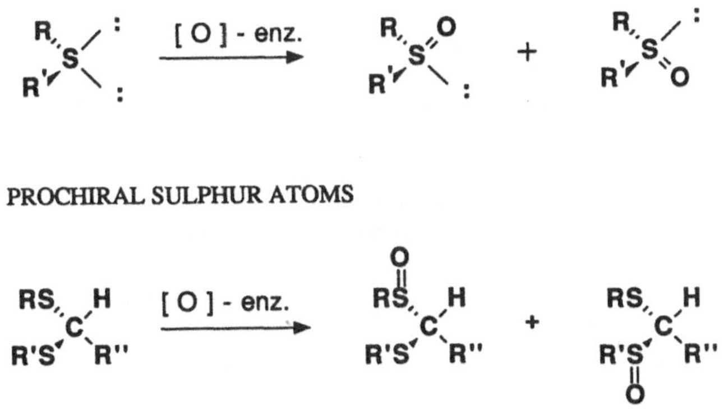

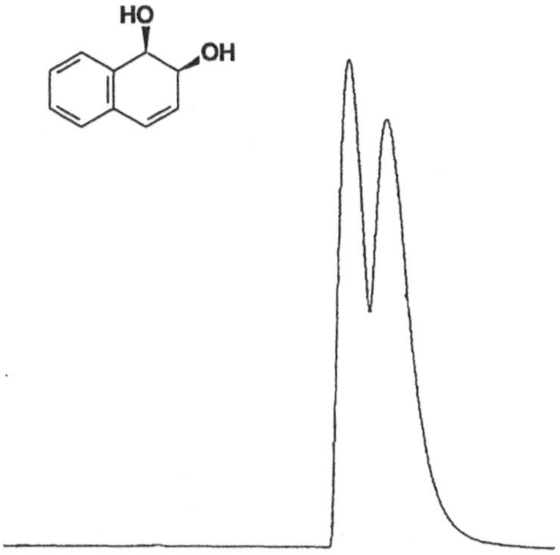

PROCHIRAL SULPHUR ATOMS

Fig. 7. Separation of [1R,2S] and [1S,2R] enantiomers of *cis*-1,2-dihydroxy-1,2-dihydronaphthalene on a Chiralcel OC column. Chromatographic conditions were 4% isopropyl alcohol in hexane as eluent at a flow rate of 0.5 ml min^{-1}. An α value of 1.11 was obtained.

respectively. These microbial systems are known to have produced acyclic homochiral sulphoxides in previous work from these laboratories [10].

It should be strongly emphasised that the absolute configuration of (+)- and (-)-1,3-benzodithiole-1-oxide has not yet been unequivocally established. On the basis of empirical methods (e.g., using [¹H]-NMR spectral data in the presence of optically pure 1,1,1-trifluoromethylphenyl carbinol) the tentative assignment shown in Scheme 9 and Figure 9 has been adopted.

Using the resolution and alkylation sequence shown in the right-hand portion of Scheme 8 it has now been possible to stereochemically correlate the parent sulphoxide enantiomers (1,3-benzodithiole-1-oxide) to a range of *cis*- and *trans*-2-alkyl-1,3-benzodithiole-1-oxides (alkyl = methyl, ethyl, 2-propyl). Using electronic circular dichroism spectral analysis, the sulphoxide isomers formed on oxidation of 2-(tert-butyl)-1,3-benzodithiole have also been stereochemically correlated to the other sulphoxides produced from the alkylation work. The validity of this correlation of absolute configuration is dependent upon a correct assignment for the parent sulphoxide. The sequence shown on the left hand side of Scheme 8 has been carried out but a suitable crystal for X-ray crystallography has not yet been produced. A successful conclusion to the latter study will thus provide an anchor structure and an unequivocal stereochemical correlation to sixteen 2-alkyl-1,3-benzodithiole-1-oxide enantiomers which have recently been produced in these laboratories [11].

Scheme 8 STEREOCHEMICAL CORRELATION OF SULPHOXIDES

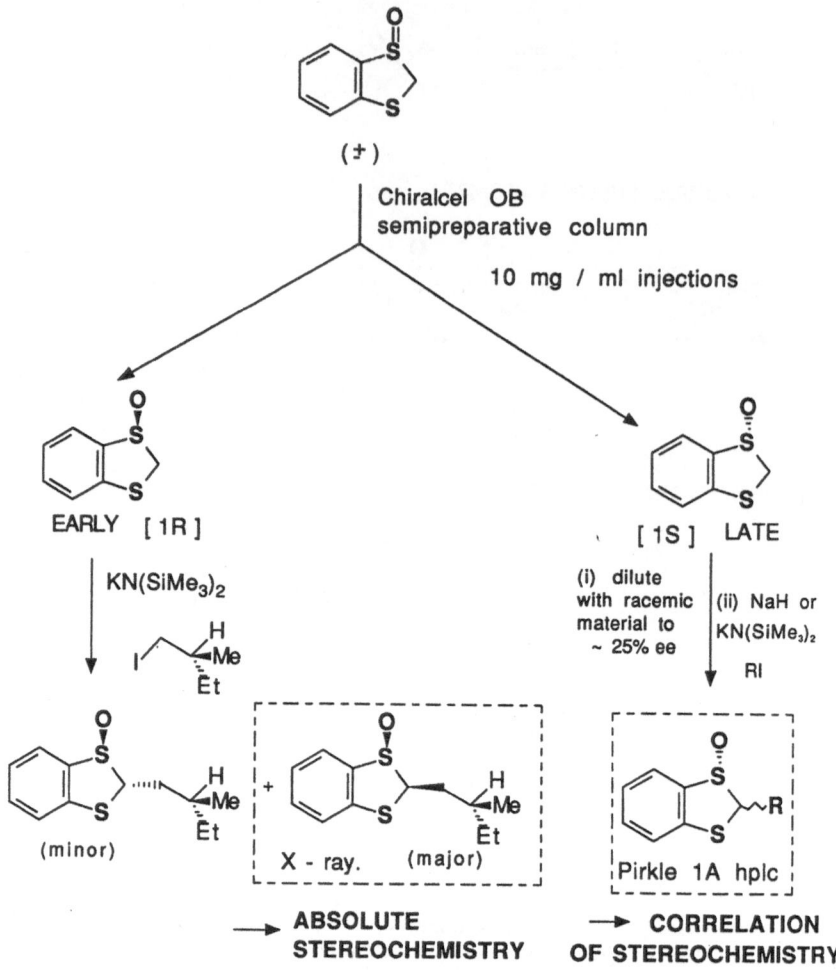

Using 2-methyl-1,3-benzodithiole as substrate, addition of an oxygen atom to the pro-R sulphur atom gave the *cis*- (1S,2R) and *trans*- (1R,2R) isomers (Table 4 and Scheme 9). Conversely, monooxygenation at the pro-S sulphur atom gave the *cis*- (1R,2S) and *trans*-(1S,2S) isomers.

Scheme 9 PROCHIRAL LONE PAIRS AND PROCHIRAL SULPHUR ATOMS

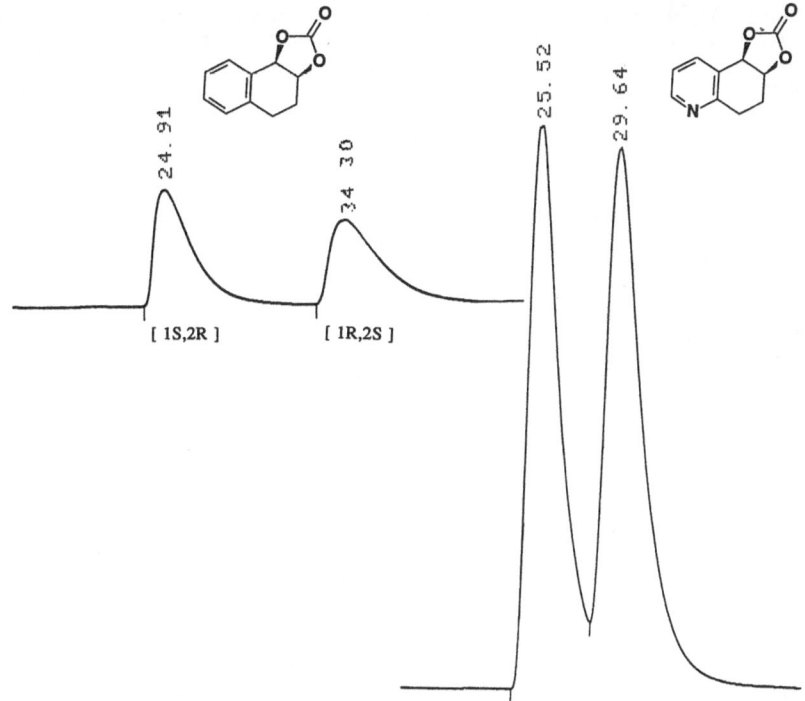

Fig. 8.   Separation of [1R,2S] and [1S,2R] enantiomers of the carbonates of *cis*-1,2-dihydroxy-1,2,3,4-tetrahydronaphthalene and *cis*-5,6-dihydroxy-5,6,7,8-tetrahydroquinoline on a Chiralcel OB column. Chromatographic conditions were 50% isopropyl alcohol in hexane as eluent at a flow rate of 0.5 ml min^{-1}. Under these conditions α values of 1.6 and 1.3, respectively, were obtained.

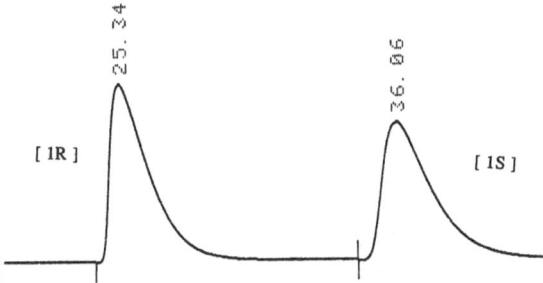

Fig. 9.   Separation of [1R] and [1S] enantiomers of 1,3-benzodithiole-1-oxide on a Chiralcel OB column using 50% isopropyl alcohol in hexane as eluent at a flow rate of 0.5 ml min^{-1}. An α value of 1.38 was obtained.

Using a Pirkle-type 1A (Regis Chemical Co. Ltd.) ionically bonded column and a solvent system of 10% isopropyl alcohol in hexane it was possible to simultaneously separate both *cis* and *trans* isomers and enantiomers of each of the 2-alkyl-1,3-benzodithiole-1-oxides (Figure 10). Using this CSP-HPLC method a range of chiral oxidants has been examined with 2-methyl-1,3-benzodithiole as substrate (Table 4). This method for probing the stereoselectivity of chiral oxidants for pro-R or pro-S lone pairs on a sulphur atom (and pro-R or pro-S sulphur atoms on a carbon atom) has been applied to both chemical and enzymatic oxidants. A strong preference for the pro-R sulphur atom (ca. 80%) was observed with the two chemical oxidants (Table 4). The intact growing fungal [1] and bacterial [1] systems appeared to show a strong preference for the pro-S sulphur atoms (> 90%). The latter estimate may, however, be affected by a small degree of asymmetric destruction of one sulphoxide enantiomer. This analytical technique is ideally suited for

Table 4. Relative Yields and Optical Yields of *cis* and *trans* 2-Methyl-1,3-benzodithiole-1-oxide and Stereoselectivity for the pro-R Sulphur Atom

Chiral Oxidant	% *cis*	(% ee)	%*trans*	(% ee)	% pro-R
Fungus (*Aspergillus foetidus*)	88	(90)	12	(5)	10
Bacteria (*Pseudomonas putida*)	96	(≥ 98)	4	(≥ 98)	4
Kagan's modification of the Sharpless asymmetric oxidations system	14	(47)	86	(78)	80
Bovine serum albumin/NalO$_4$	28	(23)	72	(70)	78

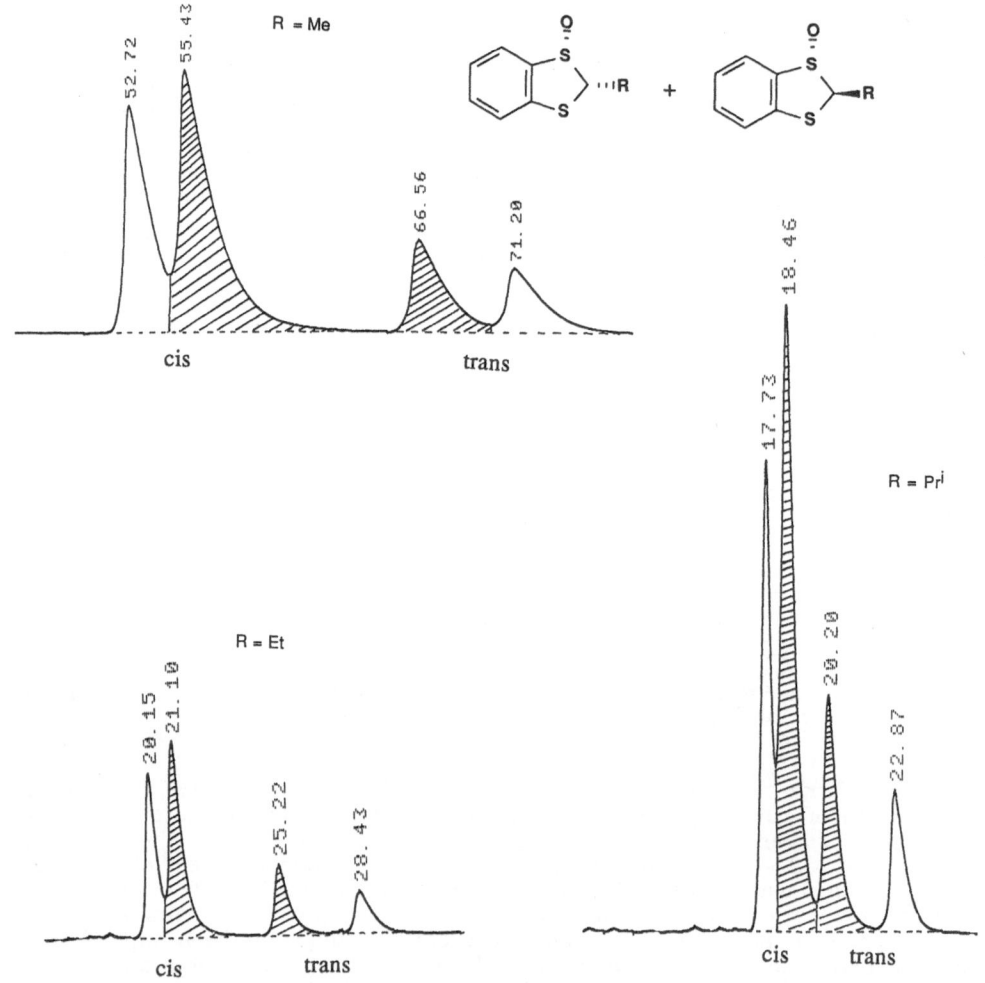

Fig. 10. Separation of enantiomers of *cis*- and *trans*-2-alkyl-1,3-benzodithiole-1-oxides using two Pirkle 1A columns in series with 10% isopropyl alcohol in hexane as eluent at a flow rate of 2.0 ml min^{-1}.

small scale oxidation reactions thus enabling results to be obtained using pure monooxygenase enzyme systems [11].

CONCLUSIONS

We have demonstrated the power of CSP-HPLC analysis in estimating the ability of enzyme systems to stereodifferentiate between prochiral hydrogen atoms (benzylic and allylic monohydroxylation), prochiral faces of an arene (arene oxide or cis-dihydrodiol formation), prochiral lone pairs (monosulphoxidation of a thioether) and prochiral sulphur atoms (monosulphoxidation of a 1,3-thioacetal). This method has proved to be particularly useful for the analysis of (1) very small quantities of metabolites, e.g., from pure enzyme reactions, and (2) rather unstable and transient metabolites, e.g., arene oxides and arene hydrates. Using the semipreparative CSP-HPLC method it has been possible to obtain significant quantities of pure enantiomers for chemical studies which are not readily available by other methods.

ACKNOWLEDGEMENTS

While all of the CSP-HPLC work described has been carried out at The Queen's University of Belfast, the wider programmes of enzyme-catalysed oxidation are the result of collaborative work with the groups of Professor H. Dalton (bacteria metabolism, University of Warwick, UK, Dr. P. J. van Bladeren (liver enzyme metabolism, TNO-CIVO Institutes, Zeist, The Netherlands) and Professor J. R. Cashman (organosulphur oxidation, University of California, San Franciso, USA). We wish to thank these groups and other research colleagues in Belfast who provided samples for CSP-HPLC analysis and assistance with the separations (Drs R. Dunlop, J. J. McCullough and N. D. Sharma). We are grateful to Professor D. Armstrong, University of Missouri-Rolla, for providing preliminary results on the resolutions of β-tetralol by CSP-GLC.

Finally, we wish to thank the Department of Education for Northern Ireland for a Postgraduate Studentship (to R. A. S. McM).

REFERENCES

1. R. Agarwal, D. R. Boyd, R. A. S. McMordie, G. A. O'Kane, P. Porter, N. D. Sharma, H. Dalton, and D. J. Gray, manuscript submitted.
2. H. L. Holland, E. J. Bergen, P. C. Chenchaiah, S. H. Khn, B. Munoz, R. W. Ninniss and D. Richards, *Can. J. Chem.*, 65:502-507 (1987).
3. D. R. Boyd, R. A. S. McMordie, N. D. Sharma, H. Dalton, P. Williams and R. O. Jenkins, *J. Chem. Soc., Chem. Commun.*, 339-340 (1989).
4. D. R. Boyd, R. A. S. McMordie, N. D. Sharma, R. Agarwal, G. M. Bessems, B. van Ommen, and P. J. van Bladeren, manuscript submitted.
5. D. R. Boyd, M. R. J. Dorrity, J. F. Malone, R. A. S. McMordie, N. Sharma, H. Dalton and P. Williams, *J. Chem. Soc. Perkin Trans. 1*, 489-494 (1990).
6. D. R. Boyd and D. M. Jerina, in: "The Chemistry of Heterocyclic Compounds", Vol. 42, A. Weissberger and E. C. Taylor, eds., Part 3, "Small Ring Heterocycles", A. Hassner, ed., Wiley-Interscience, New York, 197 (1985).
7. S. K. Agarwal, D. R. Boyd, H. P. Porter, W. B. Jennings, S. J. Grossman and D. M. Jerina, *Tetrahedron Lett.*, 26:4253-4256 (1986).
8. D. R. Boyd and D. M. Jerina, manuscript in preparation.
9. D. R. Boyd, R. A. S. McMordie, H. P. Porter, H. Dalton, R. O. Jenkins and O. W. Howarth, *J. Chem. Soc., Chem. Commun.*, 1722-1724 (1987).
10. D. R. Boyd, C. T. Walsh and Y. C. Chen, in: "Sulphur Drugs and Related organic Chemicals: Chemistry, Biochemistry and Toxicology", Vol. 2, Part A, L. A. Damani, ed., Ellis Horwood, Chichester, UK, 67 (1989).
11. D. R. Boyd, R. A. S. McMordie, J. R. Cashman and L. D. Olsen, manuscript in preparation.

SEPARATION AND ASSIGNMENT OF THE STEREOISOMERS OF

β-(p-CHLOROPHENYL) GLUTAMIC ACID BY LIGAND EXCHANGE CHROMATOGRAPHY

P. M. Udvarhelyi, D. C. Sunter and J. C. Watkins

Department of Pharmacology
Medical School, University of Bristol
Bristol BS8 1TD, UK

SUMMARY

The resolution of β-(p-chlorophenyl) glutamic acid and β-phenylglutamic acid into two pairs of enantiomers was accomplished by chiral ligand exchange chromatography (LEC). The configurations of the four separated stereoisomers of β-(p-chlorophenyl) glutamic acid at the α-centre were determined by chiroptical spectroscopy. From these measurements it was also possible to establish which of the stereoisomers were enantiomers.

INTRODUCTION

The resolution of amino acid enantiomers is perhaps the most frequently pursued of the chiral separations using liquid chromatography. One very successful technique, capable of resolving a wide range of both proteinogenic and non-proteinogenic amino acids and derivatives, is ligand exchange chromatography (LEC). Pioneered by Davankov [1], LEC has recently achieved a high level of interest. Numerous review articles have discussed both the theory and practise of LEC [2-5]. Briefly, the technique relies on the formation of labile ternary mixed-ligand transition metal complexes, which, being diastereomeric, may be separated in a chromatographic system as shown schematically below for a proline (pro)-based system. Frequently LEC entails the use of cupric ions together with a chiral chelating amino acid. The chiral chelating amino acid may be present as a mobile phase additive or may be attached to the stationary phase material (a discussion concerning the differences between these two chromatographic strategies is presented in [6]).

In our experiments cupric ions and S-proline were added to the mobile phase, thus:

$$(S\text{-}Pro)Cu(S\text{-}Pro) + S\text{-}aa = (S\text{-}Pro)Cu(S\text{-}aa) + S\text{-}Pro$$

$$(S\text{-}Pro)Cu(S\text{-}Pro) + R\text{-}aa = (S\text{-}Pro)Cu(R\text{-}aa) + S\text{-}Pro.$$

S-aa and R-aa are the two amino acid enantiomers under study which in this case were β-(p-chlorophenyl) glutamic acid (A) and β-phenylglutamic acid (B).

$$NH_2\text{-}C^*H\text{-}COOH$$
$$|$$
$$X\text{-}Ph\text{-}C^*H\text{-}CH_2\text{-}COOH$$

where X = p-chloro = β-(p-chlorophenyl glutamic acid (A) and X = H = β-phenylglutamic acid (B).

Recent Advances in Chiral Separations, Edited by D. Stevenson and
I. D. Wilson, Plenum Press, New York, 1991

Compounds A and B were selected for study by LEC because of (1) the presence of two centres of chirality, hence the existence of four stereoisomers; (2) the presence of an aromatic group aiding detection by UV; and (3) neurobiological interest in determining the stereoisomer(s) in which biological activity resides, and whether enantiomeric masking occurs.

EXPERIMENTAL

Both analytical and preparative separations were carried out using 5 μm C_{18} reversed-phase stationary phase material packed into stainless steel columns (250 mm x 4 mm id and 250 mm x 10 mm id).

A range of mobile phase mixtures were found to effect an adequate separation, but the most frequently used mobile phase composition consisted of 4.0 mM copper (II) sulphate, 8.0 mM S-proline in 10% methanol. The pH was adjusted to a value of 7.4. (It was observed that lower pH values resulted in increased retentions.) Sample solubilisation was assisted by dissolving the amino acid in 0.25 M sodium hydroxide and titrating to approximate neutrality. Sample concentrations of up to 15 mg ml^{-1} could be achieved, which allowed for the separation of approximately 0.75 mg of A per run on a preparative column.

The instrumentation consisted of an LDC ConstaMetric III pump and a Spectro-Monitor UV/visible detector connected to a Perkin-Elmer R100 chart recorder. Sample injection was via a Rheodyne 7125 with either a 20 or 50 μl loop.

Solutions of the four chromatographically separated isomers of A were collected, pooled and reduced to approximately one-fifth their original volumes. Each of these solutions was then applied to a BioRad AG 1 hydroxide ion exchange resin (200-400 mesh) in a small column. This was washed with aqueous 0.25 M acetic acid resulting in the passage down the column of S-proline and cupric ions. The migration of cupric ions was clearly visible and, when completed, 1.5 M acetic acid in 25% acetonitrile was added to the head of the column and the subsequent second fraction collected. Paper electrophoresis with ninhydrin detection confirmed that the 0.25 M acetic acid fraction contained proline, while the second fraction contained the separated amino acid isomer (also in the presence of cupric ions). Higher ionic strength acetic acid fractions contained no ninhydrin-positive components.

The separated isomers of A were analysed on a Chiralplate (Macherey-Nagel) by ligand exchange thin-layer chromatography (TLC) [7]. The solvent system consisted of acetonitrile-methanol-water (100:25:25 v/v). Detection was by spraying with 0.5% ninhydrin in ethanol-water and heating at 110°C for 5 min.

Circular dichroism (CD) spectra on the separated stereoisomers of A were obtained using a Jasco J-40 CS spectrometer. The amino acid isomers were proline free, but still complexed with cupric ions (in fact the d-d transition of Cu^{2+} was utilised for the stereochemical assignment of one of the chiral centres).

Finally cupric ions could be removed from solution by passing the solution through a small chelite (Serva) ion-exchange column.

RESULTS AND DISCUSSION

Ligand Exchange

The separation achieved for A and B using S-proline as the mobile phase additive is shown in Table 1. Four well separated peaks could be discerned and collected. The separation detailed in Table 1 was optimised for the rapid collection of A; the separation factors could be increased by reducing the methanol content of the mobile phase. The retention, and hence the separation factor, of B was much improved when no methanol was present in the mobile phase and the pH was reduced. Variation of the monitoring wavelength between 190 and 265 nm produced no change in peak area ratios implying that stereoisomers were being observed. However, as the mobile phase did contain an absorbing species, copper (II) bis-proline, negative system peaks were also observed. System peaks are a feature of chromatographic separations that utilise absorbing species in the mobile phase [8]. They are a result of the stationary phase re-adsorbing copper (II) bis-proline from the mobile phase and thereby producing a temporary reduction in the "background" absorbance level.

The result of substituting racemic proline in place of S-proline are detailed in Table 2. In this achiral separation the enantiomers were coeluted and only the diastereomers separated. Under the same injection conditions the peaks were approximately double the intensity when compared to the chiral experiment. The retentions recorded in the racemic chromatogram did not allow us to determine which of the four separated isomers of Table 1 were enantiomers. The principal of reciprocity in chiral recognition [6] will only be observed when the R-amino acid antipode is used as the chiral selector.

A separation based on R-proline is presented in Table 3. Within experimental variability this chromatogram was identical to Table 1, with the major exception the elution order of the isomers was expected to be reversed.

Ligand Exchange TLC

Chiral TLC of unresolved A produced three well separated spots (Figure 1). The most mobile spot was substantially more intense than the lower two spots, which were of equal intensity. Clearly in the more intense spot, two of the isomers of A had run together. Chiral TLC confirmed that a change in the chirality of the mobile phase selector from S- to R-proline did reverse the elution order of the isomers (see Figure 1). The elution order of the eluted isomers, with respect to time, in the S-proline (SP) experiment were designated SP1 < SP2 < SP3 < SP4; while in the R-proline (RP) experiment the fractions may be designated RP1 < RP2 < RP3 < RP4, where SP1 was the first fraction obtain in the S-proline experiment SP2 the second, and so on. From Figure 1 it may be seen:

$$\text{SP1 } (R_f, 0.58) = \text{RP3 } (R_f, 0.61)$$
$$\text{SP2 } (R_f, 0.51) = \text{RP4 } (R_f, 0.54)$$
$$\begin{bmatrix} \text{SP3 } (R_f, 0.70) = \text{RP1 } (R_f, 0.71) \\ \text{SP4 } (R_f, 0.71) = \text{RP2 } (R_f, 0.71) \end{bmatrix}$$

Table 1. Chiral Separation of A and B Isomers by Ligand Exchange Chromatography (LEC)

	$*k'$ SP1	k' SP2	k'SP3	k'SP4	$\dagger\alpha'$
A	1.27	1.93	3.57	5.17	1.52
B	0.37	0.47	0.80	1.33	1.27

Mobile phase: 4.0 mM Cu^{2+}, 8.0 mM S-proline in 10% methanol,pH 7.4 - see Experimental for details.
* k'_{SP1} = Capacity factor of isomer SP1, the least retained isomer when S-proline is the mobile phase additive.
† α' = Separation factor between SP2 and SP1, the two peaks of closest proximity.

Table 2. Achiral Separation of A Isomers by Ligand Exchange Chromatography (LEC)

	$*k'$ RAC1	k' RAC2	α'
A	2.27	3.10	1.37

Mobile phase: as in Table 1 except S-proline was replaced by racemic proline.
*k'_{RAC1} = Capacity factor of the least retained isomer in the racemic experiment.

	k' RP1	k' RP2	k' RP3	k' RP4	α'
A	1.17	1.70	3.40	5.07	1.45
B	0.40	0.53	0.99	1.51	1.33

Mobile phase: as in Table 1 except S-proline was replaced by R-proline.
RP1 was the least retained isomer when R-proline is the mobile phase additive.

These results indicated that isomers SP1 and SP3, and SP2 and SP4 were enantiomers. However, TLC did not allow us to make any statement concerning the absolute configurations of the separated isomers apart from noting that other experimenters [7] have observed that generally the R-α carbon configuration has a lower R_f value compared to the S-α carbon configuration. This would tentatively indicate that SP2 has an R configuration at the α-carbon centre. For a more rigorous and general assignment it was necessary to utilise chiroptical spectroscopy.

Circular Dichroism Spectroscopy

The CD spectra of SP1-4 complex with copper (II) are shown in Figures 2 and 3. The CD band in the red region of the spectrum originates from the d-d transition of Cu^{2+}, and is an induced band resulting from the attachment of a chiral ligand to Cu^{2+}. The sign of the d-d transition has been used previously to assign the configuration of the α-centre of amino

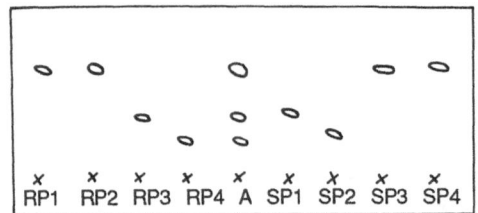

Fig. 1. Chromatogram of the TLC of β-(p-chlorophenyl) glutamic acid isomers on Chiralplate. RP1 is the first eluted isomer from LEC using R-proline as the chiral selector - see text for details. A is unresolved β-(p-chlorophenyl) glutamic acid.

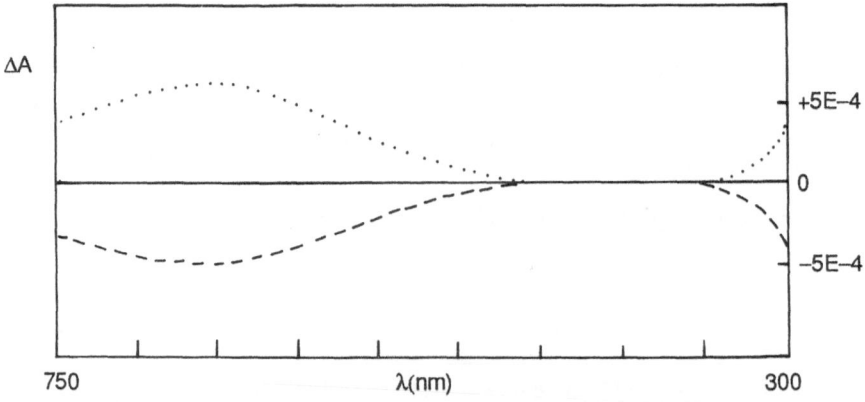

Fig. 2. CD Spectra of cupric complexes of SP1 and SP3. Key: ... SP1, --- SP3.

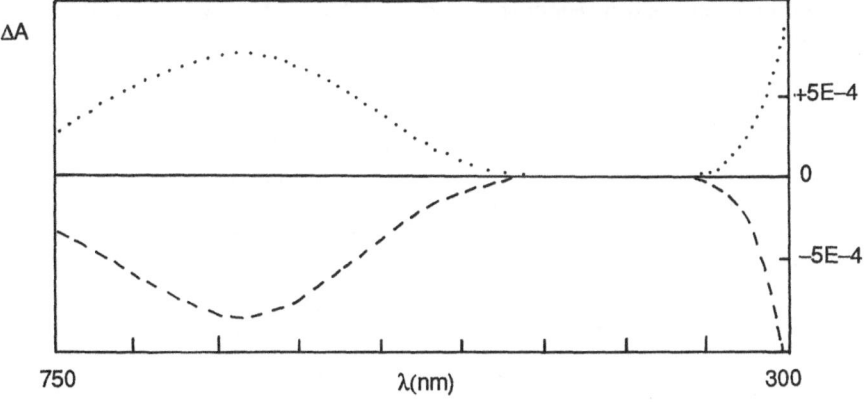

Fig. 3. CD Spectra of cupric complexes of SP2 and SP4. Key: ... SP2, --- SP4.

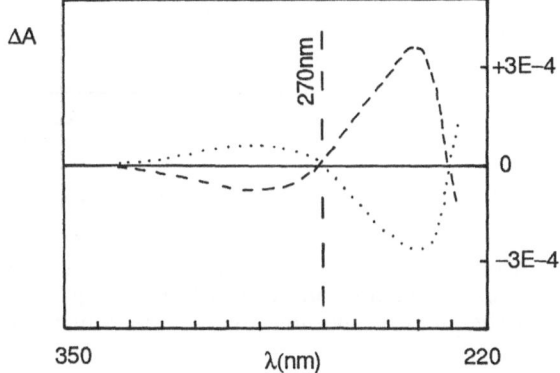

Fig. 4. CD Spectra of cupric complexes of SP1 and SP3. Key: ... SP1, --- SP3.

acids [9]. All S-amino acids, when complexed to Cu^{2+}, give rise to a negative highest energy d-d transition (ca. 600 nm) [A. F. Drake, personal communication]. This allows for the assignment of the configuration of isomer SP2 as R on the basis that the above mentioned band is positive (Figure 3). This result concurs with the tentative assignment from chiral TLC. Likewise the isomer SP1 may be assigned as R at the α-centre because its CD spectrum was also positive in the relevant region (Figure 2). Hence isomers SP1 and SP2 differed in their respective configurations at the α-centre. In the same way isomers SP3 and SP4 may be assigned as S (Figures 2 and 3).

In order to exclude the possibility that the presence of a second chiral centre may invalidate this assignment protocol, the CD spectra of the four recently synthesised configurational isomers of 3-benzylglutamic acid [10] were measured in the presence of Cu^{2+}. As expected the results show that the sign of the CD band at ca. 600 nm is dependent on the configuration at the α-centre, and is positive for both the α-R configurations (Table 4). Therefore the sign of the CD band is independent of the β-centre configuration.

CD spectra of isomers SP1-4 in the UV region (Figures 4 and 5) allow us to state that isomers SP1 and SP3 and SP2 and SP4 were mirror image sets rather than any other alternative pairing. This is the pairing also indicated by the TLC experiment.

Table 4. Circular Dichroism of Highest Energy d-d Transition of Cu^{2+}/3-Benzylglutamic Acid

3-Benzylglutamic Acid Isomer	CD (600 nm)
2R, 3S	Positive
2R, 3R	Positive
2S, 3R	Negative
2S, 3S	Negative

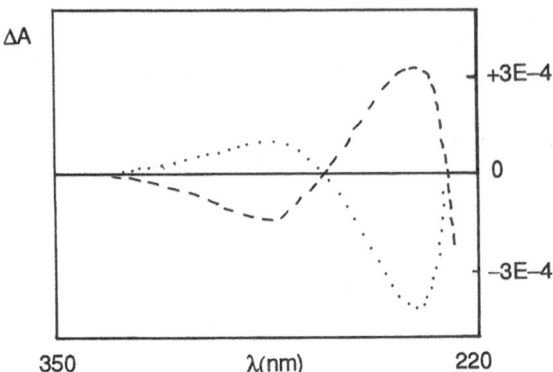

Fig. 5. CD Spectra of cupric complexes of SP2 and SP4. Key: ... SP2, --- SP4.

The determination of the enantiomeric pairs means that the elucidation by X-ray crystallography of the β-centre of just one of the four configurational isomers of A will allow for assignment of the three remaining β-centre configurations. This work will be undertaken presently.

ACKNOWLEDGEMENTS

The authors would like to thank Professor Shirahama, Department of Chemistry, Hokkaido University, Sapporo, Japan for kindly supplying samples of the configurational isomers of 3-benzylglutamic acid. We thank Dr. Alex F. Drake, Department of Chemistry, Birkbeck College, London for valuable spectroscopic discussions.

REFERENCES

1. V. A. Davankov and S. V. Rogozhin, *J. Chromatogr.*, 60:280-283 (1971).
2. S. Lam, *J. Chromatogr. Sci.*, 22:416-423 (1984).
3. V. A. Davankov, *Adv. Chromatogr. (NY)*, 18:139 (1980).
4. H. Burckner, *Chromatographia*, 24:725 -738(1987).
5. T. Takenchi, H. Asai and D. Ishii, *J. Chromatogr.*, 407:151-158 (1987).
6. V. A. Davankov, A. A. Kurganov and T. M. Ponomareva, *J. Chromatogr.*, 452:309-316 (1988).
7. K. Gunther, *J. Chromatogr.*, 448:11-30 (1988).
8. P. K. Gupta and J. G. Nikelly, *J. High Resolut. Chromatogr.*, 9:572-576 (1986).
9. J. P. Greenstein and M. Winitz, "Chemistry of the Amino Acids", Vol. 1, John Wiley, New York (1961).
10. M. Yanagida, K. Hashimoto, M. Ishida, H. Shinozaki and H. Shirahama, *Tetrehedron Lett.*, 30:3799-3802 (1989).

A *NOTE ON* DIRECT SEPARATION OF MEFLOQUINE ENANTIOMERS

BY LIQUID CHROMATOGRAPHY ON A UREA-LINKED CHIRAL STATIONARY PHASE

F. Gimenez*†, F. Bertrand‡, M. Bouley†, A. Thuillier†,
G. Hazebroucq* and R. Farinotti*‡

*Laboratoire de Pharmacie Clinique
Université Paris XI Chatenay-Malabry, France
†Service de Pharmacie-Pharmacocinétique
Hôpital Pitié-Salpétrière, Paris, France, and
‡ Service de Pharmacie Clinique et des Biomatériaux
Hôpital Bichat, Paris, France

INTRODUCTION

(2-Piperidyl)-2,8-bis(trifluoromethyl)-4-quinolinemethanol (structure shown in Figure 1) has two asymmetric carbons. Both erythro- and threo-diastereoisomers have been synthesised and the erythro-isomer (mefloquine) is now used as an antimalarial agent. Here we report some preliminary results on the separation of the enantiomers of mefloquine by liquid chromatography.

MATERIALS AND METHODS

Apparatus

The liquid chromatography system consisted of a Schimadzu LC6A pump, SPD6A variable wavelength UV detector (at $\lambda = 285$ nm) and a CR6A integrator. An (S)-napthylurea column (25 cm x 4.6 mm id) was used for chromatography. Solvent was delivered at 1.5 ml min^{-1}.

Chemicals

Racemic mefloquine was kindly supplied by Hoffman la Roche (Basel, Switzerland). (+)-3-Bromocamphor-8-sulphonic acid (ammonium salt) was purchased from Aldrich-Chemie (FRG).

Mobile Phases

Three binary mobile phases were used consisting of different proportions of hexane-2-propanol, hexane-dichloromethane and hexane-chloroform.

Ternary phases were prepared by the addition of modifiers such as methanol and acetonitrile to the binary mobile phases listed above.

Standards

(+)- and (-)-Mefloquine were prepared using (+)-3-bromo-8-camphorsulphonic acid, ammonium salt according to the method of Carroll and Blackwell [1]. Identity of the samples was verified by mass spectrometry and determination of melting points.

Recent Advances in Chiral Separations, Edited by D. Stevenson and
I. D. Wilson, Plenum Press, New York, 1991

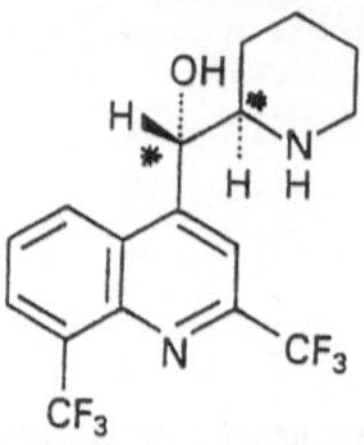

Fig. 1. The structure of mefloquine.

RESULTS AND DISCUSSION

Influence of the Mobile Phase Polarity on Capacity Factors

The three binary mobile phases listed in Materials and Methods were tested with different proportions of the more eluotropic solvent (from 90:10 to 10:90 v/v) for the separation of mefloquine on the napthylurea column. The use of dichloromethane, a dipolar molecule, interacting with the stationary phase via the carbonyl group through a dipole-dipole interaction, or chloroform, a hydrogen donor, interacting with the amide group through non-reciprocal hydrogen bonding at the carbonyl oxygen, did not result in the enantiomers being eluted. However, the hexane-2-propanol (80:20 v/v) mobile phase did allow isomer resolution. In this case, an increase in the propanol concentration produced a decrease of k'.

Influence of Methanol and Acetonitrile

In order to study the influence of modifiers such as methanol and acetonitrile, the effect of addition of up to 16% of these solvents to a hexane-2-propanol mobile phase mixture (80:20 v/v) was studied. These results are summarised in Table 1.

At a methanol concentration of more than 10%, we observed a slight increase of selectivity and a decrease of resolution due to the reduction of k'; however, efficiency was increased. The effect of solvent composition on selectivity and resolution is shown in Figure 2. In order to optimise the resolution, we maintained a concentration of 14% of methanol and decreased the strength of mobile phase by reducing the 2-propanol

Table 1. Influence of Methanol and Acetonitrile on Capacity Factors

Volume of Modifier Added to 100 ml of Hexane-2-Propanol (80:20 v/v)	Acetonitrile		Methanol	
	$k'(+)$	$k'(-)$	$k'(+)$	$k'(-)$
0	8.46	10.60	8.46	10.60
1	4.98	6.44	6.36	8.19
2	4.64	6.14	5.57	7.25
3	3.95	5.28	4.66	6.12
5	3.06	4.12	3.71	4.95
10	2.36	3.20	2.78	3.80
16	-	-	1.71	2.38

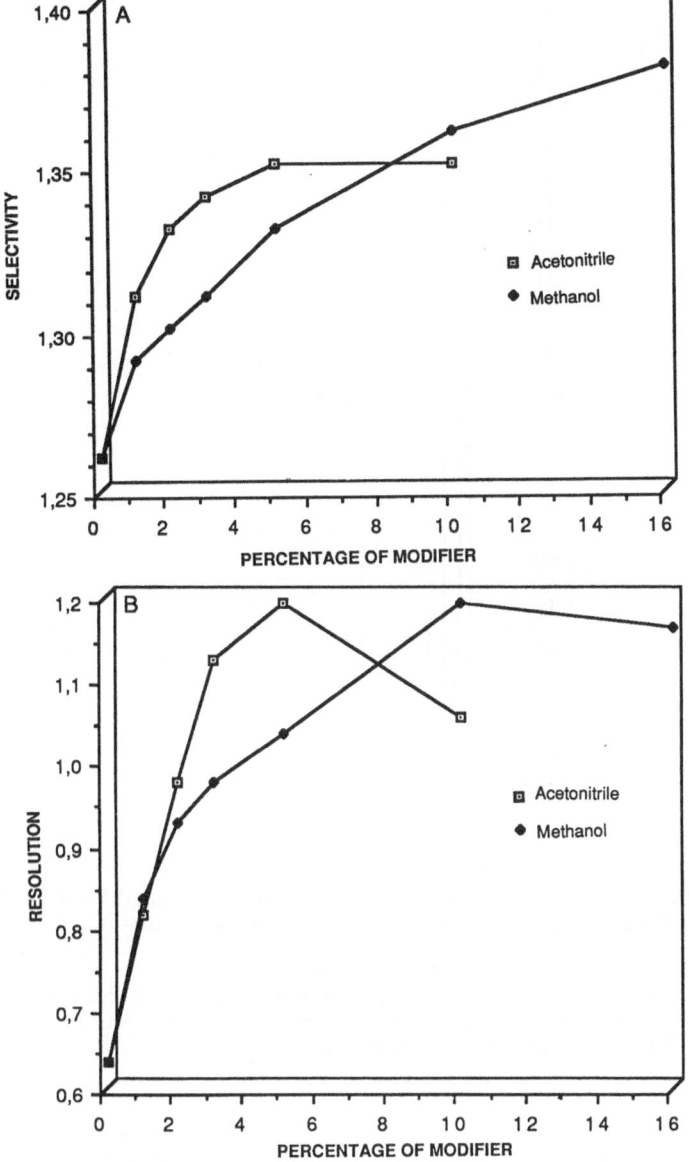

Fig. 2.   Influence of methanol and acetonitrile on (A) selectivity and (B) resolution.

concentration. Finally, a hexane-2-propanol-methanol (82:4:14 v/v) mixture was obtained which enabled the resolution of mefloquine enantiomers (Figure 3) with an α of 1.29. Injection of the individual enantiomers showed that (+)-mefloquine was eluted before (-)-mefloquine.

CONCLUSION

In our studies on the separation of the enantiomers of mefloquine we have tested a number of stationary phases including (S)-naphthylurea, α-1-glycoprotein, (R)-phenyl-glycine and polyacrylamide. However, of these the (S)-naphthylurea phase was the only one which allowed the separation of the mefloquine enantiomers. These results show the influence of modifiers such as methanol or acetonitrile on retention, selectivity and

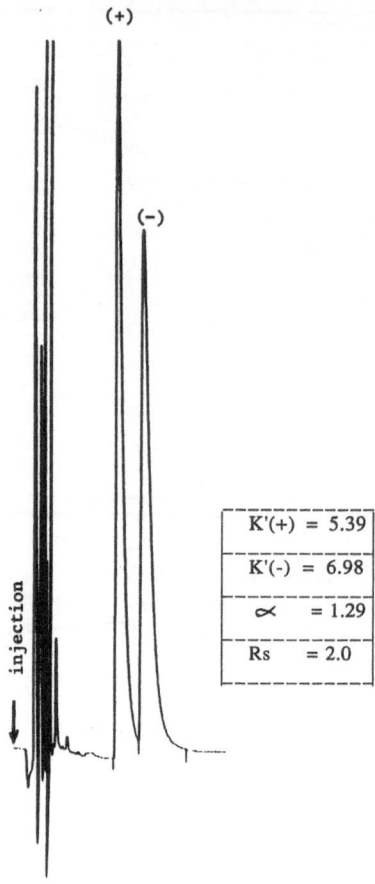

Fig. 3.   Resolution of mefloquine enantiomers on a (S)-napthylurea column (25 cm x 4.6 mm) with a mobile phase of hexane-2-propanol-methanol (82:4:14 v/v) and a flow rate of 1.5 ml min^{-1}.

resolution, probably due to a competition of methanol and 2-propanol for the active sites of the chiral stationary phase. The lower viscosity of methanol or acetonitrile compared to 2-propanol may also have contributed to these effects.

REFERENCE

1.   F. I. Carroll and J. T. Blackwell, *J. Med. Chem.*, 17:210-219 (1974).

AN EVALUATION OF SOME CHIRAL STATIONARY PHASES

FOR THE SEPARATION OF β-BLOCKER DRUGS

G. A. Kingston and D. Stevenson

Robens Institute of Health and Safety
University of Surrey
Guildford, Surrey GU2 5XH, UK

SUMMARY

Three different chiral stationary phases have been evaluated for their ability to resolve racemic propranolol and other β-blocker drugs. Cyclobond 1 was not successful in separating underivatised propranolol. α_1-Acid glycoprotein successfully separated the drugs atenolol, alprenolol, metoprolol, oxprenolol, practolol, pronethalol and propranolol into their enantiomers. Separations for propranolol, pronethalol, oxprenolol, metoprolol and alprenolol as their 1- or 2-naphthamide derivatives were achieved on a covalent 3,5-dinitrobenzoyl phenylglycine phase. The derivatisation reaction was rapid and quantitive.

INTRODUCTION

In common with a large number of cardiovascular drugs, propranolol and other β-blocking aminoalcohols contain an asymmetric centre and, although evidence exists to show their activity is highly stereoselective, they are administered as racemic mixtures. The β-receptor antagonist properties of propranolol, for example, are largely attributable to the (S)-isomer [1] and this is further complicated by the preferential metabolism of the (R) form which promotes higher plasma concentrations of the more active (S)-isomer [2,3].

The enantiomers of pronethalol and metoprolol have been successfully resolved by high-performance liquid chromatography (HPLC) on a cyclodextrin column without prior derivatisation [4]. Better separations have been achieved for these and other β-blockers using an α1-acid glycoprotein phase [5] although in some cases this required the formation of oxazolidone derivatives [6]. Propranolol has been resolved on a cellulose tris(3,5-dimethylphenyl-carbamate) phase without derivatisation [7] and on a (3,5-dinitrobenzoyl)-phenylglycine column as an amide [8] or an oxazolidone derivative [9,10]. The oxazolidone derivatives have also been resolved by capillary gas-liquid chromatography (GLC) using an XE-60-L-valine-(R)-α-phenylethylamide phase [11].

Other enantioselective separations have been achieved through the use of both mobile phase additives [12-13] and chiral derivatising reagents, with the derivatives resolved by either HPLC [14-16] or thin-layer chromatography (TLC) [17].

This study involved the evaluation of three types of commercially available HPLC columns with respect to their use in the separation of β-blockers. According to the classification system devised by Wainer [18], the three phases selected were a Type 1 3,5-dinitrobenzoyl phenylglycine column, a Type 3 β-cyclodextrin column and a Type 5 α1-acid glycoprotein (AGP) column.

A method was developed to separate the enantiomers of propranolol and other β-blockers on the Type 1 phase via the formation of naphthamide derivatives. This method originally involved the use of an ionic 3,5-dinitrobenzoyl phenylglycine column but, as this proved unsuccessful, was transferred to a covalently bound column of the same phase. The elution order of the enantiomers was established where possible by derivatising samples of the pure enantiomers in a similar fashion to the racemic mixture. In each case the (+)-isomer was found to elute prior to the (-)-isomer.

Separations were also achieved using the Type 5 protein phase, and retention studies were carried out to assess the effect of variations in mobile phase composition on resolution and retention.

EXPERIMENTAL

Apparatus

Liquid chromatography was performed using either a Beckman 110 or a Severn Analytical SA6410B solvent delivery system, a Rheodyne 7120 sample injection valve, a Pye Unicam LC-UV detector and a JJ Lloyd CR652S chart recorder.

Chiral Stationary Phases (CSPs)

The three different phases employed here were: (1) β-cyclodextrin bonded phase (Cyclobond 1, Advanced Separation Technology), 25 cm x 4.6 mm id; (2) (R)-3,5-dinitrobenzoyl phenylglycine phase (Hichrom, Reading) 25 cm x 4.6 mm id; and (3) AGP column (ChromTech, Sweden), 15 cm x 4.6 mm id.

Mobile Phases

All solvents used for mobile phases were of HPLC grade or equivalent and reagents used for buffers were of a comparable quality.

The cyclodextrin column was used in conjunction with eluents of methanol in either water or 1% triethylamine acetate buffer, pH 4.1. The 3,5-dinitrobenzoyl phenylglycine column was used exclusively with mobile phases of isopropanol in hexane, while the AGP column was used with a 0.01 M pH 7.0 phosphate buffer eluent, modified with up to 10% of either propan-2-ol or acetonitrile.

Model Compounds

Racemic atenolol, alprenolol, oxprenolol, practolol and propranolol were obtained, mainly as salts, from ICI Pharmaceuticals Division, Macclesfield, UK. The (+)- and (-)-isomers of propranolol (Inderal) and pronethalol (Alderlin) were obtained from the same source. Racemic metoprolol tartrate was obtained from Geigy Pharmaceuticals, Macclesfield, UK.

Formation of Naphthamide Derivatives

Naphthamide derivatives of the β-blockers were formed as shown in Figure 1 in order to reduce the polarity of the sample molecules and also to provide additional sites for possible interactions with the 3,5-dinitrobenzoyl phenylglycine HPLC stationary phase. An acid anhydride would normally have been the first choice of derivatising reagent since any excess anhydride is relatively easy to remove. In this case, due to the availability of reagents, 1-naphthoylchloride was used to derivatise the samples according to the procedure described below.

A fixed volume of a 2 M sodium hydroxide solution was added to 200 µl of an aqueous or methanolic solution of the sample, and a suitable volume of the derivatising reagent was added. The reaction mixture was vortexed for 2 min then the derivatives were extracted by vortex mixing for a further 2 min with an appropriate solvent such as chloroform. The samples were centrifuged for 5 min at 2000 rpm and a known volume of the organic extract was transformed to a fresh tube and taken to dryness under nitrogen. The residue was then dissolved in a suitable solvent.

In several cases, pyridine was added to the reaction mixture in an attempt to increase the rate and yield of the reaction since it acts as a scavanger for the excess hydrochloric acid produced by the reaction. Additionally, a methanolic solution of 2-naphthoyl chloride was occasionally substituted for 1-naphthoyl chloride as the derivatising reagent.

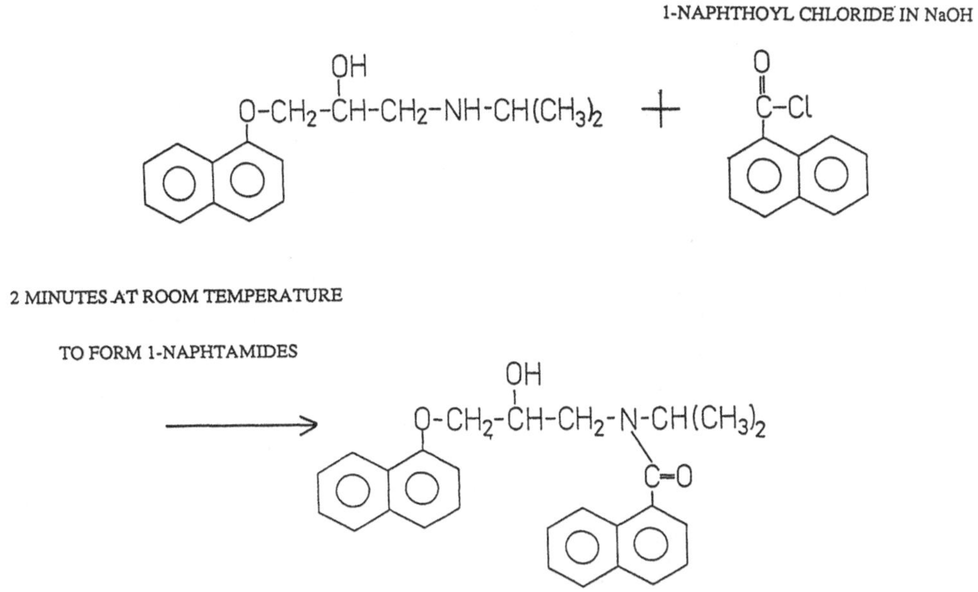

OH
|
O-CH₂-CH-CH₂-NH-CH(CH₃)₂   +   C-Cl

2 MINUTES AT ROOM TEMPERATURE

TO FORM 1-NAPHTAMIDES

OH
|
O-CH₂-CH-CH₂-N-CH(CH₃)₂
|
C=O

Fig. 1. Formation of naphthamide derivatives.

RESULTS AND DISCUSSION

Cyclobond 1 Phase

The cyclodextrin phase is compatible with the polar eluents used in reversed-phase chromatography and, as such, the column initially appeared to be the best choice for the resolution of the polar β-blocking drugs. However, our success with this column was limited and, to date, no separations have been achieved by us for any of the aminoalcohols under investigation.

An assessment of column performance was performed on receipt of the columns by measuring the efficiency of the phase under the recommended test conditions with a sample solution of m-nitroaniline. In each case the efficiency was considerably below that claimed in the specifications. In communication with the suppliers it was established that since the mass transfer characteristics of the cyclodextrin phase were relatively poor, an improvement in efficiency could possibly be acheived by a decrease in both the injection volume and the concentration of the sample solution. Upon investigation this was found to be the case and the plate number was increased from about 2000 to 10,000 by injecting 1 μl of a 0.5 μg ml^{-1} solution, although further improvements to this did not appear to be possible. It has also been suggested that decreasing the flow rate will cause an increase in the efficiency [19], but in this case no such improvement was experienced in our laboratories.

Although such measures succeeded in increasing the efficiency of the separation to a more acceptable value, very low injection volumes are not readily suited to quantitative analysis unless specialised injection valves are used since syringe injection is not sufficiently precise without the presence of an internal standard. It was found that the relationship between the sample concentration and the retention time was not constant [20], and that low amounts of injected sample caused retention time increases.

An attempt to resolve the enantiomers of propranolol was initially thought to be successful with the peak of interest showing a marked shoulder (Figure 2). Further experimentation to optimise the chromatography indicated that this was due more to column overloading than to any sort of enantiomeric separation. Resolution of both propranolol and metoprolol enantiomers have, however, been achieved using similar conditions by Armstrong et al. [4], who found it necessary to use two 25 cm cyclobond columns in series to increase the efficiency.

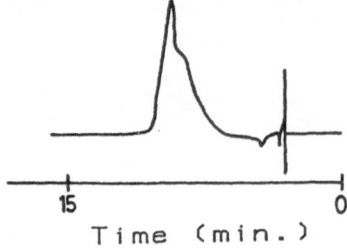

Time (min.)

Fig. 2. Resolution of propranolol on a Cyclobond 1 Column. Eluent: 25% methanol in 1% triethylamine acetate buffer pH 4.1 at 1.0 ml min^{-1}; detection: UV 288 nm; injection volume: 10 µl.

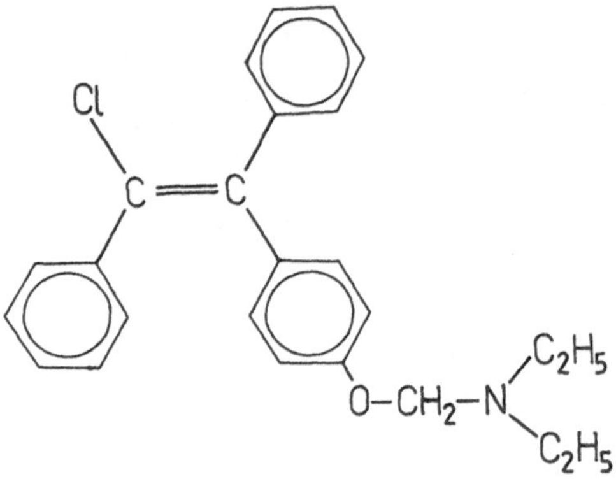

Fig. 3. The structure of clomiphene.

Although the β-cyclodextrin phase did not readily lend itself to the resolution of enantiomeric β-blockers, the separation of structural isomers was also investigated. For this, the resolution of the *cis*- and *trans*-isomers of clomiphene (Figure 3) was studied. As would have been expected, it was found that such geometric isomers were well separated, again using an eluent of 1% triethylamine acetate buffer at pH 4.1 in methanol, and that baseline resolutions could be achieved at methanol concentrations of up to 70% (Figure 4). The column thus seemed to be working successfully, despite its inability to resolve propranolol isomers.

(R)-3,5-Dinitrobenzoyl Phenylglycine Phase

Literature studies have revealed that, of all the commercially available CSPs, the "Pirkle" columns have the widest range of applications. Hence the "Pirkle" phase was selected to investigate the possibility of enantiomeric resolution of the β-blocking drugs.

Such phases work on the basis of having a minimum of three points of interaction between the solute and the phase. If such interaction sites are not present in the sample a derivatisation reaction can be performed to add a suitable group to the molecule. However, there are a number of disadvantages inherent with derivatisation reactions, not least of which is the increase in sample preparation time and also in often unavoidable sample losses.

Attempts therefore were initially made to resolve a selection of compounds without prior derivatisation, but these proved to be unsuccessful. This was due to both the polarity of the solute molecules and, in a number of cases, to the lack of sufficient suitable functional groups to interact with the phase. Since the aminoalcohols are highly polar molecules, as is the phase itself, they are not readily eluted from the phase, with eluents of low polarity. Yet increasing the polarity of the mobile phase caused the solutes to be eluted

70

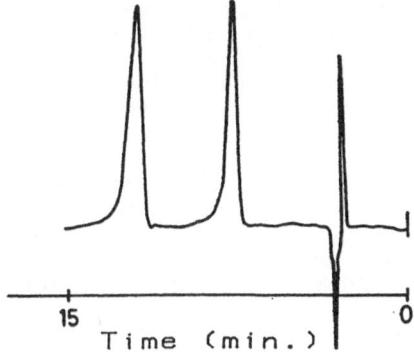

Fig. 4.   Separation of *cis*- and *trans*-clomiphene on a Cyclobond 1 phase. Eluent: 50% methanol in 1% trethylamine acetate buffer pH 6.0 at 1.0 ml min^{-1}; detection: UV 245 nm; injection volume: 10 µl.

from the column too rapidly for enantioseparation to occur and thus some form of derivatisation of the samples was deemed to be necessary.

As the point of performing a derivatisation reaction was to introduce a functional group which would provide a suitable site for interaction with the phase when necessry, the choice of derivative was critical. Successful resolutions have been achieved for a number of amines [21] and aryl aminoalkanes [22] on this phase by the formation of their α-naphthamide derivatives and hence a similar approach was taken with the β-blockers. The formation of naphthamide derivatives from the secondary amine solute not only reduces the polarity of these compounds by masking the secondary amino group, but the π basic naphthyl ring system which is introduced in this way is able to form strong π-π interactions with the π acidic 3,5-dinitrobenzoyl group.

This approach was eventually found to be very successful and resolutions were achieved for five of the seven compounds under investigation (see an example in Figure 5). However, the initial attempts involved the use of an ionic phase which gave no separations at all, and it proved necessary to transfer the method to a covalently bound column of the

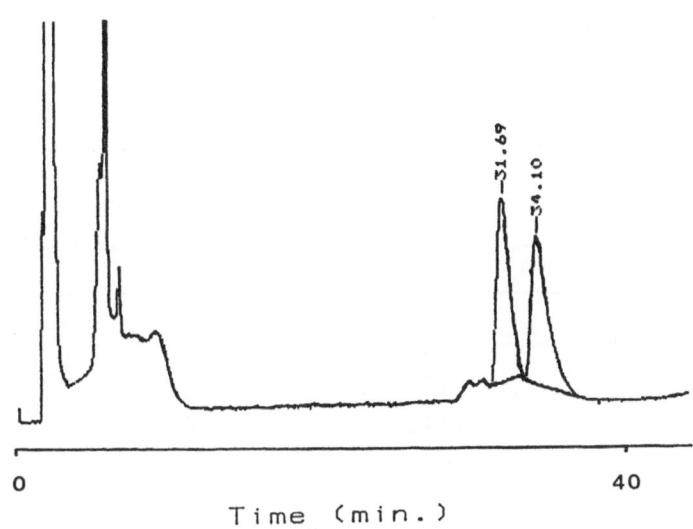

Fig. 5.   Resolution of propranolol as a naphthamide derivative on a (R)-3,5-dinitrobenzoyl phenylglycine phase. Eluent: 10% propan-2-ol in hexane at 2.0 ml min^{-1}; detection: UV 280 nm; injection volume: 20 µl.

same phase before suitable chiral separations were attained. Such covalent columns have been found to be considerably more robust and also give significantly higher efficiencies when tested with 2,2,2-trifluoro-1-(9-anthryl) ethanol under standard conditions. In this case 21,700 theoretical plates were obtained for the covalent column compared to 9000 plates for the ionic phase. All further work was carried out using covalent phenylglycine columns.

The influence of the mobile phase composition on both retention and separation was investigated using samples of the propranolol naphthyl derivative. The mobile phase was varied between 8 and 20% propan-2-ol in hexane. The retention times and separation factors for each eluent show the expected effect that increasing the polarity of the mobile phase decreased both retention and resolution of the two enantiomers (Table 1).

An attempt to assess the effect of the structure of the analyte on the retention and separation of the enantiomers was made by derivatising samples of each of the other β-blockers and subjecting them to analysis with a mobile phase of 10% propan-2-ol in hexane. Wherever possible the individual enantiomers of each compound were also derivatised and chromatographed to ensure that the two peaks were definitely due to the two isomers. In each case the (+)-isomer eluted before the (-)-isomer. The retention times and separation factors are shown in Table 2.

Attempts were also made to resolve both atenolol and practolol, but neither was separated under these conditions. A full discussion of these results would require the development of a complicated model and would be impossible without the aid of extensive computer modelling facilities. However, it is not sufficient to simply consider the required

Table 1. The Effects of Variations in Mobile Phase Composition on the Resolution of Propranolol Naphthamide Derivatives on a Covalent (R)-3,5-Dinitrobenzoyl Phenylglycine Phase

% Propan-2-ol in Hexane (v/v)	k_1'	k_2'	α
8	25.32	27.40	1.082
10	20.51	22.20	1.082
12	17.71	18.91	1.068
14	14.20	15.11	1.064
16	12.20	12.89	1.057
20	9.95	10.36	1.041

Eluent: propan-2-ol in hexane at 2.0 ml min-1; detection: UV 280 nm; injection volume: 20 μl.

Table 2. Retention of Aminoalcohols as their 1-Naphthamide Derivatives on a Covalent (R)-3,5-Dinitrobenzoyl Phenylglycine Phase

Analyte	k_1'	k_2'	α
Propranolol	22.06	23.88	1.082
Pronethalol	12.75	13.67	1.072
Oxprenolol	16.00	17.00	1.063
Alprenolol	7.00	7.40	1.058
Metoprolol	16.92	17.83	1.054

Eluent: 10% propan-2-ol in hexane at 2.0 ml min^{-1}; detection: UV 280 nm; injection volume: 20 μl

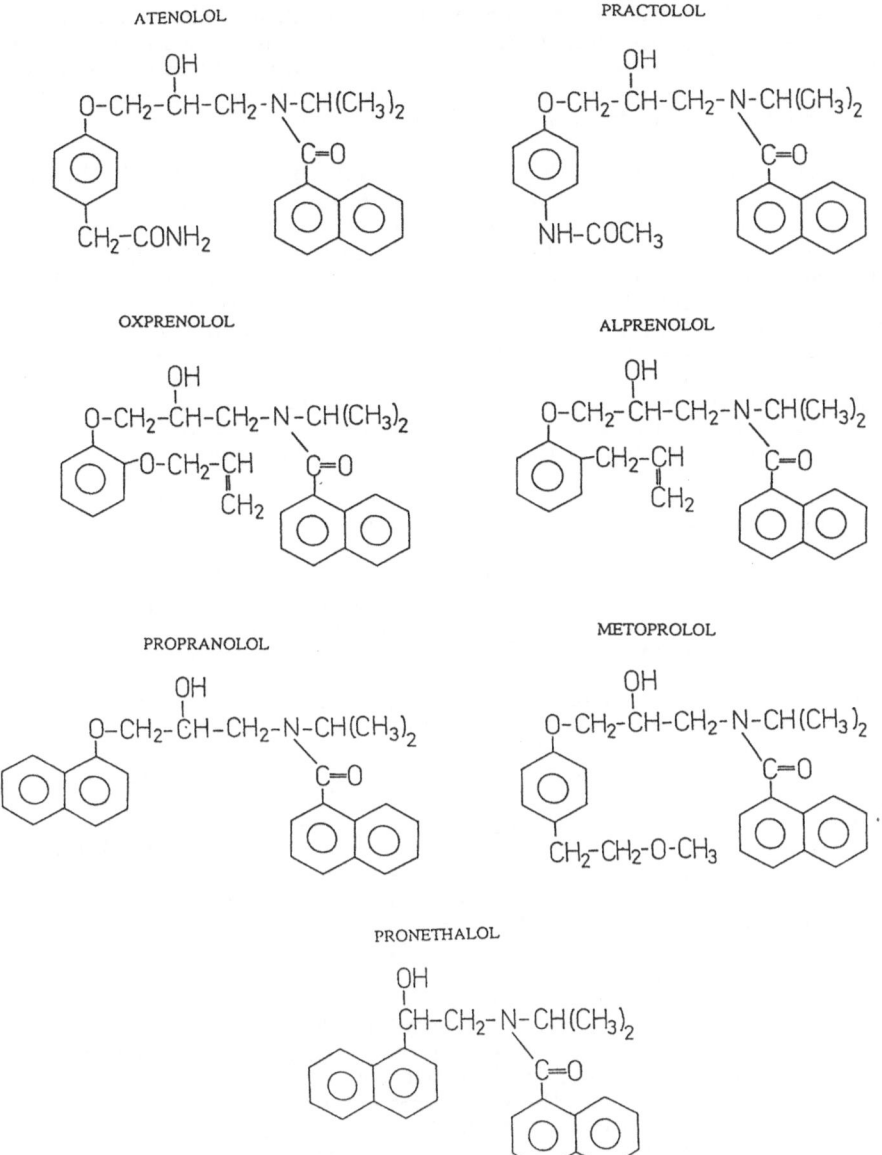

Fig. 6. Structures of the aminoalcohols as their naphthamide derivatives.

three points of interaction and certain factors must be taken into account. These include the polarity of the solute, the effect of the mobile phase (and any conformational anomalies it may cause) and the nature of the linkage between the phase and the silica support. In addition, the effect of residual silanols must be considered before any accurate predictions can be made.

From a study of the molecular structures, shown as their 1-naphthamide derivatives in Figure 6, several trends can be established. Both propranolol and pronethalol contain two naphthyl groups, either of which could potentially form the transient π-π bonds with the 3,5-dinitrobenzoyl group. However, in the case of propranolol, the naphthamide system is the least π-basic of the two, owing to the electron withdrawing power of the carbonyl group, and thus the interaction occurs at the opposite end of the molecule. This is also likely to be true for pronethalol, since again the carbonyl group is more electron withdrawing than is a hydroxyl group. The differences in retention and resolution between the two probably arises from differences in the aminoalcohol chain length. With the aromatic end secured by π-π interactions, the longer chain length of propranolol may enable the other functional groups to approach in closer proximity and thus form stronger hydrogen bonds than would be possible for pronethalol. In addition to this the π-π interactions are likely to be stronger for propranolol.

Since none of the other β-blockers have a second naphthyl system in their molecular structures, the π-π interaction between the solute and phase obviously occurs at the naphthamide group. The difference in retention and resolution between oxprenolol and alprenolol is more interesting since to all intents and purposes the major interactions of π-π and hydrogen bonding should be similar for each. The difference between the two structures lies in the substituted phenyl groups; that of oxprenolol being more electron deficient than that of alprenolol. For this to affect the retention and resolution the phenyl group must also be involved in the phase-solute interactions, although it is not yet clear in what way. It was also interesting to note that the two compounds which were not resolved under these conditions (atenolol and practolol) both had amide groups on the phenyl ring, but it is difficult to assess whether the lack of separation is due to the formation of different interactions or whether, in fact, the derivatisation has not been successful.

A further investigation of the effect of structural differences in the solute was carried out by altering the derivatising reagent. Two samples of propranolol were derivatised with either 1-naphthoyl chloride or 2-naphthoyl chloride. When chromatographed under standard conditions the results were as shown in Table 3. The increase in both retention and resolution of the 2-naphthamide derivative is possibly more due to the different orientation of the 2-naphthamide ring system causing less steric hindrance between the solute-phase complex than to differences in any of the primary interactions.

Although several stages of the derivatisation procedure were investigated, the unavailability of a pure unextracted naphthamide standard made an assessment of the yield or recovery of the method a difficult task. However, several standard curves were assayed over the range 0.01 to 1 mg ml^{-1} propranolol, and in each case the correlation coefficients were not less than 0.99, suggesting that the derivatisation procedure was, at least, consistent over this range. Problems with the method occurred at concentrations below this range since the excess derivatising reagent tended to swamp the detector response to the enantiomeric derivative. This could possibly be overcome by either solid phase extraction or column switching techniques when the method is put into routine use for monitoring enantiomeric plasma levels.

α$_1$-Acid Glycoprotein (AGP) Phase

Protein phases such as AGP appear to mimic the *in vivo* stereoselective binding processes and such phases are capable of resolving numerous enantiomeric drugs. Unfortunately, the initial phases of this type had low capacities and short lifetimes unless treated with the greatest of care. However, these columns have now been improved by a patented process by which the AGP is bonded to the silica support [22], thus increasing the phase stability. The columns are used almost exclusively with mobile phases of aqueous phosphate buffers with the addition of small amounts of organic modifiers such as methanol, ethanol, propan-1-ol, propan-2-ol and acetonitrile. For the purposes of this study the latter two were chosen to investigate the effect of changes in mobile phase composition on retention and resolution.

Table 3. Retention of Propranolol Naphthamide Derivatives on a Covalent (R)-3,5-Dinitrobenzoyl Phenylglycine Phase

Derivative	k_1'	k_2'	α
1-Naphthamide	17.4	18.2	1.046
2-Naphthamide	25.8	27.4	1.062

Eluent: 10% propan-2-ol in hexane at 2.0 ml min^{-1}; detection: UV 280 nm; injection volume: 20 μl

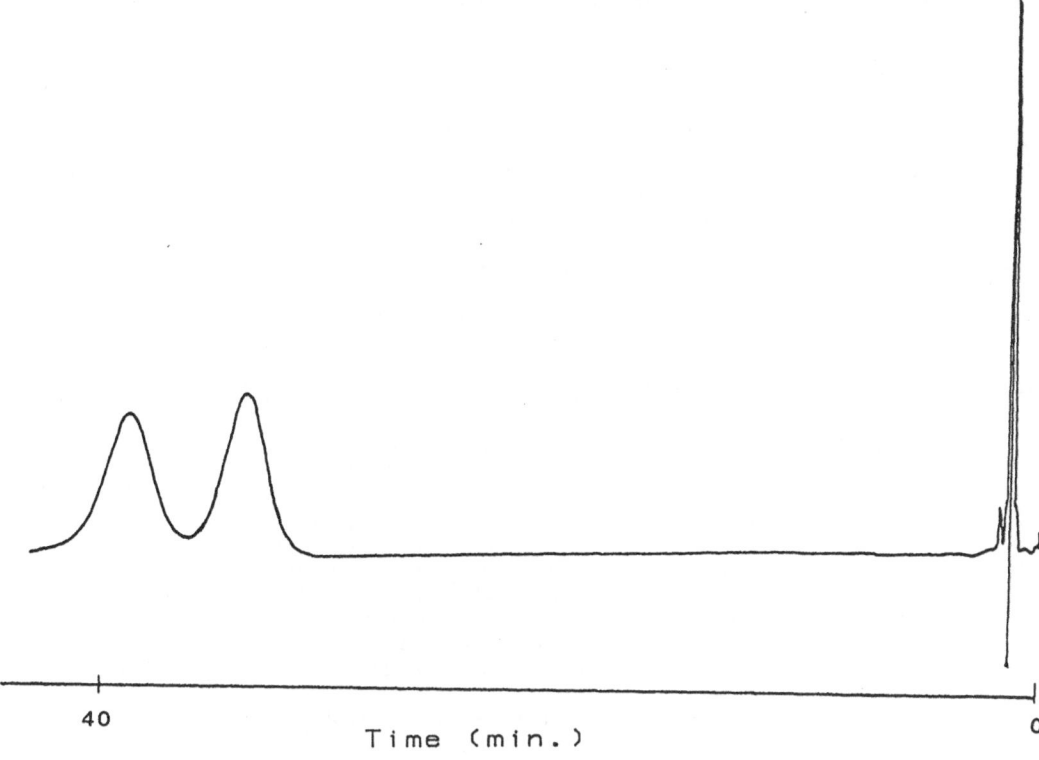

Fig. 7.   Resolution of propranolol on an AGP phase. Eluent: 5% propan-2-ol in 0.01 M phosphate buffer pH 7.0 at 0.9 ml min⁻¹; detection: UV 225 nm; injection volume: 20 μl.

The eluents were varied between 0.01 M pH 7.0 phosphate buffer, and 5 and 10% of either acetonitrile or propan-2-ol in buffer. Within this variation, separations were achieved for all the β-blockers without the need for prior derivatisation (see an example in Figure 7). The retention times followed the expected trend of decreasing with increasing concentrations of organic modifier. At similar concentrations acetonitrile gave longer retention times than propan-2-ol. In each case, longer retention times gave greater separation factors.

Since this is only a preliminary study it would be unwise to speculate about column lifetime and, more importantly, how well AGP will withstand the routine analysis of plasma samples, both of which are important considerations considering the high cost of the column.

CONCLUSIONS

There are a number of points to be drawn from this study, indicating that the choice of phase depends largely on the type of analysis required. The AGP column gave far better separations than any other phase investigated so far, without the requirement for any form of prior derivatisation. However, the cost of these phases is significantly higher than for the 3,5-dinitrobenzoyl phenylglycine phases, and as yet the column life has not been determined. On the other hand analysis with the phenylglycine phase involved derivatisation, so increasing the time, cost and possibly reducing the precision of the assay. The derivatisation reaction proposed here is however very rapid and will be investigated futher.

The cyclodextrin column was not suitable for the stereoselective analysis of the β-blockers in this study, and yet still remains one of the best approaches if only the efficiency

75

could be increased to a suitable level, since it is reasonably priced and reversed-phase eluents can be used to separate underivatised samples.

REFERENCES

1. A. M. Barrett and V. A. Cullum, *Br. J. Pharmac.*, 34:43-50 (1968).
2. B. Silber, N. H. G. Holford and S. Riegelman, *J. Pharm. Sci.*, 71:699-703 (1982).
3. L. S. Olanoff, T. Wale, T. D. Cowart, U. K. Walle, M. J. Oexmann and E. C. Conradi, *Clin. Pharm. Ther.*, 40:408-414 (1986).
4. D. W. Armstrong, T. J. Ward, R. D. Armstrong and T. E. Beesley, *Science*, 232:1131-1135 (1986).
5. G. Schill, I. W. Wainer and S. A. Barkan, *J. Chromatogr.*, 365:73-88(1986).
6. J. Hermansson, *J. Chromatogr.*, 325:379-384 (1985).
7. R. J. Straka, R. L. Lalonde and I. W. Wainer, *Pharmaceut. Res.*, 5:187-189 (1988).
8. W. H. Pirkle, J. M. Finn, J. L. Schreiner and B. C. Hamper, *J. Am. Chem. Soc.*, 103:3964-3966 (1981).
9. I. W. Wainer, T. D. Doyle, K. H. Donn nd J. R. Powell, *J. Chromatogr.*, 306:405-411 (1984).
10. K. H. Donn, J. R. Powell and I. W. Wainer, *Clin. Pharm. Ther.*, 37:191 (1985).
11. O. Gyllenhaal, W. A. Konig and J. Vesman, *J. Chromatogr.*, 350:328-331 (1985).
12. M. B. Gupta, J. W. Hubbard and K. K. Midha, *J. Chromatogr.*, 424:189-194 (1988).
13. C. Pettersson, A. Karlsson and C. Gioelli, *J. Chromatogr.*, 407:217-229 (1987).
14. G. Gubitz and S. Mihellyes, *J. Chromatogr.*, 314:462-466 (1984).
15. M. J. Wilson and T. Walle, *J. Chromatogr.*, 310:424-430 (1984).
16. T. Walle, D. D. Christ, U. K. Wale and M. J. Wilson, *J. Chromatogr.*, 341:213-216 (1985).
17. G. Pfugmann, H. Spahn and E. Mutschler, *J. Chromatogr.*, 416:331-339 (1987).
18. I. W. Wainer, *Trends in Anal. Chem.*, 6:125-134 (1987).
19. K. G. Feitsma, J. Bosman, B. F. H. Drenth and R. A. De Zeeuw, *J. Chromatogr.*, 333:59-68 (1985).
20. W. H. Pirkle and C. J. Welch, *J. Org. Chem.*, 49:138-140 (1984).
21. W. H. Pirkle, C. J. Welch and M. H. Hyun, *J. Org. Chem.*, 48:5022-5026 (1983).
22. J. Hermansson, *J. Chromatogr.*, 269:71-80 (1983).

SYNTHESIS OF AMINOALCOHOL AND OXAZOLIDONE CHIRAL STATIONARY PHASES FOR HIGH-PERFORMANCE LIQUID CHROMATOGRAPHY

W. C. Ferraz Lourenço and D. R. Taylor

Chemistry Department
UMIST, PO Box 88
Manchester M60 1QD, UK

SUMMARY

Several examples of a new family of chiral stationary phases (CSPs) have been prepared by treatment of silica gel with asymmetric optically active organosilanes derived from the reactions between epoxides and amines. When a racemic epoxide reacts regiospecifically with an optically active amine, two readily separated optically active diastereoisomers are produced. We have devised procedures for attaching the resultant aminoalcohols to silica gel, along lines used previously by ourselves and others for synthesis of CSPs via organosilane precursors. Protection of the reactive aminoalcohol function was essential to avoid polymerisation prior to bonding.

INTRODUCTION

The resolution of enantiomers by high-performance liquid chromatography (HPLC) has received much attention over the past few years [1,2] with numerous novel chiral stationary phases (CSPs) being reported in the literature (see for example Refs [3,4]). In this study, we have prepared several examples of a new family of CSPs by treatment of silica gel with optically active organosilanes derived from reactions between epoxides and amines.

When a racemic epoxide reacts regiospecifically with an optically active amine, the two diastereoisomeric aminoalcohols so formed are readily separated as their oxazolidone derivatives [5]. The two optically active aminoalcohols are then easily regenerated by hydrolysis in a basic medium [6].

We have devised procedures for attaching the resultant aminoalcohols and oxazolidones to silica gel, along lines used previously by ourselves [7,8] and others [9,10] for the synthesis of CSPs via organosilane procursors. Protection of the reactive aminoalcohol function is essential to avoid polymerisation prior to bonding.

EXPERIMENTAL

Compounds are identified by name or number. The number in bold type refers to the structures shown in the appropriate Reaction Scheme or Figure.

*Preparation of the Aminoalcohol **2** (Reaction Scheme 1)*

A solution of R(+)-α-methylbenzylamine (3.6 g; 30 mmol) in chloroform (10 ml) was added to allylglycidyl ether (2.3 g; 20 mmol) and the mixture refluxed for 6 hr. Evaporation of the solvent gave a slightly yellow liquid, which was heated to 120°C 5 mm Hg in a kugelrohr distillation apparatus to remove any unreacted starting material. The residue, a colourless oil, showed two spots on TLC (R_f = 0.41 and 0.57; using ethylacetate-isopropyl alcohol 9:1 v/v, saturated with ammonia. Flash chromatography (silica gel 40-63 μm; ethylacetate) afforded two components: component A, a colourless oil (0.31 g) was identified

Recent Advances in Chiral Separations, Edited by D. Stevenson and I. D. Wilson, Plenum Press, New York, 1991

by NMR and IR as the tertiary amine **1** (Scheme 1) $[\alpha]_D^{20} = -8.9$ (c = 2.375; chloroform), (found: C, 69.2; H, 9.3; N, 4.5%. $C_{20}H_{31}NO_4$ requires C, 68.8; H, 8.9; N, 4.0%). Component B, a colourless oil (2.51 g), was a mixture of two diastereoisomers identified spectroscopically as the aminoalcohol **2**, $[\alpha]_D^{20} = +40.0°$ (c = 2.225; chloroform), (found: C, 71.5; H, 9.2; N, 6.2%. $C_{14}H_{21}NO_2$ requires C, 71.5; H, 8.9; N, 6.0%).

*Preparation of the (R,R)(+) and (S,R)(+) Oxazolidones **4** and **5** (Reaction Scheme 2)[1]*

Oxazolidones were prepared following a procedure described by Hyne [5]. Thus, the aminoalcohol **2** gave a diastereoisomeric mixture of oxazolidones **3** as shown in Scheme 2 in 70% yield, $[\alpha]_D^{20} = +50.4°$ (c = 1.850; chloroform), shown by [^{1}H]-NMR and TLC (R_f = 0.32 and 0.34; diethyl ether) to consist of two compounds which were isolated by flash chromatography (silica gel 40-63 μm; 4:1 diethyl ether/30:40 petroleum ether). The less retained component (on silica) was a slightly yellow oil identified spectroscopically as the (R,R)(+)-oxazolidone **4**, (0.37 g; 1.4 mmol), $[\alpha]_D^{20} = +12.8°$ (c = 0.960; chloroform) (found: C, 68.8; H, 7.6; N, 5.7%. $C_{15}H_{19}NO_3$ requires C, 69.0; H, 7.3; N, 5.4%). The more retained component (on silica) was also a slightly yellow oil which was identified by NMR and IR as the (S,R)(+)-oxazolidone **5** (0.35 g; 1.5 mmol), $[\alpha]_D^{20} = +80.6°$ (c = 1.060; chloroform), (found: C, 68.7; H, 7.6; N, 5.5%).

Preparation of Organosilanes from Oxazolidones

The (R,R)(+)-oxazolidone derivative **4** (0.83 g; 3.2 mmol) was dissolved in chloroform (10 ml), and this solution was added to a solution of excess 3-mercaptopropyl-trimethoxysilane (1.9 g; 9.5 mmol) and azobisisobutyronitrile (0.05 g; 0.3 mmol) in chloroform (5 ml). The mixture was heated to reflux for 6 hr in an inert atmosphere (argon). Then the solvent and excess of mercaptosilane were distilled off leaving a yellow oil identified spectroscopically as **6** (Scheme 3) (1.3 g; 2.8 mmol; 89%), $[\alpha]_D^{20} = +8.6°$ (c = 2.280; chloroform), (found: C, 55.3; H, 7.7; N, 2.9; S, 7.3%. $C_{21}H_{35}SiSNO_6$ requires C, 55.1; H, 7.7; N, 3.1; S, 7.0%).

Similarly, the diastereoisomeric (S,R)(+)-oxazolidone **5** (0.84 g; 3.2 mmol) in chloroform (15 ml), and a solution of 3-mercaptopropyltrimethoxysilane (1.9 g; 9.5 mmol) and azobisisobutyronitrile (0.05 g; 0.3 mmol) in chloroform (5 ml) gave a yellow oil identified spectroscopically as **7** (Scheme 3) (1.3 g; 2.9 mmol; 90%); $[\alpha]_D^{20} = +45.2$ (c = 2.380; chloroform), (found: C, 55.0; H, 7.6; N, 3.3; S, 7.1%).

*Preparation of the (R,R)(+) and (S,R)(+) Aminoalcohols **8** and **9** (Reaction Scheme 4)*

Oxazolidones **4** and **5** were cleaved by basic hydrolysis following a procedure described by Teng et al. [6]. Thus, the oxazolidone **4** gave, after purification by flash chromatography, a yellow oil identified by NMR and IR as the (R,R)(+)-aminoalcohol derivative **8**, (0.45 g; 1.9 mmol; 51%), $[\alpha]_D^{20} = +44.8°$ (c = 2.165; chloroform), (found: C, 71.2; H, 9.1; N, 6.2%. $C_{14}H_{21}NO_2$ requires C, 71.5; H, 8.9; N, 6.0%).

Similarly, the oxazolidone **5** (1.6 g; 6.1 mmol) afforded, after flash chromatography, a yellow oil identified spectroscopically as the (S,R)(+)-aminoalcohol derivative **9**, (0.67 g; 2.9 mmol; 47%), $[\alpha]_D^{20} = +39.7°$ (c = 2.050; chloroform), (found: C, 71.2; H, 9.2; N, 6.3%).

Preparation of O,N-Trifluoroacetates from Aminoalcohols (Reaction Scheme 4)

A mixture of (R,R)(+)-aminoalcohol **8** (0.7 g; 2.9 mmol) and trifluoroacetic anhydride (2.5 g; 1.7 ml; 12 mmol) in chloroform (10 ml) was cooled to 0°C with stirring. A solution of triethylamine (1.1 g; 1.5 ml; 10.3 mmol) in chloroform (4 ml) was added dropwise and the mixture was stirred at room temperature for 1 hr. The mixture was then heated to reflux for 4 hr, cooled to room temperature and stirred overnight. The mixture was diluted with chloroform (50 ml), washed with 10% hydrochloric acid (2 x 30 ml), 5% aqueous sodium bicarbonate (2 x 50 ml), dried magnesium sulphate, and evaporated to afford a yellow liquid which, after flash chromatography (silica gel 40-63 μm; 1:1 diethyl ether/30:40 petroleum ether), gave a yellow oil identified spectroscopically as the (R,R)(+)-O,N-trifluoroacetate aminoalcohol **10** (Scheme 4) (0.63 g; 1.5 mmol; 50%), $[\alpha]_D^{20}$ = + 3.9° (c = 1.000; chloroform), (found: C, 50.4; H, 4.5; N, 3.5; F, 26.7%. $C_{18}H_{19}F_6NO_4$ requires C, 50.6; H, 4.4; N, 3.3; F, 26.7%).

Similarly, a solution of triethylamine (2.5 g; 3.5 ml; 25 mmol) was added dropwise to a solution of (S,R)(+)-aminoalcohol **9** (1.6 g; 6.7 mmol) and trifluoroacetic anhydride (6.0 g; 4.0 ml; 29 mmol) in chloroform (10 ml). Evaporation of the solvent gave a yellow liquid which, after purification by flash chromatography, gave a yellow oil identified spectroscopically as the (S,R)(+)-O,N-trifluoroacetate aminoalcohol **11** (Scheme 4), $[\alpha]_D^{20}$ = + 28.1° (c = 1.420; chloroform), (found: C, 50.5; H, 4.4; N, 3.6; F, 26.8%).

Preparation of Organosilanes from Protected Aminoalcohols (Reaction Scheme 4)

The same procedure used in the preparation of compounds **6** and **7** was followed here. Thus, the (R,R)(+)-O,N-trifluoroacetate aminoalcohol **10** (0.44 g; 1.03 mmol), 3-mercapto-propylsilane (1.0 g; 5.2 mmol) and azobisisobutyronitrile (0.05 g; 0.3 mmol) in chloroform (5 ml) gave a yellow oil identified by NMR and IR as **12** (Scheme 4) (0.58 g; 0.93 mmol; 90%), $[\alpha]_D^{20}$ = + 27.4 (c = 1.680; chloroform), (found: C, 46.0; H, 5.9; N, 2.3; S, 5.0; F, 17.9; Si, 4.7%. $C_{24}H_{35}SiSF_6NO_7$ requires C, 46.2; H, 5.6; N, 2.2; S, 5.1; F, 18.3; Si, 4.5%).

Similarly, the (S,R)(+)-O,N-trifluoroacetate aminoalcohol **11** (Scheme 4) (0.50 g; 1.2 mmol), 3-mercaptopropyltrimethoxysilane (1.0 g; 5.2 mmol) and azobisisobutyronitrile (0.5 g; 0.3 mmol) in chloroform (5 ml) gave a yellow oil identified spectroscopically as **13** (Scheme 4) (0.69 g; 1.1 mmol; 90%), $[\alpha]_D^{20}$ = + 18.3° (c = 1.135; chloroform), (found: C, 46.3; H, 5.9; N, 2.3; S, 5.2; F, 17.8; Si, 4.6%).

Attachment of Organosilanes to Silica Gel (Reaction Schemes 3 and 4)

The following general procedure was used in the preparation of CSPs **14**-**17**. The silylating reagent (2.0 mmol) was added to a suspension of silica gel (5.0 g; 5 μm Spherisorb, which had been dried at 180°C over phosphorus pentoxide *in vacuo* overnight, in toluene (50 ml) and the mixture was then stirred at 90°C for 72 hr under an argon atmosphere. On cooling the bonded phase was washed successively with toluene, ethyl acetate and diethyl ether (70 ml each) on a microfiltration apparatus and then dried over phosphorous pentoxide *in vacuo* for 24 hr.

RESULTS AND DISCUSSION

The reaction of an optically active amine, α-methylbenzylamine, with racemic allylglycidyl ether was investigated. TLC analysis of the products indicated that a mixture of two compounds was formed. The first product eluted during purification by flash chromatography was proved by elemental analysis to be, not a 1:1 adduct generated by ring opening at the more substituted side of the epoxide, but instead a tertiary amine identified spectroscopically as the 2:1 adduct (Compound **1** in Scheme 1, R = allyl, R^1 = CHMePh).

The second component, which was the major product, was identified as a 1:1 adduct, presumably a mixture of two diastereoisomeric aminoalcohols **2** (Scheme 1).

Reaction Scheme 1

Aminoacohols react rapidly with phosgene in toluene resulting in oxazolidones [5], which have also been prepared by reaction of epoxides with isocyanates [11,12]. The diastereoisomeric mixture of aminoalcohols, **2**, when reacted with phosgene in toluene at 0°C afforded compound **3** (Scheme 2), also as a mixture of diastereoisomers. These were readily separable by flash chromatography on silica gel.

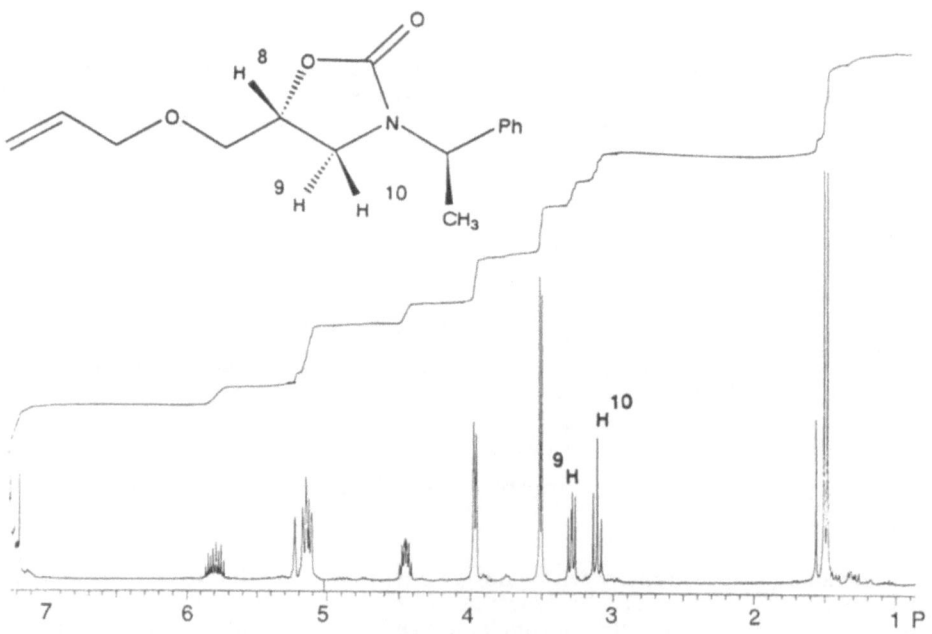

Reaction Scheme 2

The NMR spectra of the pure oxazolidones **4** and **5** (Figures 1 and 2) showed certain distinctive features: (i) non-equivalence of ring CH_9H_{10} protons (AB pattern) and (ii) characteristic H_9H_{10} chemical shift differences:

Isomer 4 (less retained) $\delta_{AB} = 42$ Hz
Isomer 5 (more retained) $\delta_{AB} = 114$ Hz.

Fig. 1. Proton NMR of the less-retained (R,R)-oxazolidone **4**.

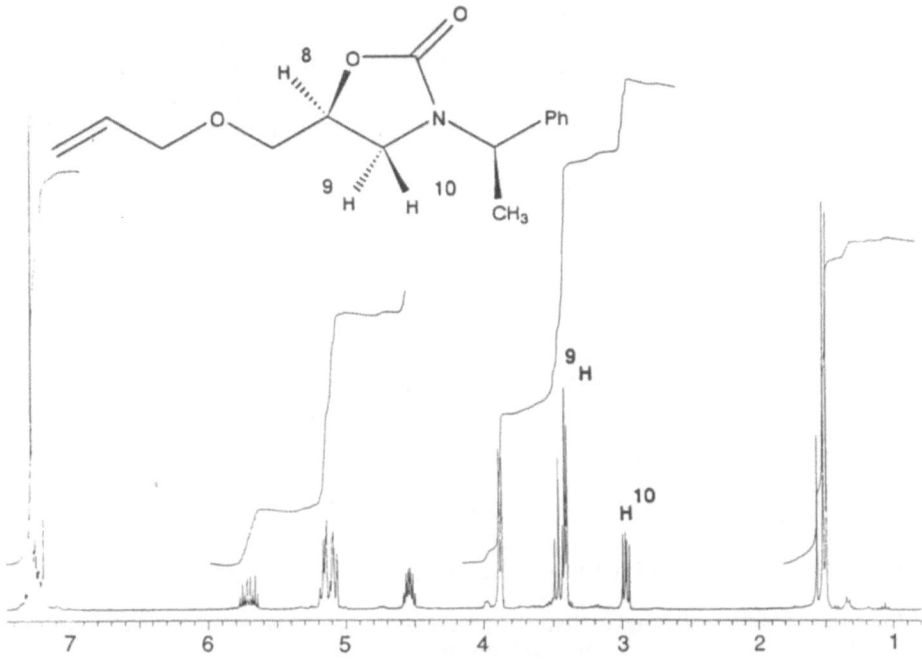

Fig. 2. Proton NMR of the more-retained (S,R)-oxazolidone **5**.

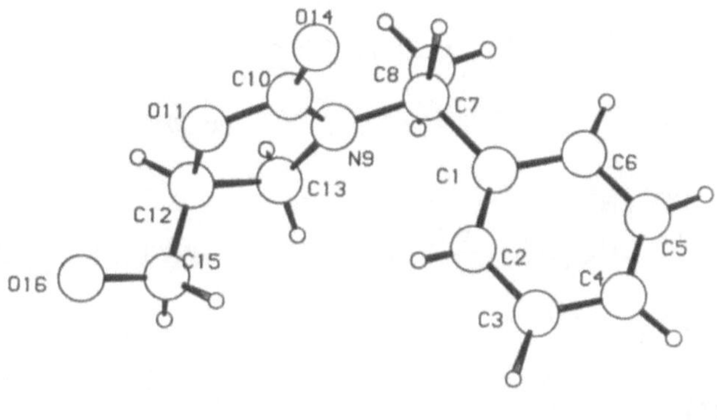

Fig. 3. X-Ray crystallographic picture of alcohol derived from the oxazolidone **5**.

The structure of oxazolidones **4** and **5** were correlated by NMR and by X-ray analysis of an alcohol derivative of **5** (Figure 3). The X-ray analysis showed that compound **5** had the following features: (i) S-configuration at C12 and (ii) R-configuration at C7.

It can be concluded that the isomer which is more retained on silica gel generally has the (7R,12S)-configuration, (C7 is known to be R, since the asymmetric centre derives from the amine). This isomer was the one which had the larger chemical shift difference (δ_{AB} = 114 Hz) between protons 9 and 10.

The two pure diastereoisomeric oxazolidones **4** and **5** were converted to optically active silylating reagents **6** and **7** and then attached to silica gel resulting in CSPs **14** and **15** (scheme 3, R = allyl).

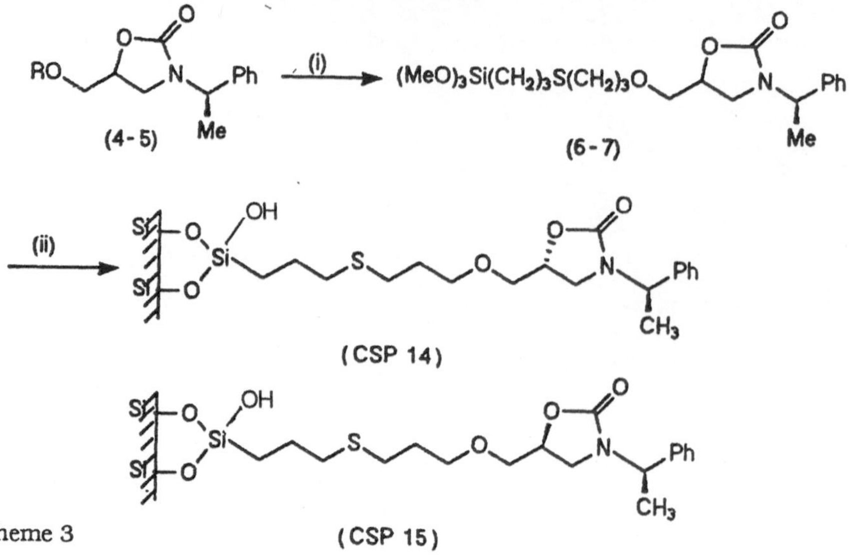

(4-5) Me

(6-7)

(CSP 14)

(CSP 15)

Reaction Scheme 3

(i) (Me)$_3$Si(CH$_2$)$_3$SH/AIBN/CHCl$_3$/reflux, (ii) Silica gel/toluene/90°C/72 hr

The pure diastereoisomeric oxazolidones **4** and **5** were then hydrolysed in basic medium to the corresponding aminoalcohols **8** and **9** (Scheme 4). In order to avoid polymerisation during the preparation of the organosilane precursors **12** and **13**, protection of the reactive aminoalcohol function by trifluoroacetylation was carried out. The organosilanes were then attached to silica gel resulting in CSPs **16** and **17** (Scheme 4).

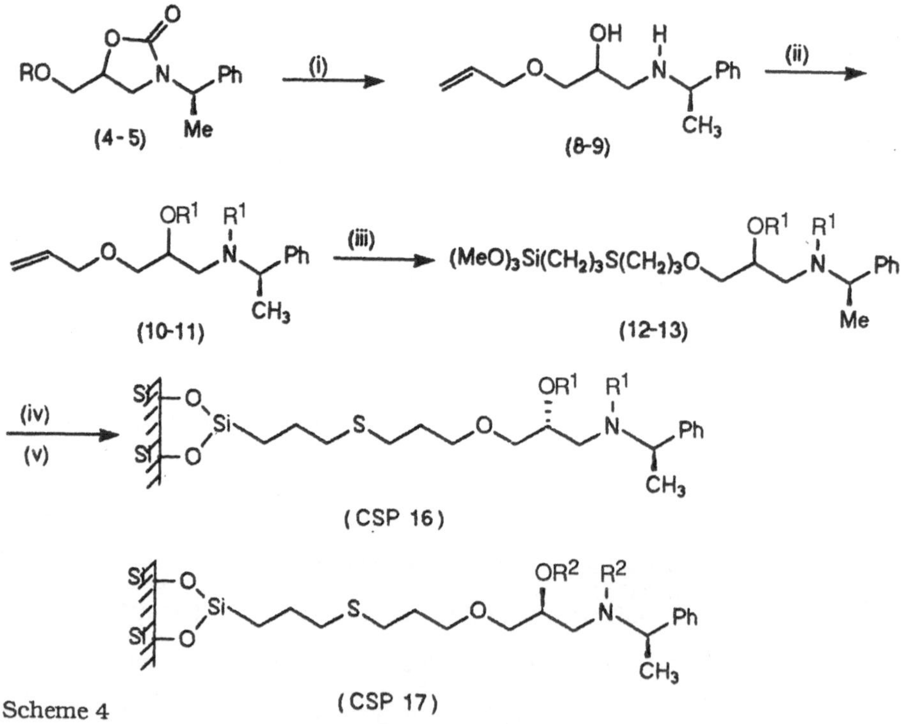

(4-5) Me

(8-9)

(10-11)

(12-13)

(CSP 16)

(CSP 17)

Reaction Scheme 4

R = allyl, R^1 = COCF$_3$; R^2 = H
(i) 3.5% w/v KOH/methanol/reflux, (ii) (CF$_3$CO)$_2$O/N(Et)$_3$/CHCl$_3$/reflux
(iii) (MeO)$_3$Si(CH$_2$)$_3$SH/AIBN/CHCl$_3$/reflux, (iv) Silica gel/toluene/90°C/72 hr
(v) 0.35 M NH$_3$ in methanol/16 hr/0°C

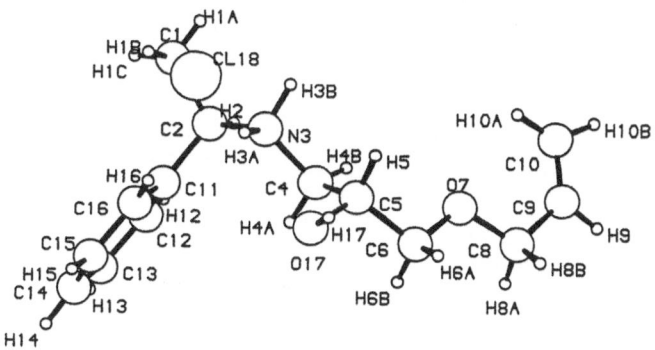

Fig. 4. X-Ray crystallographic picture of the aminoalcohol hydrochloride derived from **8**.

The hydrochloride of the aminoalcohol **6**, derived from the less retained (on silica) oxazolidone **4** was prepared, and an X-ray analysis (Figure 4) showed that compound **6** possesses the following features: (i) R-configuration at C2 and (ii) R-configuration at C5.

This is in agreement with the expected result from hydrolysis of the oxazolidone derivative **4** and it can be concluded that ether cleavage (alcohol derived from oxazolidone **5**) and hydrolysis of oxazolidones derivatives proceeded without racemisation.

ACKNOWLEDGEMENTS

The authors wish to thank Conselho Nacional de Desenvolvimento Científico e Tecnologico (CNPq) from Brazil and ICI Plant Protection Division, Jealott's Hill Research Station, Bracknell, Berkshire, UK for their financial support during this work.

REFERENCES

1. W. H. Pirkle and J. M. Finn, *J. Chromatogr.*, 192:143-158 (1980).
2. W. H. Pirkle, M. H. Hyun and B. Bank, *J. Chromatogr.*, 316:585-604 (1984).
3. W. H. Pirkle and J. E. McCune, *J. Chromatogr.*, 441:311-322 (1988).
4. Y. Dobashi and S. Hara, *J. Org. Chem.*, 52:2490-2496 (1987).
5. J. B. Hyne, *J. Am. Chem. Soc.*, 81:6058-6061 (1959).
6. L. Teng and R. B. Bruce, *J. Labelled Compd. Radiopharm.*, 15:321-328 (1978).
7. D. R. Taylor, "International Symposium on Chiral Separations", Guildford, UK, unpublished results (1987).
8. G. J. Bridger, PhD Thesis, UMIST (1987).
9. W. H. Pirkle, M. Hyun, A. Tsipouras, B. C. Hamper and B. Banks, *J. Pharm. Biomed. Anal.*, 2:173-181 (1984).
10. C. Rosini, C. Bertucci, D. Pini, P. Altemura and P. Salvadori, *Tetrahedron Lett.*, 26:3361-3364 (1985).
11. I. Shibata, A. Baba, H. Iwasaki and H. Matsuda, *J. Org. Chem.*, 51:2177-2184 (1986).
12. B. M. Trost and A. R. Sudhaker, *J. Am. Chem. Soc.*, 109:3792-3794 (1987).

THE ROLE OF MOBILE PHASE ADDITIVES IN DEVELOPING
AND OPTIMISING SEPARATIONS OF WATER-SOLUBLE ENANTIOMERS
BY HIGH-PERFORMANCE LIQUID CHROMATOGRAPHY

R. M. Gaskell and B. Crooks

Physical Chemistry Section
ICI Pharmaceuticals
Mereside, Alderley Park
Macclesfield, Cheshire SK10 4TG, UK

SUMMARY

Chiral separations are presented which demonstrate the role of mobile phase additives for enantiomer separations by high-performance liquid chromatography (HPLC).

Salts of relatively hydrophobic (log $P > 2.2$) aminoalcohol racemates are often resolved as free bases on chiral stationary phases (CSPs) using organic eluents modified with a base. However, efficient chromatographic separations for a number of more hydrophilic molecules of this type have been precluded by their low solubilities in organic solvents. Using a CSP we have developed acidic eluents which generate organically soluble ion pairs. The optimum acid has been identified. For a range of compounds the selectivity of acidic and basic eluents was compared and related to log P. Acidic eluents offered significant advantages for polar aminoalcohols.

In a second example a substituted 1,3-dioxan was resolved using a Cyclobond I (β) CSP. The efficiency, selectivity and capacity of this separation were measured and a comparison with results obtained using a nitrile (CN) column with 'β'-cyclodextrin as an eluent additive was made. Data are presented to illustrate eluent/column selection. Enhanced parameters were obtained using the modified eluent.

Finally the application of several CSPs for the optical purity measurement of a methotrexate analogue is discussed. In this case it was also perceived that the solute anion would bind to serum proteins. A separation was achieved using bovine serum albumin (BSA) in a buffered eluent on a Diol column. The effects of pH, bonded phase and organic modifier are presented.

The use of eluent additives has proved to be a flexible technique for the separation and isolation of polar water-soluble enantiomers which are poorly resolved on CSPs.

INTRODUCTION

In HPLC enantiomers can be separated indirectly as diastereomeric derivatives or directly via diastereomeric interaction with a chiral selector used as a CSP or dissolved in the mobile phase. Problems (e.g., racemisation) can arise from the indirect approach and direct methods on CSPs are often preferred. However, polar multifunctional molecules require polar eluents for solvation and elution, and weak solute-CSP interactions (e.g., H-bond), critical for chiral recognition, are diminished. Enantioselectivity is often poor.

Commercial CSPs are expensive, often with a low preparative capacity and a short lifetime. A variety of phases are available but difficulties arise in predicting the application of a specific CSP to a particular separation problem. In our experience the use of mobile phase modifiers has provided a flexible alternative with several advantages, e.g., enhanced solubility. If chiral modifiers are used, conventional columns (e.g., silica) which generate

Recent Advances in Chiral Separations, Edited by D. Stevenson and
I. D. Wilson, Plenum Press, New York, 1991

higher efficiencies/capacities than CSPs, can be utilised to separate the diasteromeric species formed. From the five main types of chiral separation mechanism in the literature using additives [1], we consider that methods using ion pair, inclusion and protein modifiers are the most flexible for analytical and preparative work. We have previously demonstrated a highly selective ion pair procedure for aminoalcohols using tartaric acid as the counter ion, and an eluent based on propan-2-ol with a silica column for analytical and preparative applications [2].

Here we illustrate how the limited range of CSPs at our disposal has been initially applied to three chiral separation problems. Our intention has been to show the practical limitations of these separations and how the use of different mobile phase modifiers/separation mechanisms has proved advantageous. Specifically we have highlighted:

1. How solubility can be enhanced via non-chiral ion pair modifiers, and how this can be used to extend the range of a CSP.
2. The benefits arising from increased column efficiency and capacity obtained by using cyclodextrin additives.
3. The use of protein additives to separate a racemate which could not be achieved using several commercial CSPs.

EXPERIMENTAL

Compounds 1-7 correspond to a range of racemic β-blockers of varying log P from the general series $R'OCH_2CH(OH)CH_2NHR''$. Compound 8 (below) is a racemic thromboxane antagonist and compound 9 (below) is a methotrexate analogue. All of these compounds were synthesised at ICI Pharmaceuticals, Alderley Park.

Solvents and Reagents

The solvents used included hexane, propan-2-ol, methanol (Fison's HPLC Grade) and ethanol (rerectified absolute). The following chemicals were also used: trifluoroacetic acid (Rathburn Chemicals Gas Phase Sequencer Grade), diethylamine (Aldrich), sodium dihydrogen phosphate and disodium hydrogen phosphate were "Analar" grade (BDH Ltd., Poole, Dorset, UK), β-cyclodextrin (Aldrich) bovine serum albumin (BSA) (Sigma Fr.V). Water was double-distilled from glass.

Columns

The columns used in this study included a 25 cm x 4.6 mm Daicel OD (Anachem, UK), 25 cm x 4.6 mm id Cyclobond I (Tecnicol, UK), 25 cm x 4.6 mm Spherisorb 5 μ C_{18}, cyanopropyl (CN) and 25 cm x 4.6 mm Lichrosorb 10 μ Diol (Hichrom, UK).

HPLC Equipment

The majority of the work was carried out using a Varian 5560 ternary HPLC system fitted with a variable wavelength UV detector (UV 200 Varian). Injections were made via a Varian 8085 autosampler and a Varian 402 Data System was used to collect and measure the raw data. Other experiments were carried out using an Altex 110A pump fitted with a Rheodyne 7125 loop injector and a Perkin Elmer LC75 variable wavelength UV detector.

RESULTS AND DISCUSSION

The Use of Non-chiral "Ion Pair" Modifiers

One of our objectives has been to develop HPLC methods to measure 0.5% of a minor enantiomer in resolved β-blockers. Indirect methods were available but efforts to establish

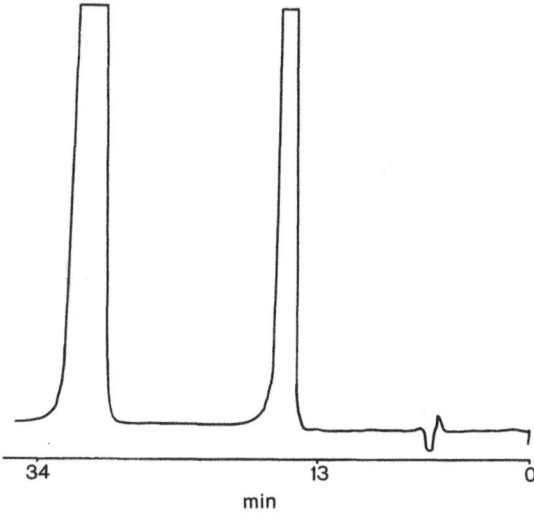

Fig. 1.   Separation of the enantiomers of a model "hydrophobic" β-blocker with a log P > 2.2. Chromatographic conditions: column, 25 cm x 4.6 mm id Chiralcel ID; eluent, propan-2-ol-hexane-diethylamine (50:50:0.1 v/v); flow rate, 1 ml min[-1].

direct chromatographic methods for certain hydrophilic blockers were hindered by their low log P which resulted in poor solubilities in organic solvents and strong, non-specific interactions with CSPs.

When salts of relatively hydrophobic blockers are chromatographed in non-aqueous eluents a basic modifier is often used to free the more lipophilic base [3] and to interact preferentially with specific column sites, e.g., silanols, which may cause peak "tailing". Many "hydrophobic" β-blocking compounds of interest to us could be separated directly on a Chiralcel OD column using propan-2-ol-hexane (1:1) modified with diethylamine as illustrated by the example shown in Figure 1. However, the low solubility of the hydrophilic β-blockers in this system generated broad, inefficient chromatographic peaks (Figure 2). It was reasoned that the solubility of these compounds in non-aqueous media would be increased in eluents modified with monobasic acids.

HPLC systems were developed for the hydrophilic β-blockers using a 25 cm x 4.6 mm Chiralcel OD column and mixtures of ethanol and hexane containing various "ion pair" modifiers.   The separations were sufficiently selective to assign minimum levels of detection of 0.2% for the minor enantiomer for a range of aminoalcohols (Figure 3).   The results were tabulated for a model compound (Table 1).   Trifluoroacetic acid (TFA) was selected as the optimum modifier and the effect of concentration was studied (Table 2).

The selection of TFA was based on: (1) the long retention of the (R)-enantiomer relative to the unretained peak of fumaric acid also present in the sample. This feature of long retention is important when measuring the quantity of the (R)-enantiomer (least retained) in the presence of a large excess of the (S)-enantiomer; (2) the higher α and R_s values generated for the enantiomeric separation when TFA was used compared to the other ion pairs.

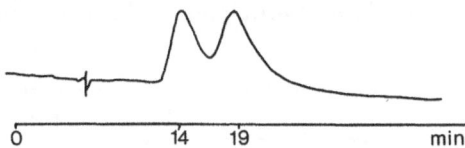

Fig. 2.   Separation of the enantiomers of a model "hydrophilic" β-blocker with a log P < 2.2. Chromatographic conditions: column, 25 cm x 4.6 mm id Chiralcel OD; eluent, hexane-propan-2-ol-diethyamine (50:50:0.1 v/v); flow rate, 1 ml min[-1].

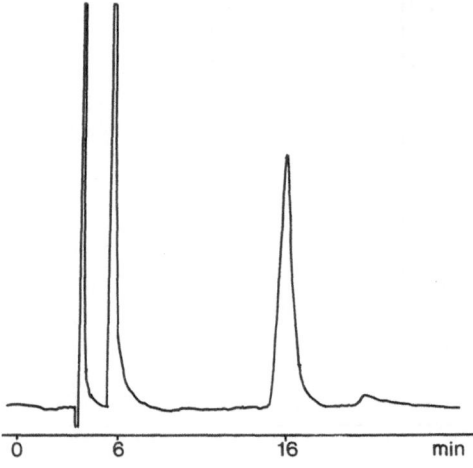

Fig. 3.   Separation of the enantiomers of the "hydrophilic" β-blocker shown in Figure 2 with an acidic eluent.   Chromatographic conditions: column, 25 cm x 4.6 mm id Chiralcel OD; eluent, hexane-ethanol-trifluoroacetic acid (70:30:0.5 v/v); flow rate, 1 ml min[-1].

Table 1.   Variation in Chromatographic Resolution with Different Counter Ions for a Model Hydrophilic β-Blocker

Counter Ion	pKa	k_1	k_2	α	R_s
Heptafluorobutyric	-0.4	0.59	4.22	7.2	5.1
Trifluoroacetic	0.2	0.78	4.92	6.3	3.9
Trichloroacetic	0.6	0.62	2.37	3.8	2.3
Formic	3.8	0.78	3.73	4.8	3.2
Acetic	4.8	0.60	3.61	6.0	3.3
Propanoic	4.9	0.96	5.06	5.3	4.0

The eluent used in these studies was hexane-ethanol-counter-ion 65:30:0.5 v/v.

Table 2.   Effect of Variation of Trifluoroacetic Acid (TFA) Concentration on the Chromatographic Resolution of the Enantiomers of a Model Hydrophilic β-Blocker

Concentration of TFA (% v/v)	k_1	k_2	α
0.2	0.566	3.464	6.120
0.4	0.552	3.451	6.252
0.6	0.559	3.482	6.229
0.8	0.586	3.734	6.372
1.0	1.014	5.243	5.171

The solvent used in this study was hexane-ethanol 63:35 v/v modified with TFA.

The use of acidic ion pair systems to achieve high retention and selectivity for low log P compounds was extended to methods where the use of a competing base had been previously used. A comparison was made for seven of these compounds (Table 3) against optimised systems modified with diethylamine on the basis of chromatographic parameters and log P.

For the compounds separated in both basic and acidic systems a plot of α versus log P was made (Figure 4) for each eluent. This indicates that at log P 2.2 - 2.6 the compounds were sufficiently soluble as free bases in the mobile phase to achieve efficient, selective chromatography. However, when log P < 2.2 the solubility as the free base became lower and solvation could be enhanced via ion pair formation with a counter ion and the acidic mobile phases became appropriate. It should be noted, however, that although some α

Table 3.   Comparison of Acidic and Basic Eluents: Correlation with log P for Seven β-Blockers (General Structure R'OCH$_2$CH(OH) CH$_2$NHR'')

Compound	log P	α Basic	α Acidic
1	2.6	2.5	2.4
2	2.1	3.8	3.7
3	2.0	3.4	5.4
4	1.3	1.2	2.1
5	0	1.6	3.7
6	-0.29	1.6	6.3
7	-0.50	2.1	3.7

values generated in basic systems appear large, this effect was largely negated by the excessive width of the peaks, a feature which was overcome using the acidic systems. It is anticipated that the correlation of log P with α for acidic and basic eluents will enable chromatographic conditions to be quickly predicted for related compounds in this series. The use of acidic eluents has considerably extended the range of the Chiralcel OD column.

The Use of a β-Cyclodextrin Additive

Our objective was to obtain a direct separation of compound 8, a substituted 1,3-dioxan, which could be used to measure 0.2% of a minor enantiomer in resolved material, and to assess the potential of the method for preparative separations. The majority of our CSPs had proved non-selective for the separation of compound 8 and related compounds where our aim had been to use ion-suppression systems on cellulose derivatives (Daicel OD, OB) and on a Pirkle column ((R)-3,5-dinitrobenzoyphenylglycine). Inefficient, unresolved peaks resulted, probably due to the modest solubility of the compounds in low polarity eluents.

A successful preliminary separation of compound 8 was achieved using a Cyclobond I (β-cyclodextrin)-bonded phase and a buffered aqueous-organic eluent at pH 7.0 in which the compound was water soluble as the anion (pK_a COOH = 3.5). After optimisation of the buffer strength (50 mM potassium dihydrogen phosphate (Table 4) and the choice of organic modifier (methanol), a selective separation of compound 8 was achieved (Figure 5). The efficiency, selectivity and resolution were measured for a loading of 100 µg.

Efforts were also made to develop a similar separation using β-cyclodextrin as the mobile phase additive on an achiral column. The optimum polarity of the column bonded

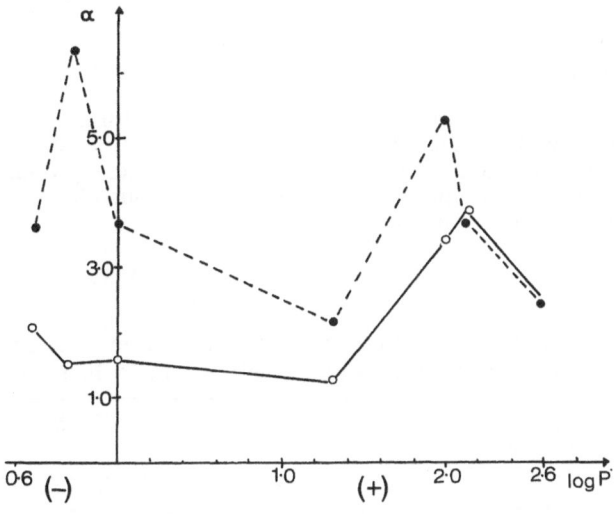

Fig. 4.   Correlation of α with log P for acidic and basic eluents for the separation of the enantiomers of seven β-blockers on Chiralcel OD. Key: ●---●, acidic; o——o, basic.

Table 4. Optimisation of Buffer Strength for the Separation of the Enantiomers of a Substituted 1,3-Dioxan on a Cyclobond I Column

Buffer Strength (mM)	k_1	k_2	α
50	3.78	4.79	1.27
10	2.66	3.32	1.25
1	6.08	7.44	1.22

The solvent used for these studies was phosphate buffer (pH 7.0)-methanol (75:25 v/v).

phase was established using Diol, CN and C_{18} and mixtures of phosphate buffer (pH 7.0) and methanol. Retention times were noted for each combination and a CN column was selected for further experiments because it achieved high solute retention (with elution) in the water-rich eluent which was necessary to dissolve the added β-cyclodextrin.

To the selected eluent (80% buffer:20% methanol) β-cyclodextrin was added at a concentration close to the maximum at room temperature (12.5 mM). A good separation was achieved (Figure 6) and the chromatography parameters were measured and compared with the results using the Cyclobond I column (Table 5).

From Table 5 it was apparent that the modified eluent technique was superior for the measurement of trace amounts of a minor enantiomer because of the higher efficiency achieved compared to the Cyclobond column. The high resolution maintained at a 100 μg loading also made this the method of choice for preparative work.

Use of a Protein Additive

In the final example we attempted to devise a separation for the enantiomers of a methotrexate analogue (compound 9). Our aim was to develop a method to measure the optical purity of the (S)-enantiomer. "Racemised" material was supplied for references. Several separation techniques were evaluated:

1. Non-aqueous ion-suppression systems (mixtures of hexane, ethanol and TFA) with cellulose and Pirkle phases. Chromatographic peaks were broad due to low solute solubility in the eluent and no enantioselectivity was achieved.

2. Solvating aqueous buffered eluents (pH 7.0) with a β-cyclodextrin-bonded phase. No enantioselectivity was demonstrated.

3. Cyclodextrin (β) and ion pair additives (quinine) were evaluated in aqueous and non-aqueous (dichloromethane) eluents respectively, without any success.

We therefore examined a further alternative, namely chiral separations on proteins. The high affinity of various anionic species for proteins, e.g., BSA, has been reported in the literature [4]. It was anticipated that commercial bonded-phase protein columns, e.g., α_1-acid glycoprotein, which are compatible with solvating aqueous eluents, might demonstrate enantioselectivity, but such columns are expensive.

Table 5. Comparison of the Separation of a 1,3-substituted Dioxan on a Cyclobond I Column and on a CN Column with β-Cyclodextrin in the Mobile Phase

Technique	Load (μg)	k_1	k_2	α	R_S	N_1	N_2
Bonded Phase	100	3.8	4.8	1.3	0.9	1250	1223
Modified Eluent	100	1.3	1.8	1.4	1.7	2843	2492

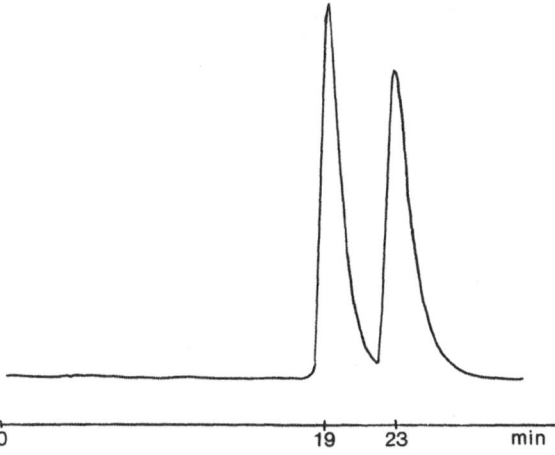

Fig. 5.   Separation of the enantiomers of the 1,3-disubstituted dioxan (compound 8). Chromatographic conditions: column, 25 cm x 4.6 mm id Cyclobond I; eluent, potassium phosphate buffer (50 mM, pH 7.0)-methanol (75:25 v/v); flow rate, 1 ml min^{-1}.

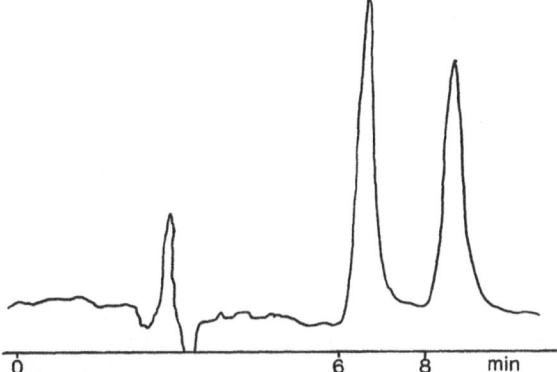

Fig. 6.   Separation of the enantiomers of the 1,3-disubstituted dioxan (compound 8). Chromatographic conditions: column, 25 cm x 4.6 mm id Spherisorb S5CN; eluent, potassium phosphate buffer (50 mM, pH 7.0)-methanol (80:20 v/v) containing 12.5 mM β-cyclodextrin; flow rate 1 ml min^{-1}.

As an alternative BSA was added to an aqueous (pH 7.0) phosphate buffer (3 g l^{-1}). This pH was used to generate the dianion of compound 9, thus enhancing its water solubility. Using this eluent a preliminary separation of the enantiomers was demonstrated on a Diol column (Figure 7). Other bonded-phase polarities were evaluated and the order of retention was C_{18} > CN > Diol (Table 6). The effect of pH on retention and separation was also studied (Table 7). As the pH was decreased to 3.0 the separation was lost. This indicates that the nature of the binding of compound 9 in its free-acid form is altered and enantioselectivity is lost. The addition of an organic modifier (propan-2-ol) reduced the retention times of the enantiomers allowing their elution from the ODS column on which they had previously been retained. Thus a solvent of 50 mM pH 4.0 sodium phosphate buffer modified with 10% (v/v) of propan-2-ol gave k_1 and k_2 values of 9.2 and 12.1, respectively, with an α of 1.3.

The method was used to measure < 0.2% of (R)-enantiomer in compound 9. This is preliminary work and the effects of various factors, e.g., protein concentration, still require detailed study. However, it is concluded that use of BSA as an additive provided a successful alternative to commercial bonded-phase columns. It must also be recognised that a successful separation would not be guaranteed with the available commercial protein columns.

CONCLUSION

The use of additives in the mobile phase has been exploited to achieve separations of polar, water-soluble racemates which were poorly resolved on CSPs. Solute solubility,

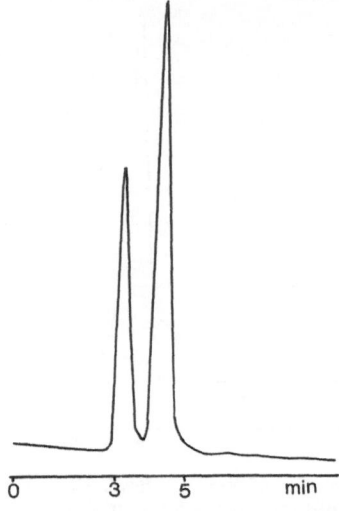

Fig. 7.   Separation of the enantiomers of a methotrexate analogue (compound 9). Chromatographic conditions: column, 25 cm x 4.6 mm id Lichrosorb 10 μ Diol; eluent, sodium phosphate buffer (50 mM, pH 7.0) containing 3 g l^{-1} of BSA; flow rate, 1 ml min^{-1}.

Table 6.   Variation of Bonded Phase Polarity on the Retention and Separation of Methotrexate Analogue Enantiomers with Bovine Serum Albumin (BSA) in the Mobile Phase

Column	k_1	k_2	α
Diol	0.13	0.61	4.7
CN	7.6	21.4	2.8
ODS	Retained	-	

The solvent was 50 mM sodium phosphate pH 7.0 containing BSA at 3 g l^{-1}.

Table 7.   Effect of pH on the Retention and Separation of Methotrexate Analogue Enantiomers on a Diol Column with Bovine Serum Albumin in the Mobile Phase

pH	k_1	k_2	α
7.0	0.13	0.61	4.7
4.9	0.34	0.66	1.9
4.0	2.32	2.67	1.2
3.0	2.42	2.42	1.0

The solvent was 50 mM sodium phosphate pH 7.0 containing BSA at 3 g.l^{-1}.

column efficiency and capacity have been increased and the useful range of a CSP has been extended.

REFERENCES

1.   C. Pettersson, *Trends Anal. Chem.*, 7:209-217 (1988).
2.   R. Gaskell and B. Crooks, *in* "Chiral Separations", D. Stevenson and I. D. Wilson, eds, Plenum, New York, 65-70 (1988).
3.   Y. Okamoto, M. Kawashima, R. Aburatani, K. Hatada, T. Nishiyama and M. Masuda, *Chem. Letters*, 1237-1240 (1986).
4.   U. Kragh-Hensen, *Pharm. Rev.*, 33:17-53 (1981).

A NOTE ON CHIRAL ION-PAIR CHROMATOGRAPHY OF

NOVEL BASIC ANTIHYPERTENSIVE AGENTS

V. de Biasi, M. B. Evans* and W. J. Lough†

SmithKline Beecham, Coldharbour Road
Harlow, Essex CM19 5AD, UK
*Division of Chemical Sciences, Hatfield
Polytechnic, Hatfield, Hertfordshire, UK, and
†School of Pharmaceutical and Chemical Sciences
Sunderland Polytechnic, Sunderland
Tyne and Wear, UK

INTRODUCTION

Since Schill et al. first reported chiral ion-pair high-performance liquid chromatography (HPLC) in 1981 [1], much work has been carried out using this technique [2,3]. However, apart from the use of L-quinine to resolve the racemates of some acidic drugs [5], it has been used almost exclusively for the enantiomeric resolution of β-blockers. This is due to this class of compound having the necessry features for chiral recognition by the acidic chiral ion-pairing reagents used, D-10-camphorsulphonic acid and N-benzyloxy-carbonyl-glycyl-proline. The chiral centre is very close to the basic centre of the molecule and also contains a functional group capable of hydrogen-bonding interaction with the carbonyl groups on the acidic chiral ion-pair reagents [3].

It was noted that these features were also present in some members of a series of substituted *trans*-3,4-dihydro-2,2-dimethyl-2H-1-benzopyrans which have been developed for use as antihypertensive agents [4]. Although there are two asymmetric carbon atoms, their fixed stereochemistry with respect to each other implies the existence of a single enantiomeric pair (Figure 1). It has been shown that there is a marked difference in biological activity between the individual enantiomers of this class of compounds and as such it was important to have a method for enantiomeric determinations for members of this series. The use of chiral ion-pair chromatography for this work was thought appropriate as the distance between the basic centre and the hydrogen bonding centre is of the same order of magnitude as in the β-blockers. Moreover, these two centres lie on a relatively rigid structure, regarded as important in other examples to obtain good enantio-selectivity [1,5].

EXPERIMENTAL

The column used throughout this work was Apex WP Diol, 100 x 1 mm packed by Capital HPLC (Edinburgh, UK). The HPLC system consisted of an electronically modified Waters 6000 A pump (Milford, MA, USA), a Kratos SF770 variable wavelength detector (ABI, NJ, USA) fitted with a microbore flowcell of 0.5 µl volume. Injection was via a Rheodyne 7450 injector (Cotati, CA, USA) and the data was recorded using a Linseis chart recorder (Linseis GMBH, FRG).

All solvents used were of HPLC grade and obtained from Rathburn Chemical Co. (Walkerburn, UK). N,N-Dimethyloctylamine was obtained from Aldrich Chemicals (Gillingham, Dorset, UK). D,L-Alprenolol and D,L-propranolol were obtained from Sigma Chemicals (Poole, Dorset, UK) and the substituted benzopyrans from SmithKline Beecham (Harlow, Essex, UK). The chiral ion-pair reagent, N-benzyloxycarbonyl-glycyl-proline (ZGP), was synthesised at SmithKline Beecham (Harlow, Essex, UK). All samples were

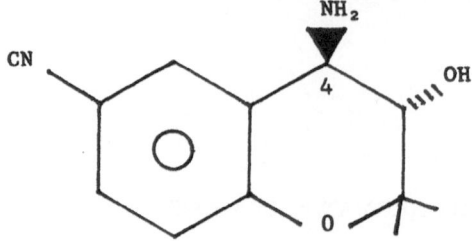

Fig. 1. Structure of *trans*-3,4-dihydro-2,2-dimethyl-2H-1-benzopyran.

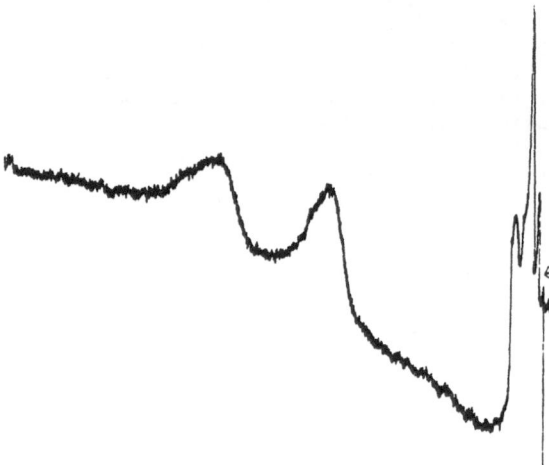

Fig. 2.   Separation of *trans*-3,4-dihydro-2,2-dimethyl-2H-1-benzopyran enantiomers with a hydroxyl group *trans* to a primary amine.

chromatographed at room temperature at a flow rate of 100 µl min^{-1} and at an absorbance of 0.05 Aufs.

RESULTS AND DISCUSSION

The application of chiral ion-pair reagents proved successful for the separation of a number of biologically active substituted benzopyrans. It appears that the chiral recognition is critically dependent on the distance between the hydroxyl group and the chiral centre.   Separation of enantiomers occurred with the hydroxyl group *trans* to a primary amino function (Figure 2) whereas with *cis* stereochemistry separation was not achieved.

Although good enantioselectivity was obtained with compounds having a primary amino group in the 4-position, no examples were obtained for equivalent compounds containing a secondary or tertiary amine (Table 1).

Despite the slight loss in enantioselectivity in going from the 6-methyl to the electron withdrawing 6-nitro substituent, changes to the substituent in the 6-position did not have a marked effect on the retention or resolution of the enantiomers (Table 2). However, the possibility cannot be discounted that the aromatic ring and its substituents had a minor role in contributing to the three points of interaction which are deemed necessary for chiral recognition.

Table 1. Effect of the Substituent on the 4-Position on the Separation of Enantiomers

Substituent	α^*
	1.30
	Single peak ($k' = 6.0$)
	Single peak ($k' = 3.0$)
	Single peak ($k' = 0.5$)

* $\alpha = k'_2/k'_2$, where k'_1 = retention index for the fastest eluting enantiomers and k'_2 = retention index for the slowest eluting enantiomers.

Table 2. Effect of the Substituent on the 6-Position on the Separation of Enantiomers

Substituent	α^*
(Fig. 2)	1.30
	1.32
	1.56
	1.25

* $\alpha = k'_2/k'_2$, where k'_1 = retention index for the fastest eluting enantiomers and k'_2 = retention index for the slowest eluting enantiomers.

The use of microbore HPLC did not compromise the degree of resolution that could be obtained, resolutions of β-blockers obtained on this system (de Biasi et al., in preparation) being comparable with literature values obtained on conventional columns. Microbore HPLC use was advantageous in minimising the consumption of the chiral ion-pair reagent.

This will be even more important in studies currently underway in which less readily available chiral ion-pair reagents are being studied.

CONCLUSIONS

Chiral ion-pair chromatography using the acidic ion-pairing agent, N-benzyloxy-carbonyl-glycyl-proline (ZGP), can be used to separate the enantiomers of *trans*-3,4-dihydro-2,2-dimethyl-2H-1-benzopyrans having a primary amino group in the 4-position. The enantiomeric separation of this class of compounds may be conveniently carried out using microbore HPLC.

REFERENCES

1.   C. Petterson and G. Schill, *J. Chromatogr.*, 204:179-183(1981).
2.   C. Petterson and G. Schill, *Chromatographia*, 16:192-197 (1982).
3.   C. Petterson and G. Schill, *J. Liq. Chromatogr.*, 9:269-290 (1986).
4.   J. M. Evans, C. S. Fake, T. C. Hamilton et al., *J. Med. Chem.*, 26:1582-1589 (1983).
5.   C. Petterson and K. No, *J. Chromatogr.*, 282:671-684 (1983).
6.   V. de Biasi, M. B. Evans and W. J. Lough, manuscript in preparation.

OPTIMISATION OF CHIRAL SEPARATION OF DOXAZOSIN ENANTIOMERS
BY HIGH-PERFORMANCE LIQUID CHROMATOGRAPHY ON A
SECOND GENERATION α_1-ACID GLYCOPROTEIN COLUMN

G. W. Ley, A. F. Fell and B. Kaye*

Department of Pharmaceutical Chemistry
School of Pharmacy, University of Bradford
Bradford BD7 1DP, UK, and
*Pfizer Central Research, Sandwich
Kent CT13 9NT, UK

SUMMARY

The separation of the enantiomers of doxazosin (1-(4-amino-6,7-dimethyoxy-2-quinazolinyl)-4-(1,4-bendioxan-2-yl-carbonyl) piperazine on a second generation α_1-acid glycoprotein column is described. Following optimisation of pH and organic modifier concentration using a modified simplex procedure, an optimum mobile phase composition of phosphate buffer (0.035 M) at pH 7.25 and acetonitrile (87:13 v/v) was established. The method was linear over the range 2.6 to 520 ng (on column) with relative standard deviations of 4.0 and 4.9% for the first and second eluting enantiomers, respectively. The limit of detection for the first and second eluting enantiomers, using UV detection at 254 nm, was 1.3 and 1.7 ng, respectively, on column.

INTRODUCTION

Doxazosin (Figure 1), 1-(4-amino-6,7-dimethoxy-2-quinazolinyl)-4-(1,4-benzdioxan-2-yl-carbonyl) piperazine, is a postsynaptic α_1-adrenoceptor antagonist [1]. The aim of the work reported here was to develop a chiral assay with a view to investigating the drug's chiral metabolism *in vitro* using high-performance liquid chromatography (HPLC).

Chiral columns using α_1-acid glycoprotein (AGP) have been shown to resolve a wide range of different types of compounds of pharmaceutical interest [2]. A second generation Chiral-AGP column recently developed by Hermansson was used in this study. First attempts at separating the enantiomers of doxazosin on this column produced only a partial separation.

To improve upon this initial result a logical stepwise optimisation procedure was instigated as follows:

1. A range of organic modifiers was considered.
2. A factorial design was carried out to interrogate the magnitude of the effects of the chromatographic variables and their interactions.
3. A modified simplex procedure was carried out to optimise the variables.
4. The method was validated with regard to linearity, relative standard deviation and limit of detection.

EQUIPMENT AND MATERIALS

The HPLC equipment comprised: an LKB 2150 HPLC pump, an LKB 2151 variable wavelength monitor (set at 254 nm), an LKB 2210 two-channel recorder and a Rheodyne 7150 injection valve with a 20 µl loop and a Chiral-AGP column (ChromTech AB, Norsborg,

Recent Advances in Chiral Separations, Edited by D. Stevenson and
I. D. Wilson, Plenum Press, New York, 1991

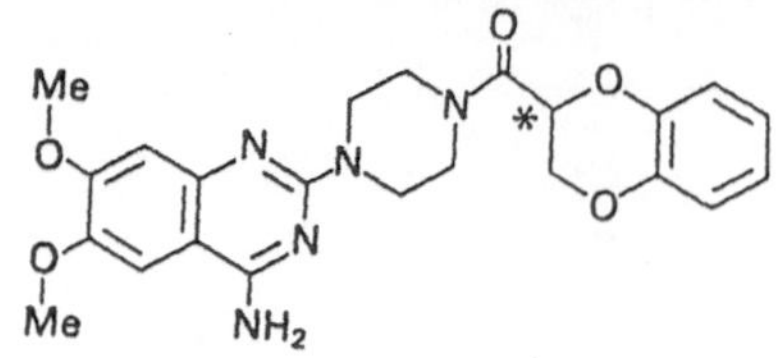

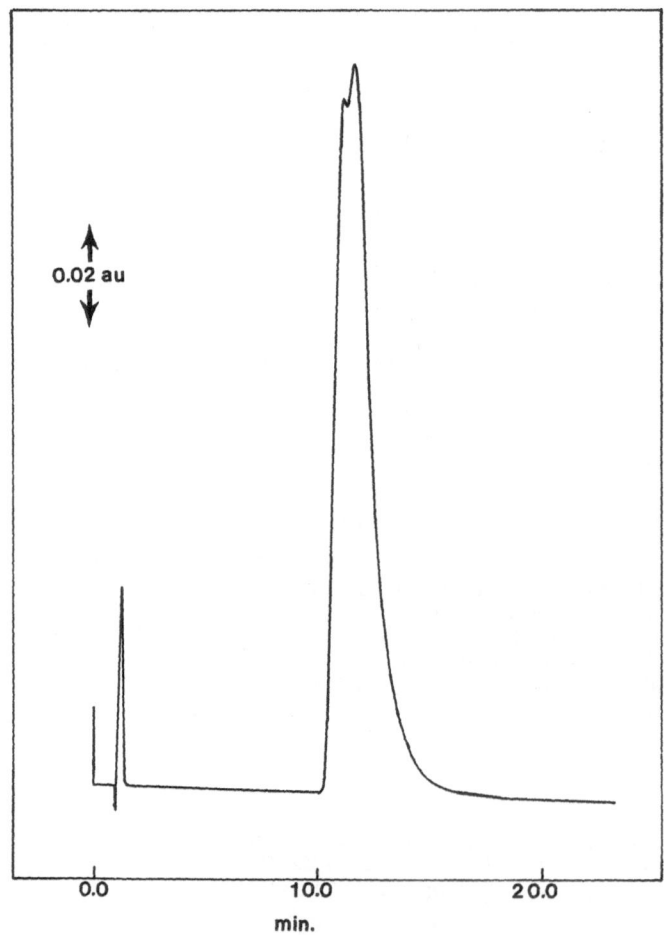

Fig. 1. Doxazosin structure. * = Chiral centre.

Sweden; 100 mm x 4.6 mm id). All solvents were HPLC grade. All mobile phases were filtered through a 0.45 µm Durapore filter and degassed by ultrasonication under reduced pressure. Doxazosin was provided by Pfizer Central Research and dissolved in buffer, and then diluted with the appropriate organic modifier.

RESULTS AND DISCUSSION

The first injection of doxazosin onto the chiral-AGP column showed only a very slight inflection in a single peak (Figure 2). A few range-finding experiments gave rise to a partial resolution. Different organic modifiers were then tried (2-propanol, 1-propanol, methanol and acetonitrile) using a pH 7 phosphate buffer. From Table 1 it can be clearly seen that, under these conditions, acetonitrile was the best of the organic modifiers investigated.

0.02 au

0.0   10.0   2 0.0

min.

Fig. 2. Mobile phase: phosphate buffer (0.01 M, pH 5.4)-2-propanol (90:10 v/v).

98

Table 1. The Values for Peak Separation Function and k' of the Last Eluting Enantiomer

Organic Modifier*		P_i	k' of (S)-Doxazosin
2-Propanol	(10%)	0.60	16.0
1-Propanol	(10%)	0.29	9.7
Methanol	(10%)	0.75	174.5
Acetonitrile	(10%)	0.80	34.6

*All experiments utilised 0.01 M phosphate buffer pH 7.0 with the flow rate set to 0.9 ml min^{-1}.

The quality of separation was assessed by measuring the peak separation function (P_i), defined by Kaiser as the ratio of the valley depth between two peaks and the average peak height, giving rise to values of $P_i = 0$ for total overlap and $P_i = 1$ for baseline separation.

The flow rate for all these experiments was set to 0.9 ml min^{-1}. All the experiments were carried out at ambient temperature (ca. 25°C).

FACTORIAL DESIGN

The effects and interactions of organic modifier percentage, buffer pH and buffer concentration were studied using a three-dimension, two-level factorial design [3] (Figure 3). A, B and C represent the three different variables under consideration; + and - represent high and low values of these variables, respectively. The result, Y, could in the experiment undertaken be expressed in terms of the capacity factors or P_i for each enantiomer.

The magnitude of each result from these equations was used to assess the relative importance of each variable. The factorial design procedure can also be used to establish the number of significant variables; this in turn can be used to select the optimisation rationale applicable to the system under scrutiny.

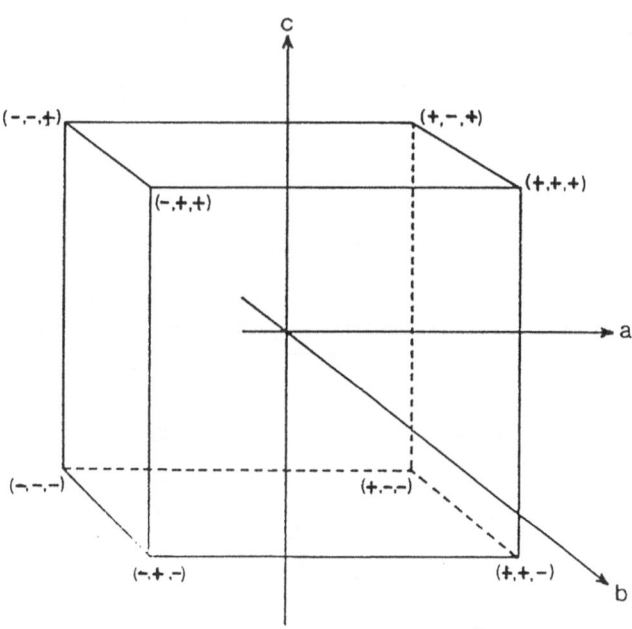

Fig. 3.   Factorial design for three variables (A, B and C): experiments were carried out at each corner of the cube.

Taking into account the data obtained from the consideration of different organic modifiers (Table 1) and the allowable range of pH (3.5 to 7.5) and buffer concentration (0.01 to 0.035 M) for the Chiral-AGP column, the factorial design was carried out for pH, buffer concentration and organic modifier (acetonitrile) percentage (Table 2).

The capacity factors (k') observed for the first and second enantiomers, respectively, and the respective α values are summarised in Table 3, which indicates that pH has the strongest influence on the enantioselectivity.

The effects and interactions of the selected variables were calculated from the k' data above, the results being tabulated in Table 4. From Table 4 it can be clearly seen for both enantiomers that only pH and organic modifier percentage interact significantly. They also have the greatest individual effect.

The results of the factorial design justify the selection of a two-variable simplex design for optimising pH and organic modifier. The concentration of phosphate buffer was 0.035 M.

Table 2.   Experimental Design Used in the Factorial Design for the Doxazosin Analysis

Experiment No.*		pH	Buffer Concentration (M)	Acetonitrile (%)
1	+++	7.0	0.035	10
2	++-	7.0	0.035	5
3	+-+	7.0	0.010	10
4	+--	7.0	0.010	5
5	-++	4.5	0.035	10
6	-+-	4.5	0.035	5
7	--+	4.5	0.010	10
8	---	4.5	0.010	5

*See also Figure 3.

Table 3.   Calculated Chromatographic Parameters for the Factorial Design for Doxazosin Analysis

Experiment No.	k' (First Enantiomer)	k' (Second Enantiomer)	α
1	12.7	17.4	1.37
2	63.2	88.9	1.41
3	24.5	34.5	1.41
4	66.2	93.7	1.42
5	6.3	8.1	1.29
6	4.1	5.2	1.27
7	1.5	1.5	1.00
8	3.6	4.6	1.28

Table 4. Modulus of the Values of the Effects and Interactions of the Variables Considered in the Factorial Design, Calculated Using Capacity Factors

Variable	Enantiomer 1	Enantiomer 2
Effect of pH	37.8	53.8
Effect of buffer concentration	2.4	3.7
Effect of organic modifier	23.0	32.7
Interaction pH x buffer concentration	5.0	7.3
Interaction pH x organic modifier percentage	23.1	32.6
Interaction buffer concentration x organic modifier %	1.1	1.6

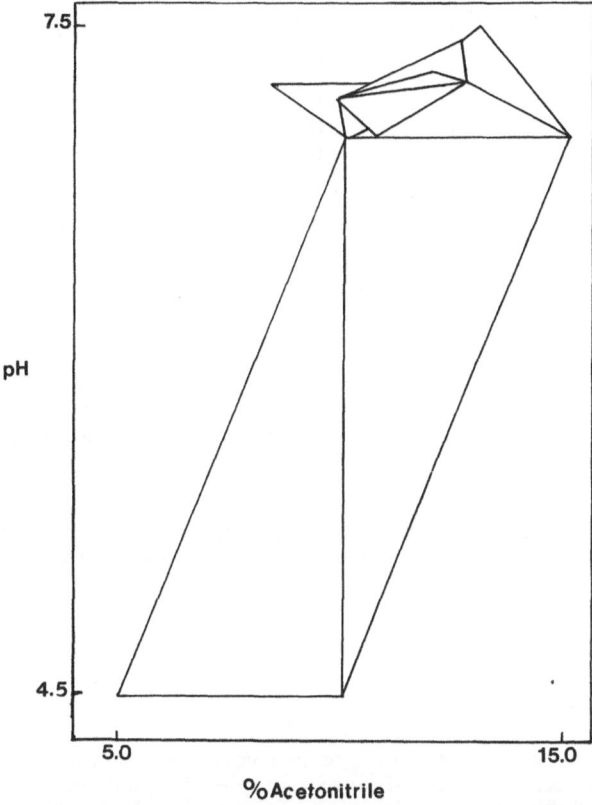

Fig. 4. Movements of the modified sequential simplex method in the location of the optimum conditions of pH and acetonitrile concentration for the resolution of doxazosin on the Chiral-AGP column.

MODIFIED SIMPLEX

The modified simplex procedure [4] makes it necessary to define a chromatographic response function (CRF) which characterises the chromatogram with respect to resolution and analysis time. Conventionally, the CRF is often defined in terms of a parameter related to separation (e.g., R_s), and two parameters related to the minimum and maximum required retention times, respectively, for example:

$$CRF = (R_s)^n + T_1 + T_2.$$

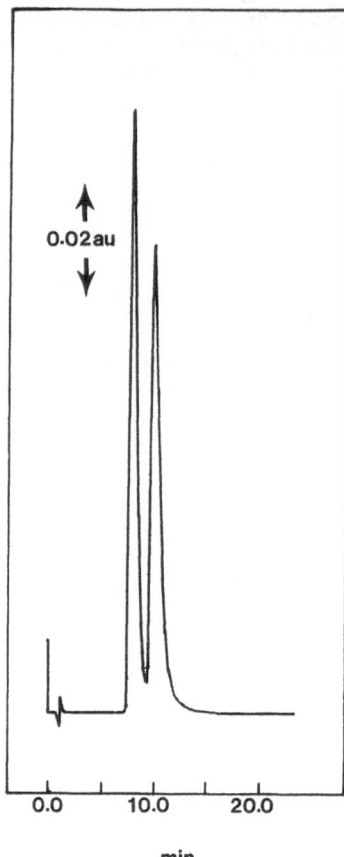

Fig. 5.   Chromatogram obtained by injection of 50 ng doxazosin into the optimised eluent system described.

The aim of the present work was to develop an optimisation procedure to obtain the best possible resolution in the shortest possible time. To this end a new empirical CRF was proposed, of the form:

$$CRF = P_i^5/\log_{10}t_r,$$

where P_i is the peak separation function (for peak pair i) and t_r is the retention time of the second enantiomer peak in minutes.

If the simultaneous optimisation of two variables is considered (typically pH and organic modifier percentage), the standard simplex method involves starting with three experiments using different values for each of the two variables (forming a triangle on the response surface). The experiments are assessed as "best" (B), "next best" (N) and "worst" (W), on the basis of the CRF values. Then the point W is reflected through the middle of B and N to determine the next pair of values for the two parameters, giving a new simplex BNW. The process is then repeated sequentially until an apparent optimum is found.

In the modified simplex procedure, two new operations, expansion and contraction, are incorporated. Thus the simplex can accelerate towards the apparent optimum region. Having approximately located this region, the search area is progressively reduced until the apparent optimum is located.

The criterion used for stopping the modified simplex in the study undertaken was that the change in the value for the CRF resulting from each experiment should be less than 5%. In fact this was obtained after eleven experiments. The final simplex plot is shown in Figure 4.

102

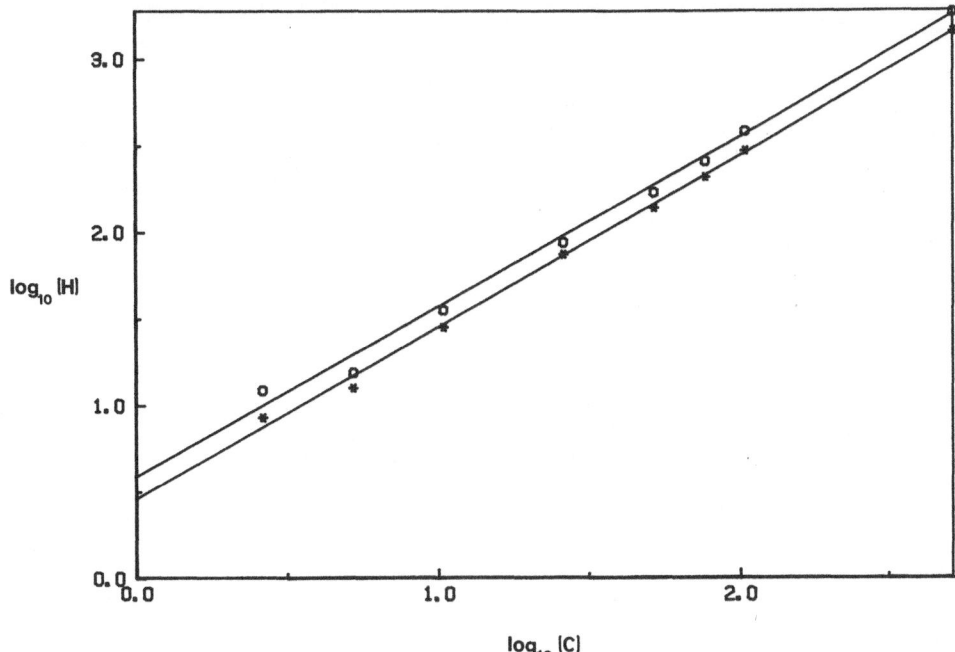

Fig. 6.   Calibration plot for doxazosin separated on a Chiral-AGP column: o, first eluting
enantiomer; *, second eluting enantiomer; C, on-column concentration (ng); H, peak height in
absorbance units (x 100).

Thus the optimum mobile phase conditions were established as: phosphate buffer
(0.035 M pH 7.25)-acetonitrile (87:13 v/v). The optimum chromatogram under these conditions is shown in Figure 5.

The linearity of the optimised method was assessed in the range 2.6 to 520 ng on-
column of each enantiomer (n = 8; Figure 6). The method was linear over this concentration,
range with a correlation coefficient for the first peak of 0.997 and for the second peak of
0.999. The curves obtained for each enantiomer also had similar slope (first enantiomer =
0.978, second enantiomer = 0.989). The relative standard deviation (RSD) was calculated by
making five injections at 52.0 ng of each enantiomer on column to give RSD values of 4.0%
for the first and 4.9% for the second eluting enantiomer. The enantiomeric concentration
corresponding to the limit of detection (signal to noise ratio = 2) was found to be 1.3 and
1.7 ng on column, for the first and second eluting enantiomers, respectively.

CONCLUSION

At this stage of development, the method developed is considered to be sufficiently
sensitive and precise for use in *in vitro* experiments on the metabolism of doxazosin. This
work forms part of a continuing series of experiments [5] designed to explore the application
of systematic optimisation in chiral HPLC.

REFERENCES

1.   P. Timmermans, H. Kwa, A. Karamat and P. van Zwieten, *Arch. Int. Pharmacodyn.
          Ther.*, 245:218-235 (1980).
2.   J. Hermansson, K. Strom and R. Sandberg, *Chromatographia*, 24:520-526 (1987).
3.   J. C. Berridge, "Techniques for the Automated Optimisation of HPLC Separations",
          John Wiley, New York, Ch. 4.3, pp 62-69 (1985).
4.   J. C. Berridge, "Techniques for the Automated Optimisation of HPLC Separations",
          John Wiley, New York, 4.3:62-69; 6.2:125-151 (1985).
5.   A. Fell, T. Noctor, J. Mama and B. Clark, *J. Chromatogr.*, 434:377-382 (1988).

ATTEMPTS TO QUANTIFY THE EFFECT OF SELECTED STRUCTURAL
PARAMETERS ON THE SEPARATION OF SOME N-ARYLTHIAZOLINE-2-THIONE
AND N-ARYLTHIAZOLIN-2-ONE ATROPISOMERS ON THE CHIRAL STATIONARY
PHASE, CELLULOSE TRIACETATE (CTA1)

C. Roussel*, J.-L. Stein, M. Sergent
and R. Phan Tan Luu

URA 1410 CNRS, ENSSPICAM
Faculte Sciences St Jerôme
13397 Marseille Cedex 13, France

SUMMARY

Molecular recognition mechanisms for cooperative chiral stationary phases such as cellulose triacetate are still poorly understood. Here studies on the effect of the position and type of the subtituent on the separation of N-arylthiazoline-2-thione and N-arylthiazolin-2-one atropisomers on cellulose triacetate (CTA1) are described.

INTRODUCTION

The molecular recognition mechanisms for cooperative chiral stationary phases (CSPs) are not clearly understood since the active sites are not easily identified in contrast to the situation for molecular CSPs in which models for chiral recognition have been proposed. A typical cooperative chiral stationary phase is microcrystalline cellulose triacetate which has been used successfully for many analytical and preparative chiral separations [1-3].

We have applied, for the first time, a two-level full factorial design to quantify the effect of three selected structural parameters (factors) on the separation and capacity factors (responses) of atropisomers in some N-arylsubstituted heterocycles (general structures shown in Figure 1) on cellulose triacetate (CTA1). This methodology allows a quantitative approach to the main effect of variables which affect the retention and separation according to the postulated model, thus greatly facilitating the discussion of the data.

MATERIALS AND METHODS

The synthesis of compounds 1 to 8 has been already described, as well as their barriers to rotation around the pivot bond [4,5]. The barriers to enantiomerisation are larger than 29 kcal mol^{-1} and thus the atropisomers are particularly stable at room temperature.

Liquid Chromatography on Microcrystalline Cellulose Triacetate

Cellulose triacetate (15-25 μm from Merck) was packed in a thermostated 200 x 25 mm glass column equipped with a 5 cm^3 injection loop. Both UV (LKB 2138 UVICORD) and polarimetric (Perkin Elmer 241) detectors were used to monitor the eluent. In all cases the elution solvent was ethanol-water (96:4), at a flow rate of 138 ml hr^{-1}, and pressure drop of ca. 1.7 bar, at a temperature of 25°C. For analytical runs, 2-3 mg of racemate were used to

---

* Author to whom correspondence should be addressed.

Recent Advances in Chiral Separations, Edited by D. Stevenson and
I. D. Wilson, Plenum Press, New York, 1991

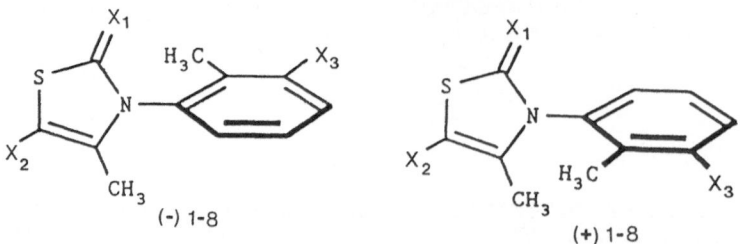

Fig. 1.　Structure of the N-arylthiazoline-2-thiones and N-arylthiazolin-2-ones used in these studies. X1 = Oxygen or sulphur, X2 = hydrogen or methyl, X3 = hydrogen or methyl.

determine k'_1, k'_2 and α. 1,3,5-Tri-t-butylbenzene was used as reference [6]. When the separation was particularly poor the capacity factors (given in Table 1) were determined by separate injection of each enantiomer. Semipreparative injections were performed on 80-100 mg of samples in 5-10 ml ethanol depending on the solubility of the sample. The flow rate used for preparative injections was 98 ml hr^{-1}. After three to five cycles, samples of high enantiomeric purity suitable for the determination of the capacity factors were obtained.

RESULTS AND DISCUSSION

　　The capacity factors for compounds 1-8 on cellulose triacetate are listed in Table 1, and are arranged according to the sign of the eluted enantiomer and not according to the actual order of elution. This arrangement corresponds to the absolute configuration since we have shown chemically and by substituent effect on rotatory power that all the positive atropisomers have the same absolute known configuration [5,7]. Misleading results would have been obtained in ranging the atropisomers in a series of "first-eluted enantiomers" and in a series of "second-eluted enantiomers" since we would have compared compounds with opposite spatial arrangement around the pivot bond.

　　We were interested in three structural modifications which may affect the spatial steric requirement of the atropisomer and (or) the dipole moment and basicity of the heterocyclic and aryl parts. These three structural modifications are denoted X_1, X_2 and X_3: X_1 is oxygen (level -1) or sulphur (level +1), X_2 is hydrogen (level -1) or methyl (level +1), X_3 is hydrogen (level -1) or methyl (level +1). The combination of these three structural modifications (factors) leads to the design of eight (2^3) compounds which were

Table 1.　Compounds, Design Levels, Responses and Column of Signs in the 2^3 Factorial Design

Compound	Design Levels			Responses			Signs for Interactions			
	X_1	X_2	X_3	$k'(+)$	$k'(-)$	k'_2/k'_1	X_1X_2	X_1X_3	X_2X_3	$X_1X_2X_3$
1	-	-	-	2.05	1.66	1.24	+	+	+	-
2	+	-	-	1.52	2.30	1.51	-	-	+	+
3	-	+	-	0.66	0.55	1.21	-	+	-	+
4	+	+	-	0.87	0.78	1.12	+	-	-	-
5	-	-	+	0.81	1.92	2.36	+	-	-	+
6	+	-	+	0.96	3.09	3.20	-	+	-	-
7	-	+	+	0.35	0.35	1.00	-	-	+	-
8	+	+	+	0.64	0.61	1.04	+	+	+	+

synthesised. In Table 1, X_1, X_2 and X_3 are the factors and + and - indicate a measure at the higher or the lower level, respectively, and define the precise structure of the compound under study. For instance, the measurement with X_1, X_2 and $X_3 = -1$ (first line of Table 1) corresponds to 3-(2'-methylphenyl)-4-methyl-4-thiazolin-2-one.

The formalism of the full factorial design according to a quadratic model indicates that an observable (response) Y can be expressed according to Eq. 1 [8]:

$$Y = b_0 + b_1X_1 + b_2X_2 + b_3X_3 + b_{12}X_1X_2 + b_{13}X_1X_3 + b_{23}X_2X_3 + b_{123}X_1X_2X_3. \tag{1}$$

The coefficients in Eq. 1 are obtained by solving the height equations obtained by replacing in the equation the factors X_i by -1 or +1 according to the experimental design 1 (Table 1). The result of this is that the response $k'(+)$ (capacity factor of the dextrorotatory atropisomer) can be expressed by Eq. 2:

$$k'(+) = 0.98 + 0.015\,X_1 - 0.35\,X_2 - 0.29\,X_3 + 0.11\,X_1X_2 + 0.09\,X_1X_3 + 0.15\,X_2X_3 - 0.075\,X_1X_2X_3. \tag{2}$$

Inspection of the coefficients in Eq. 2 points out that two steric structural modifications (X_2 and X_3) have a drastic main effect on retention: namely the introduction of a methyl in position 5 of the heterocycle or introduction of a methyl group in position 3' of the aryl part. In both cases such a modification results in a decrease of the capacity factor which is not compensated by the positive interaction term X_2X_3. To our surprise the large electronic change obtained on going from an oxygen to a sulphur in position 2 (X_1) has a minor effect on the retention of the positive atropisomer. A valuable property of the factorial design is that it makes possible not only the calculation of the main effect (average effect) but also interaction effects between variables as well. Inspection of the interaction diagrams between variables (Scheme 1) indicates that X_2 has only a slight interaction with X_1 and with X_3, respectively, whereas the interaction between variables X_1 and X_3 is very weak. On a molecular basis this might indicate that the structural change X_2 modifies to some extent the interaction of variable X_1 and X_3 with the chiral support.

The results are completely different in Eq. 3 which expresses the capacity factor of the laevorotatory atropisomers:

$$k'(-) = 1.40 + 0.29\,X_1 - 0.83\,X_2 + 0.08\,X_3 - 0.16\,X_1X_2 + 0.07\,X_1X_3 - 0.17\,X_2X_3 - 0.06\,X_1X_2X_3. \tag{3}$$

For such a spatial arrangement, two structural modifications are important (X_1 and X_2). On going from an oxygen to a sulphur, a strong positive effect is observed (coefficient 0.29), whereas on going from hydrogen to methyl in position 5 of the heterocycle a very strong negative effect is observed. The structural modification of hydrogen for methyl in position 3' of the aryl ring (X_3) has a minor effect on the retention. Inspection of the interaction diagrams (Scheme 2) shows that the interactions between variables X_1 and X_2 or X_3 are weak. X_2 and X_3 show a slight interaction.

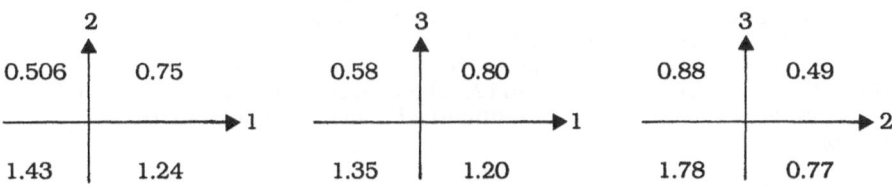

Scheme 1. Interactions between variables for the response $k'(+)$.

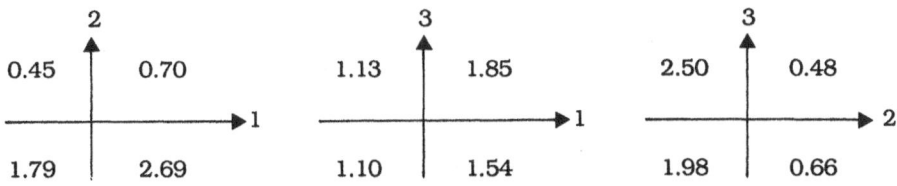

Scheme 2. Interactions between variables for the response $k'(-)$.

It emerges from these results that position 5 of the heterocycle is a very sensitive area for the interaction of both atropisomers with the cellulose triacetate chiral "receptors". This is clearly shown by the negative sign and the magnitude of both coefficients for X_2 in Eqs 2 and 3. It is worth noting that for the same heterocyclic framework in thiazoline-2-thione derivatives of alanine and 2-phenethyl amine, the introduction of a methyl group in position 5 also resulted in a similar decrease of the capacity factors [9]. Conversely, both atropisomers appear to behave independently as far as structural modifications X_1 and X_3 are concerned.

On a molecular recognition basis, such an opposite behaviour might arise either from the occurrence of a very different spatial arrangement for each atropisomer within the same cellulose triacetate chiral receptor or from the occurrence of different chiral receptors as is probably the case for cooperative phases. More probably these two hypotheses are operating together.

Equation 4, which gives the variation of the separation factor k'_2/k'_1 (as it is usually defined), indicates that strong interactions are observed and (as would be expected from an analysis of Eqs 2 and 3) that all three factors are important. One has to question the physical significance of the separation factor in the case of concurrent sites as is probably the case for cellulose triacetate [10]:

$$k'_2/k'_1 = 1.58 + 0.13\,X_1 - 0.49\,X_2 + 0.31\,X_3 - 0.14\,X_1X_2 + 0.08\,X_1X_3 - 0.38\,X_2X_3 - 0.05\,X_1X_2X_3.$$

(4)

We are currently approaching the molecular recognition mechanism in the cooperative phase CTA1 using factorial design in a larger set of atropisomers in which many substitutions are possible by unequivocal synthesis. This might be a complementary approach to the theoretical approach of the modeling of chiral recognition mechanisms on CTA1 developed recently on lactones with charge density mapping [11].

Furthermore, comparison of the analytical expression of the same responses, using the same factorial design obtained on different cooperative phases, might give a rather good description of the similarities and differences in molecular recognition mechanism within the various phases.

REFERENCES

1. S. G. Allenmark, "Chromatographic Enantioseparation: Methods and Applications", Ellis Horwood, Chichester, UK (1988).
2. M. Zief and L. J. Crane, "Chromatographic Chiral Separations", Chromatographic Science Series, Vol. 40, Marcel Dekker, New York (1988).
3. A. Mannschreck, H. Koller and R. Wernicke, *Kontakte (Darmstadt)*, 1:40-48 (1985).
4. C. Roussel, M. Adjimi, A. Chemlal and A. Djafri, *J. Org. Chem.*, 53:5076-5080 (1988).
5. C. Roussel, J. L. Stein and F. Beauvais, *New J. Chem.*, 14:169-173 (1990).
6. H. Koller, K. H. Rimböck and A. Mannschreck, *J. Chromatogr.*, 282:89-94 (1983).
7. C. Roussel and A. Chemlal, *New J. Chem.*, 12:947-952 (1988).
8. G. E. P. Box and N. R. Draper, "Empirical Model-Building and Response Surfaces", John Wiley, New York, 4:105-142 (1987).
9. R. Isaksson and J. Roschester, *J. Org. Chem.*, 50:2519-2521 (1985).
10. C. Roussel, J. L. Stein, F. Beauvais and A. Chemlal, *J. Chromatogr.*, 462:95-103 (1989).
11. R. M. Wolf, E. Francotte and D. Lohmann, *J. Chem. Soc. Perkin Trans.*, 2:893-901 (1988).

A NOTE ON TEMPERATURE EFFECTS ON HIGH-PERFORMANCE LIQUID CHROMATOGRAPHIC ENANTIOMERIC RESOLUTION USING CHIRAL STATIONARY PHASES CONTAINING *TRANS*-1,2-DIAMINOCYCLOHEXANE DERIVATIVES

F. Gasparrini, D. Misiti and C. Villani

Dipartimento Studi di Chimica e Tecnologia
delle Sostanze Biologicamente Attive
Università "La Sapienza", Piazzale A. Moro 5
00185 Rome, Italy

INTRODUCTION

The explosive growth of high-performance liquid chromatography (HPLC) during the past decade has been made possible by major advances both in materials available to the chromatographer and in the techniques needed to make new classes of compounds suitable for HPLC separation.

Several types of chiral stationary phases (CSPs) have been developed and successfully employed in the direct resolution of racemic compounds [1-3]. Totally synthetic CSPs, obtained by grafting small chiral molecules onto a silica surface, exhibit the following features:

1. Each enantiomeric form, as well as the "racemic version" of the selector is available.
2. The chemical structure of the selector can be easily manipulated in order to improve and/or broaden chiral recognition ability.
3. Highly efficient columns can be obtained.

In the framework of our research on enantiomeric resolutions by HPLC, three chiral sorbents containing different R,R(-) 1,2-diaminocyclohexane derivatives have been prepared and evaluated. These phases have been obtained by reacting glycidoxypropyl silica gel (GPSG) with R,R(-) 1,2-diaminocyclohexane and then by derivatising this matrix with 3,5-dinitrobenzoyl chloride (CSP-DACH DNB) [4,5], pentafluorobenzoyl chloride (CSP-DACH PFB) or alternatively [6] with α-naphthoyl chloride (CSP-DACH NAPHT) [7], according to Reaction Scheme 1.

These synthetic σ-bonded CSPs show great chemical and thermal stability. The CSP-DACH DNB has been almost exclusively used in normal phase chromatography, whereas CSP-DACH PFB and CSP-DACH NAPHT may be utilised for both normal and reversed-phase separations. Moreover, these phases have been successfully utilised in sub- and supercritical fluid chromatography (SFC) in packed columns [8].

In general, adequate values for enantioselectivity (α) in the resolution of different racemates have been obtained for a range of different classes of compounds and chiral drugs successfully, including sulfoxides, selenoxides, phosphinoxides, phosphinates, benzodiazepines, β-blocking agents, non-steroidal antiinflammatory agents, and so on.

Column temperature is a parameter that has, up until now, received little attention with the majority of HPLC chiral separations being conducted at room temperature. In the present work we have investigated the dependence of retention and stereoselectivity on temperature for different combinations of CSP, solute and mobile phase. Variable temperature chromatographic runs provide useful information concerning the thermodynamic parameters for the CSP-solute interaction, as given by the Gibbs-Helmholtz relationship [9-11]:

Recent Advances in Chiral Separations, Edited by D. Stevenson and I. D. Wilson, Plenum Press, New York, 1991

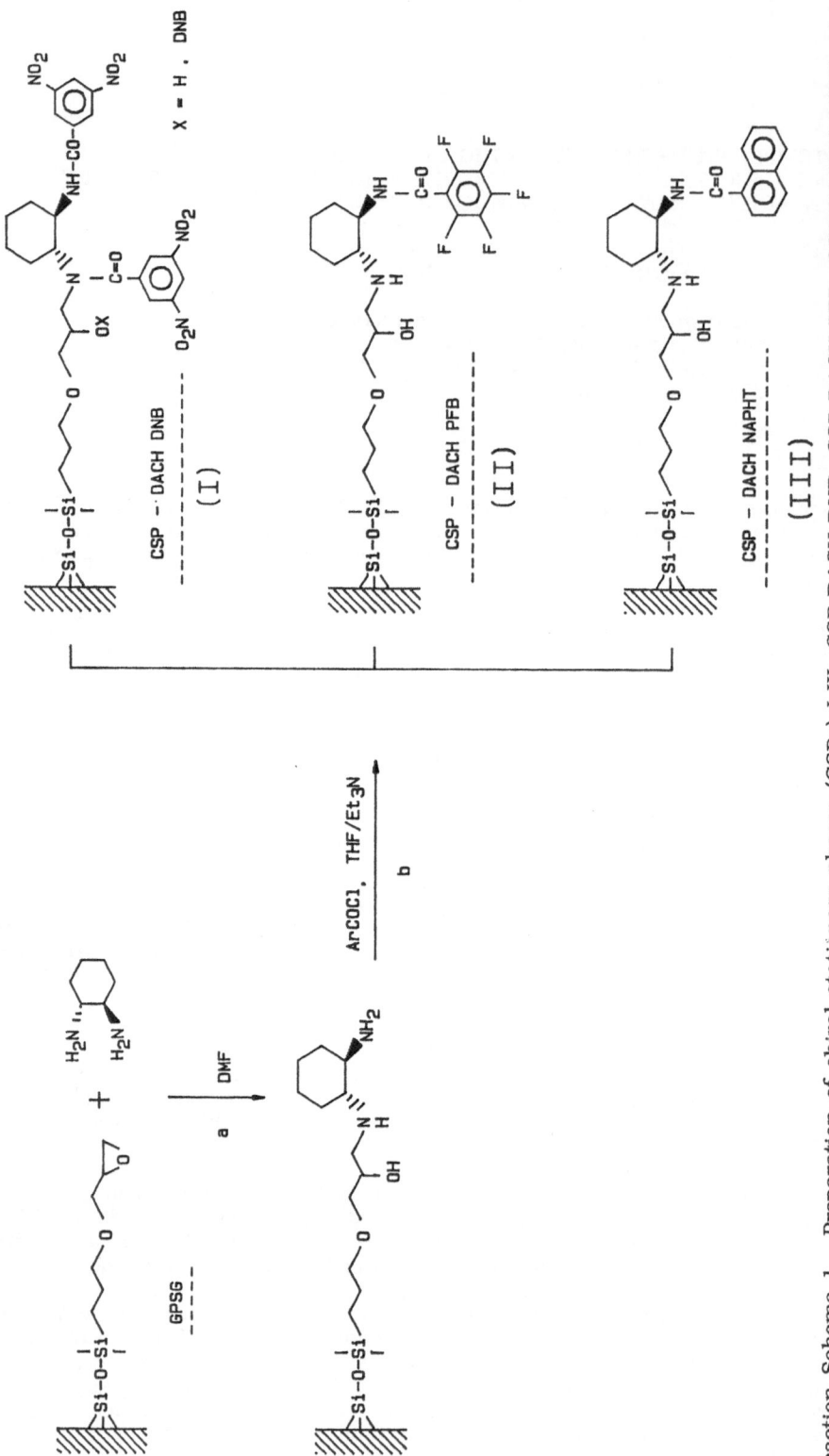

Reaction Scheme 1. Preparation of chiral stationary phases (CSPs) I-III: CSP-DACH-DNB, CSP-DACH-PFB, CSP-DACH-NAPHT. (a) Glycidoxypropylsilica gel (GPSG), R,R(-) or S,S(+)-diaminocyclohexane (DACH), DMF, 96 hr, room temperature. (b) Aroyl chloride (3,5-dinitrobenzoyl chloride (CSP-DACH DNB); pentafluorobenzoyl chloride (CSP-DACH PFB); 1-naphthoyl chloride (CSP-DACH NAPHT).

110

$$\Delta\Delta G° = \Delta\Delta H° - T\Delta\Delta S° = - RT\ln\alpha,$$

where $\Delta\Delta G°$, $\Delta\Delta H°$ and $\Delta\Delta S°$ are the differences in the standard change of free energy, enthalphy and entropy, respectively, related to the equilibria of the solutes adsorption onto the CSP, and α represents the magnitude of enantioselectivity ($\alpha = K_2'/K_1'$).

The interactions involved in the resolution processes (π-π donor acceptor, dipole-dipole and hydrogen bond interactions) have been studied through the evaluation of the relationship between enantioselectivity factor (α) and temperature. Following the same approach, the influence of the eluent composition in both normal and reversed-phase separation mechanisms on the performance of stereoselective recognition has been examined.

EXPERIMENTAL

Apparatus

Analytical liquid chromatography was performed on a Waters chromatograph (Waters Associates, Milford, MA, USA) equipped with a U6K universal injector, two M510 solvent delivery systems and a M490 programmable multi wavelength detector. A temperature control module (TMC) was employed for variable temperature runs. Chromatographic data were collected and processed on a Waters™ 840 data and chromatography control station.

Materials

LiChrosorb Si100, 5 µm and HPLC-grade solvents were from Merck (Darmstadt, FRG); 3-glycidoxypropyltrimethoxysilane (GOPTMS) was from Janssen (Beerse, Belgium); 1R,2R(-)-diaminocyclohexane (1R,2R DACH) and 1S,2S(+)-diaminocyclohexane (1S,2S DACH), 3,5-dinitrobenzoyl chloride (DNB-Cl), 2,3,4,5,6-pentafluorobenzoyl chloride (PFB-Cl), 1-naphthoyl chloride (1-NAPHT-Cl) and Lawesson's reagent were from Fluka (Buchs, Switzerland). Remaining chemicals were reagent grade and used as purchased.

Preparation of CSPs

A general procedure for the synthesis of the stationary phases is shown in Scheme 1.

Column Packing

The column was made of stainless steel (250 mm x 4.0 mm or 150 mm x 4.0 mm id) packed with LiChrosorb Si100 CSP-DACH DNB (5 µ), LiChrosorb Si100 CSP-DACH PFB (5 µ) or LiChrosorb Si100 CSP-DACH NAPHT (5 µ) using the slurry packing procedure. Grafted silica (3.5 or 1.5 g) was dispersed in chloroform (50 or 30 ml, respectively) and then treated ultrasonically for 5 min. The slurry thus obtained was packed with a Haskel Model pump DSTV-122 using n-hexane as pressurising solvent.

Analytes

Racemic compounds used in this study (Figure 1) were obtained by simple acylation (acylchloride, dichloromethane, triethylamine, room temperature) of commercially available amines; thioamide (structure 4, Figure 1) was obtained from amide (structure 3, Figure 1) by means of Lawesson's reagent [12].

Chromatographic Conditions

Normal phase separations were carried out with flow rates of 2.0 ml min^{-1}; reversed-phase separations were performed with flow rates of 1.0 ml min^{-1}; UV absorbance was monitored at 254 nm. Temperature ranges investigated were 25 to 95°C for normal phase runs and 25 to 65°C for reversed-phase runs.

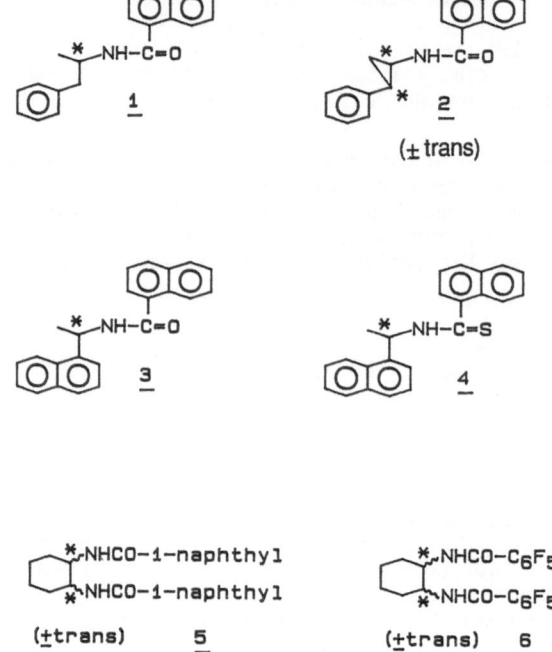

Fig. 1. Structures of racemic compounds examined.

RESULTS AND DISCUSSION

The effect of the temperature on retention and selectivity are described by Eqs 1 and 2:

$$\ln K'_1 = -\frac{\Delta H^{\circ}_i}{RT} + \frac{\Delta S^{\circ}_i}{R} + \ln\phi \tag{1}$$

$$\ln \frac{K'_i}{K'_j} = \ln\alpha = -\frac{\Delta\Delta H^{\circ}_{ij}}{RT} + \frac{\Delta\Delta S^{\circ}_{ij}}{R}, \tag{2}$$

where ΔH° and ΔS° are the enthalpy and entropy changes associated with the analyte retention process, and ϕ is the phase ratio; the subscripts i and j refer to the enantiomeric forms of a generic racemic solute. Both equations predict a linear inverse relationship between $\ln K'$ or $\ln\alpha$ and temperature.

Thermodynamic parameters obtained for analytes 1-6 (see Figure 1) in different chromatographic conditions are listed in Table 1. Typical graphs of these relationships (Van't Hoff plots) are reported in Figure 2.

Several high performance reversed-phase chromatograms of the separation of the racemic mixture of analyte 3, carried out at different temperatures on the CSP-DACH PFB column are shown in Figure 3.

The results obtained clearly show that at the lower end of the temperature range investigated, the capacity factor (K') of both isomers and the enantioselectivity factor (α) increase while a small amount of peak broadening occurs. The broadening of the peaks may have been due to a slow mass transfer process occurring in the above range of temperatures.

Table 1. Thermodynamic and Chromatographic Parameters for Analytes 1-6 and CSPs I-III in Different Chromatographic Conditions

Analyte	CSP	Eluent (v/v)		K_1 (25°C)*	α (25°C)	$-\Delta H°$† (cal mol^{-1})	$-\Delta\Delta H°$ (cal K^{-1}mol^{-1})	$-\Delta\Delta S°$ (call K^{-1}mol^{-1})
1	DACH-DNB	n-Hexane-IPA	(70:30)	12.65	1.24	4834	407	0.94
2	DACH-DNB	n-Hexane-IPA	(70:30)	30.14	2.09	5494	1282	2.86
3	DACH-DNB	n-Hexane-IPA	(60:40)	23.67	4.38	6408	2380	5.10
4	DACH-DNB	n-Hexane-IPA	(60:40)	8.21	1.39	5265	556	1.24
3	DACH-PFB	IPA-H$_2$O	(47:53)	4.05	1.80	5275	1470	3.76
3	DACH-PFB	MeOH-H$_2$O	(55:45)	3.46	1.46	4730	815	1.98
3	DACH-PFB	MeCN-H$_2$O	(45:55)	3.85	1.14	3120	258	0.61
3	DACH-PFB	MeOH-H$_2$O	(60:40)	2.12	1.97	4269	1039	2.11
6	DACH-NAPHT	MeOH-H$_2$O	(60:40)	3.24	1.33	4018	427	0.87

* K_1 = Capacity factor of the first eluted enantiomer.
† Values listed are accurate to ± 150 cal mol^{-1} (obtained from $\ln K_1$ versus $1/T$ plots).
 Values listed are accurate to ± 0.2 cal K^{-1}mol^{-1} (obtained from $\ln\alpha$ versus $1/T$ plots).
IPA = Propan-2-ol, MeOH = methanol, MeCN = acetonitrile.

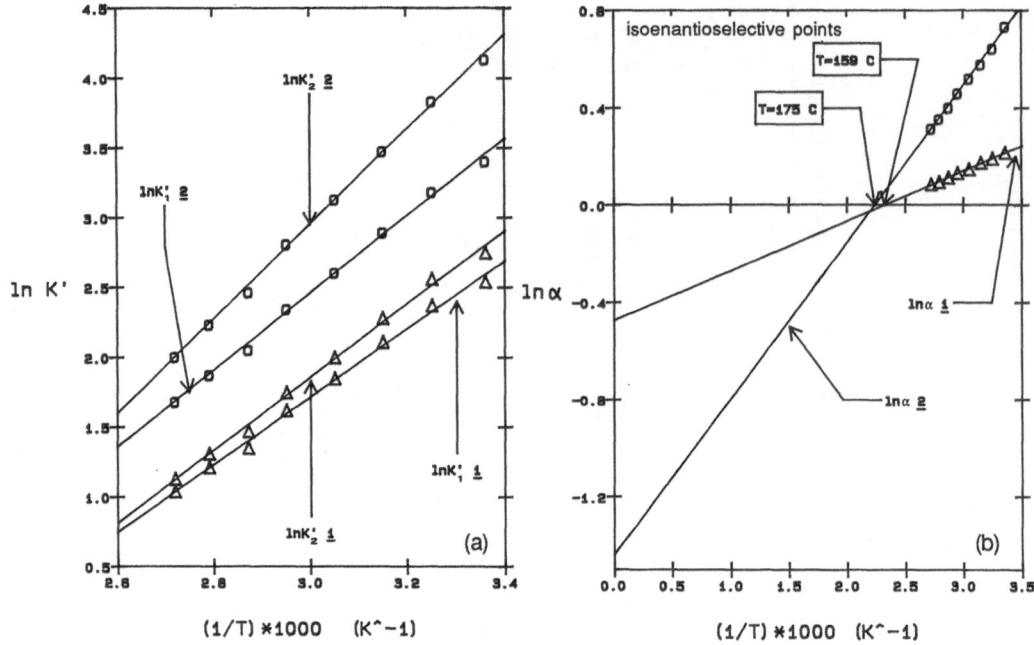

Fig. 2. The linear inverse relationship between (a) $\ln K'$ and temperature; (b) $\ln\alpha$ and temperature for analyte 1, racemic amphetamine (1-naphthoyl derivative) and analyte 2, racemic tranylcypromine (1-naphthoyl derivative). Column: 150 mm x 4 mm id CSP-DACH DNB. Eluent: n-hexane-IPA 70:30 v/v. Flow rate: 2.0 ml min^{-1}. Temperature: from 25 to 95°C.

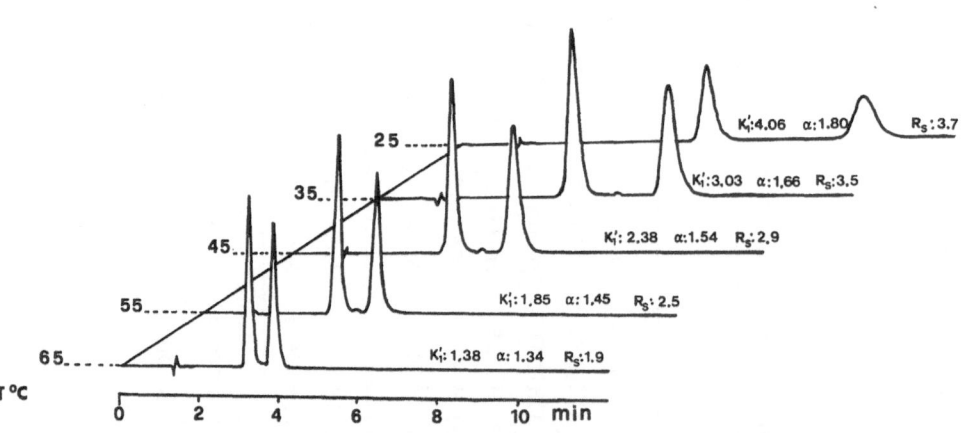

Fig. 3. Temperature influence on capacity factor (K'), enantioselectivity (α) and resolution (R_S) for analyte 3. Column: 150 mm x 4 mm id CSP-DACH PFB. Eluent: IPA-H$_2$O 47:53 v/v. Flow rate: 1.0 ml min^{-1}. Temperature: from 25 (top) to 65°C (bottom), 10°C increments.

114

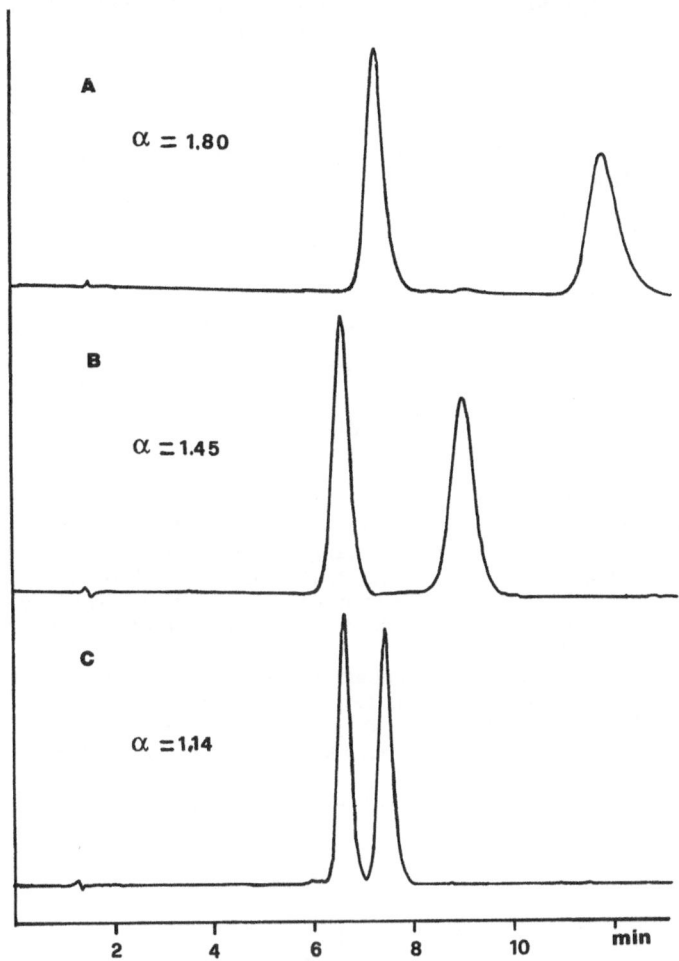

Fig. 4.   Solvent influence on enantioselectivity (α) for analyte 3. Column: 150 mm x 4 mm id CSP-DACH PFB. Eluent: (A) IPA-H_2O 47:53 v/v; (B) MeOH-H_2O 55:45 v/v; (C) MeCN-H_2O 45:55 v/v. Flow rate: 1.0 ml min^{-1}. Temperature: 25°C.

However, it is clear that decreasing the temperature caused an enhancement of the chromatographic resolution of the peaks corresponding to both enantiomers (see Figure 3). Each case should, however, be assessed individually depending on the experimental conditions such as the type of CSP and analyte.

All of the resolution processes investigated are enthalpy controlled and $\Delta\Delta S°$ values are negative, as expected on the basis of the three-points rule [13].

The results obtained for the resolution of compounds 3 and 4 on the CSP-DACH DNB clearly show the importance of H-bonding in the CSP-solute association: replacement of the amide moiety by the less polar thioamide moiety caused a drastic reduction in the stereoselectivity (the α value decreased from 4.38 to 1.39). This shows that a strong hydrogen bond between the amidic hydrogen of the CSP-DACH DNB and the oxygen of the analyte was involved in the chiral recognition process (a difference of ca. 3 Kcal mol^{-1} is found between the $\Delta H°$ values for the most retained enantiomers of compounds 3 and 4).

Both CSPs-DACH PFB and DACH-NAPHT are able to perform chiral recognition in the presence of aqueous eluents because of their marked lipophilic character. A detailed study of the effect of the mobile phase composition on the interactions between analyte 3 and CSP-DACH PFB has been carried out and from the data listed in Table 1 (CSP-DACH PFB, analyte 3) it is evident that enantioselectivity was strongly influenced by the nature of the organic modifier (Figure 4). Thus the α values at 25°C decreased as the modifier was changed from propan-2-ol ($\alpha = 1.80$) to methanol ($\alpha = 1.45$) to acetonitrile ($\alpha = 1.14$); it is worth noting that $\Delta H°$ values also decrease in the same way.

ACKNOWLEDGEMENTS

This work was financially supported by grants from Ministero della Pubblica Istruzione and Consiglio Nazionale delle Ricerche P.F. Chimica Fine II, Italy.

REFERENCES

1. W. Linder, *Chromatographia*, 24:97-107 (1987).
2. M. Zief and L. J. Crane, "Chromatographic Chiral Separations", Marcel Dekker, New York (1988).
3. W. H. Pirkle and C. Pochapsky, *in:* "Advances in Chromatography", Vol. 27, Marcel Dekker, New York, 3:73-127 (1987).
4. G. Gargaro, F. Gasparrini, D. Misiti, G. Palmieri, M. Pierini and C. Villani, *Chromatographia*, 24:505-509 (1987).
5. F. Gasparrini, D. Misiti, C. Villani, F. La Torre and M. Sinibaldi, *J. Chromatogr.*, 457:235-245 (1988).
6. F. Gasparrini, D. Misiti, G. Palmieri and C. Villani, 12th International Symposium on Column Liquid Chromatography, Washington, USA, 19-24 June 1988.
7. G. Gargaro, F. Gasparrini, D. Misiti, G. Palmieri, F. La Torre and C. Villani, 16th International Symposium on Chromatography, Paris, 21-26 September 1986.
8. F. Gasparrini, D. Misiti and C. Villani, *J. High Resolut. Chromatogr.*, 13:182-184 (1990).
9. H. Colin and G. Guiochon, *J. Chromatogr.*, 158:183-205 (1978).
10. E. Grushka and R. Leshem, *J. Chromatogr.*, 22:41-50 (1983).
11. J. N. Akanya and D. R. Taylor, *Chromatographia*, 25:639-642 (1988).
12. S. Scheibye, B. S. Pedersen and S. O. Lawesson, *Bull. Soc. Chim. Belg.*, 87:229-238 (1978).
13. W. H. Pirkle and T. C. Pochapsky, *Chem. Rev.*, 89:347-362 (1989).

ENANTIOMER SEPARATION ON DISSOLVED CYCLODEXTRIN DERIVATIVES
BY HIGH-RESOLUTION GAS CHROMATOGRAPHY: THERMODYNAMIC
DATA OF CHIRAL RECOGNITION

Volker Schurig* and Martin Jung

Institut für Organische Chemie der Universität
Auf der Morgenstelle 18, D-7400 Tübingen, FRG

SUMMARY

The analytical gas chromatographic enantiomer separation of various racemates on
dissolved, derivatised β-cyclodextrins is described. Following an overview on the properties
and use of cyclodextrins, a new phase, i.e., heptakis(2,6-O-dimethyl-3-O-heptafluoro-
butanoyl)-β-cyclodextrin, dissolved in OV-1701 or OV-225 is introduced which shows some
complementary behaviour to existing phases in enantiomer separation. For the first time,
thermodynamic data of enantiomer discrimination involving cyclodextrin derivatives are
described. The determination of the temperature dependence of enantiomer separation
shows that both the entropic and enthalpic parameters favourably contribute to chiral
recognition, and that an isoenantioselective temperature is absent.

INTRODUCTION

The unambiguous determination of enantiomeric compositions and absolute configurations is an important analytical task in contemporary research devoted to the
synthesis, characterisation and use of chiral compounds [1]. Separation of volatile
enantiomers by gas chromatography (GC) constitutes an efficient and reliable tool in
modern enantiomer analysis in regard to simplicity, reproducibility, high sensitivity and
high resolution [2,3]. The development and ready availability of high-resolution capillary
columns (WCOT = wall coated open-tubular) and the use of highly sensitive detectors permits
sample analysis without prepurification and mixtures of enantiomers may be simultaneously analysed. Hyphenated techniques such as multidimensional chromatography
and coupling to spectroscopic detection is easily feasible. Problems, typically arising in, for
example, high-performance liquid chromatography (HPLC) from gradient elution, and the
use of modifiers are absent in GC.

Three types of chiral stationary phase (CSP) have been shown to be versatile selectors
for gas chromatographic enantiomer separation:

1. Amino acid derivatives capable of forming hydrogen bonds [4], such as commercially
    available "Chirasil-Val" and "XE-60-L-Val(R)-α-pea" [5,6].
2. Optically active metal coordination compounds ("complexation gas chromatography")
    [7-12].
3. Sugar derivatives such as cyclodextrins capable of (inter alia) inclusion [13-25].

While the first two methods are now well established, the third method is still in its
infancy and its great potential for routine enantiomer analysis is only at the beginning of
its full exploitation. Cyclodextrins (cycloamyloses, Schardinger dextrins) (structures in

---

* Author to whom correspondence should be addressed.

Recent Advances in Chiral Separations, Edited by D. Stevenson and
I. D. Wilson, Plenum Press, New York, 1991

117

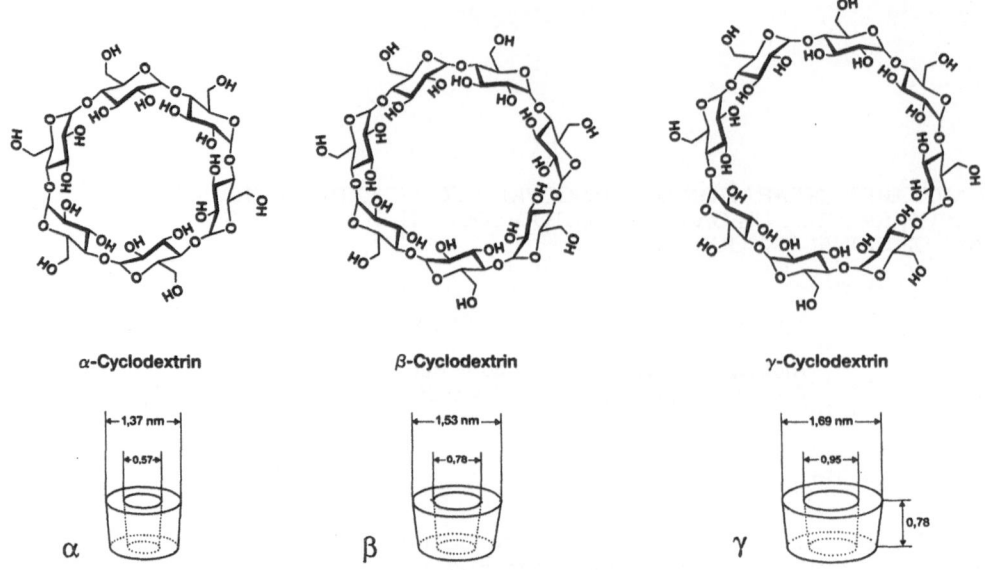

Fig. 1. The structure of α, β and γ cyclodextrin.

Figure 1) are cyclic oligomers containing 6, 7 or 8 (α, β and γ–cyclodextrin) D-glucose units which are α-1,4-linked as in starch.

In 1891, Villiers [26] isolated cyclodextrins for the first time as degradation products of starch, and Schardinger [27] prepared them in 1911 from starch using *Bacillus macerans*. Today cyclodextrins are readily commercially available.

The propensity of cyclodextrins to include organic molecules of suitable size into their hydrophobic intramolecular cavity is well established. This, together with the presence of primary and secondary hydroxyl groups that can be chemically modified, has made them interesting model substances for molecular catalysis and in mimicking enzymes [28-30]. The intramolecular cavity possesses a narrow and a wider end. The former contains primary C(6) hydroxy groups while the latter is comprised of C(2) and C(3) secondary hydroxyl groups of the respective glucose units. Of particular interest for pharmaceutical and related applications is the observation that the formation of inclusion complexes into cyclodextrins often increases the stability and/or water solubility of biologically important molecules [31].

X-Ray structure analyses of various inclusion complexes of β-cyclodextrins have been performed [32-37]. They reveal that not only 1:1 complexes, but also 2:3 or 1:2 complexes are formed between the cyclodextrin and the guest molecule. The latter cyclodextrin dimers are held together by hydrogen bonds between the secondary hydroxyl groups. Investigations on heptakis(2,6-O-dimethyl)-β-cyclodextrin showed that the removal of some hydroxy functions makes the formation of the dimers very unlikely and that, on the other hand, due to the hydrophobic groups which take their place, the space available for a hydrophobic guest molecule is enlarged [38]. This observation opens an avenue for derivatisation strategies which affect conformation and the tendency of the modified cyclodextrins to form inclusion compounds. Numerous cyclodextrin derivatives are already known, and it is possible to selectively modify the primary hydroxy groups or those attached to C(2) and C(6) [39-42].

As early as 1959, Cramer and Dietsche [43] demonstrated that cyclodextrins can be employed for the separation of racemic mixtures. Armstrong et al. have successfully applied silica-gel-immobilised cyclodextrins for enantiomer separation in HPLC [44,45].

An attempt in 1981 in our laboratory to use permethylated cyclodextrins for the enantiomer separation of saturated hydrocarbons, devoid of any chemical function - a challenge of merely academic interest - by GC proved unsuccessful (Y.H. Kim, unpublished results). After that, there appeared some remarkable reports on the use of underivatised

118

cyclodextrins for the enantiomer separation of cyclic hydrocarbons in GC using packed columns [13,14]. The importance of this finding has been fully appreciated [46] and later on, a functionalised racemic compound was resolved on a permethylated cyclodextrin using a capillary column [15]. The high melting point of cyclodextrins and their permethylated derivatives prevents their general use as an enantioselective stationary phase in GC. We have therefore employed solutions of derivatised cyclodextrins in moderately polar polysiloxanes for coating glass or fused silica capillary columns [16,22,24] while König et al. used undiluted per-n-pentylated cyclodextrins as fluids for coating glass capillary columns [17-21].

All the new cyclodextrin phases greatly extend the scope of enantiomer separation by GC. Yet, in most cases, enantiomer discrimination is quite low as compared to the use of cyclodextrins in formamide solution in GC [13,14], or of anchored cyclodextrins in HPLC [44,45], with α-values only in the range of 1.03 to 1.30. The high-resolution power of capillary columns, however, can handle such small free-energy differences for analytical purposes successfully with the concomitant merit of short analysis times.

The separation of unfunctionalised, saturated hydrocarbons [13,24] can only be explained by the formation of inclusion complexes. In general, enantiomer separation on cyclodextrin derivatives should be governed by *shape selectivity* rather than determined by *functionality selectivity* which is inherent to the first two methods cited above for homologous series of compounds. Yet for molecules possessing chemical functionalities, the participation of diastereomeric interactions, other than inclusion, should not be underestimated. This is borne out by the observation that the open-chain analogue of cyclodextrins, per-n-pentylated amylose, may separate enantiomers [23]. Per-n-pentylated amylose, which is believed to be devoid of helical structures due to the presence of large hydrophobic n-pentyl groups, may therefore also undergo interactions with the selectand by forces other than inclusion.

In our group [16,22,24] permethylated cyclodextrins and 3-O-perfluoroacylated derivatives are dissolved in a moderately polar polysiloxane such as OV-1701 (5% cyanopropyl : 7% phenyl : 88% methylpolysiloxane) or OV-225 (25% cyanopropyl : 50% phenyl : 25% methylpolysiloxane). With this strategy the cyclodextrin derivatives can be used below their melting points, and, due to the presence of the polysiloxane, the coating of deactivated vitreous capillary columns was straightforward (glass and fused silica). Furthermore, thermodynamic data of chiral recognition, $\Delta_{R,S}(\Delta G)$, $\Delta_{R,S}(\Delta H)$ and $\Delta_{R,S}(\Delta S)$, could easily be determined by measuring the retention increase R' of enantiomers on the cyclodextrin dissolved in the solvent S in comparison to an achiral reference column containing only the solvent S.

Here we describe a new phase heptakis(2,6-O-dimethyl-3-O-heptafluorobutanoyl)-β-cyclodextrin, dissolved in OV-1701 or OV-225 (Phase III) which shows some complementary behaviour to the existing phases, i.e., heptakis(2,3,6-O-trimethyl)-β-cyclodextrin (Phase I), heptakis(2,6-O-dimethyl-3-O-trifluoroacetyl)-β-cyclodextrin (Phase II) [16,22,24]. For the first time, thermodynamic data focussing on the temperature-dependence of GC separation of enantiomers with cyclodextrin derivatives are presented.

EXPERIMENTAL

Synthesis of Heptakis(2,6-di-O-methyl-3-O-heptafluorobutanoyl)-β-cyclodextrin (Phase III)

Sodium heptafluorobutanoate was prepared by adding sodium methylate in methanol to a slight excess of heptafluorobutanoic acid. The solvent was evaporated *in vacuo* and the salt was dried over phosphorus pentoxide. Heptakis(2,6-di-O-methyl)-β-cyclodextrin was prepared according to the literature [40]. Thus 12.5 ml (51 mmol) of heptafluorobutanoic acid anhydride, 6.0 g (25.9 mmol) of sodium heptafluorobutanoate, and 2.1 g (1.5 mmol) of heptakis(2,6-di-O-methyl)-β-cyclodextrin were stirred for 2 hr at 50°C and then overnight at room temperature. The excess anhydride was removed *in vacuo*. After addition of a few millilitres of anhydrous dichloromethane the viscous mass was dried *in vacuo*. The residue was refluxed with 50 ml of anhydrous dichloromethane and the solution was decanted from the sticky sodium salt. The procedure was repeated. The solvent was then removed with a rotary evaporator and the residue was dried at 60°C and 0.22 torr. The powder was dissolved in 10 ml dichloromethane and refluxed after addition of 50 ml n-hexane. The product precipitated as an oil on cooling. The product was washed with n-hexane and dried for 2 hr at 0.02 torr and 60°C and then recrystallised from dichloromethane-n-hexane. The white product was kept under nitrogen. Yield: 31% R_f (on thin-layer chromatography (TLC) silica gel) = 0.50 (benzene-ethanol 1:1 v/v), [^{13}C]-NMR (CDCl$_3$, ppm): 59.0, 59.2 (OMe); 104, 108, 112, 115, 120, 124 (CF); 156.7, 157.2, 157.6, 160.7 (CO).

Instrumentation

A Carlo-Erba gas chromatograph (Hofheim, Taunus, FRG), Fractovap 2350, equipped with a flame ionisation detector (FID) and suitable for operation with glass open-tubular columns, was used. The carrier gases were commercial high-purity nitrogen, helium or hydrogen used without further purification. The split ratio was set to 1:100. The instrument was set at its highest sensitivity at a tolerable signal-to-noise ratio.

Open-tubular Columns [47]

Duran glass tubings (obtained from Schott Ruhrglas, Mainz, FRG) were drawn to capillaries of 0.25 mm id, using a Hupe & Busch glass-drawing machine.

Acid Leaching and Rinsing

The capillary column was filled to 90% of its capacity with aqueous 6 M hydrochloric acid, sealed under vacuum and heated for 24 hr at 150°C (for the subsequent coating with OV-1701) or 175°C (for the subsequent coating with OV-225). The ends were opened and the capillary column was rinsed with approximately three volumes of 0.01 M hydrochloric acid and dehydrated for 2 hr at 250°C under a stream of dry nitrogen.

Deactivation with DPTMDS (Diphenyltetramethyldisilazane) or CPTMDS (Dicyanopropyl-tetramethyldisiloxane)

The column was dynamically coated with a solution of DPTMDS (for the subsequent coating with OV-1701) or CPTMDS (for the subsequent coating with OV-225) (Fluka, Buchs, Switzerland) in n-pentane (1:1 v/v). The column was sealed under vacuum and heated to 200°C. The temperature was then increased to 390°C (for DPTMDS), or 360°C (for CPTMDS), respectively, at a rate of 3°C min^{-1} and held there for 12 hr. After cooling to room temperature, the column ends were opened, and the capillary column was rinsed with 1 ml of n-hexane, 3 ml of methanol and 2 ml of diethyl ether.

Coating

The column was coated with a 0.05-0.15% solution of the cyclodextrin derivative and the required amount of OV-1701 or OV-225 (0.4-0.6%) in n-pentane-dichloromethane (1:1 v/v) by the static method.

Fused silica capillaries were deactivated and coated statically according to the literature [47].

RESULTS AND DISCUSSION

In comparison to the use of permethylated β-cyclodextrins in the undiluted form [15,48] the dissolution of the chiral selector in polysiloxanes offers a number of advantages:

1. There are no difficulties arising from the high melting point of permethylated β-cyclodextrin, and low temperatures, which tend to increase separation factors, can readily be employed.
2. The polarity of the stationary phase can be widely adjusted by the proper choice of the solvent.
3. The use of the polysiloxane matrix offers excellent coating properties to glass and fused silica surfaces.
4. Thermodynamic data on enantioselectivity are accessible from *relative retention data* as has previously been demonstrated in complexation gas chromatography [3,10].

In Figures 2 and 3 representative gas chromatograms, featuring enantiomer separation of compounds belonging to different classes of compounds, on the novel selector heptakis(2,6-O-dimethyl-3-O-heptafluorobutanoyl)-β-cyclodextrin (Phase III) are shown. The separation of E-5-methyl-hept-2-en-4-one, the hazelnut flavour compound Filbertone

[49], represents an example of an acyclic compound which is especially well resolved on Phase III (Fig. 2B). Acyclic halo carboxylic acid esters, such as ethyl 2-bromopropionate, can be separated into enantiomers on Phase III (Fig. 3A), but not on Phase I or II. For γ–decalactone (Fig. 3D), pulegone (Fig 2C), neomenthol (Fig. 3C) as well as a number of aromatic oxiranes, Phase III is also superior to Phases I and II. For chiral spiro acetals, which occur frequently as pheromone components of insects, Phase III is also very well suited. For example, chalcogran can be quantitatively separated into enantiomers without plateau formation (Fig. 3B) [23]. Contrary to the situation pertaining to Phase II, on which saturated hydrocarbons are not resolved, racemic cis-pinane can be separated on Phase III, albeit with a smaller separation factor as compared with Phase I.

Phases I, II and III can be employed at a wide temperature range extending from 25 to 190°C. Even temperatures up to 210°C have been used for 6 hr without detrimental effect on enantiomer separation on Phase I. For practical reasons nitrogen or helium have been employed as carrier gas, although hydrogen may also be used. As we have the facilities for preparing glass capillaries at our disposal, most separations are performed with this inexpensive material. In Figure 4 enantiomer separations with cyclodextrin derivatives, using a fused silica instead of a glass capillary column, are shown. The separation of chiral saturated hydrocarbons which are devoid of any chemical functionality is noteworthy.

In the following paragraphs the determination of the retention increase R' and of thermodynamic data of chiral recognition for representative selectands on Phase I is described. The enantiomer separation of a racemate on a cyclodextrin phase is based on the different stabilities of the diastereomeric associates formed by a rapid and reversible 1:1 equilibrium. According to Gil-Av and Herling [50], Muhs and Weiss [51], and Purnell [52], stability constants can readily be measured from retention data by GLC. This approach has later been extended, and simplified, by Schurig et al. [53] and applied to enantiomer separation [10,22].

The method is outlined as follows. First the net retention ratio of the selectand (the racemic compound) in relation to an inert reference standard (n-octane or n-decane), not interacting with the *selector*, r_0, is determined on a reference column containing the pure solvent (the polysiloxane OV-1701). Then the net retention ratio r of the same selectand in relation to the same inert standard is determined on a column containing the concentration m (molality) of the selector in the solvent. On the column containing the selector (the cyclodextrin derivative), the reversible formation of inclusion and/or association complexes is adding to the physical partition between the mobile and stationary phase and consequently leads to a retention increase R' as the result of chemical selectivity.

Unfortunately, contrary to complexation gas chromatography, we can not discern a totally *inert* reference standard being devoid of interaction with cyclodextrins. Thus, the n-alkanes used as reference standards will to some extent be included into the cavity of the cyclodextrins as the following results show. Therefore, the approach described here is affected by a systematic error, and the conclusions drawn from our investigations should only be treated in a qualitative manner.

A retention increase $R' > 0$ is a necessary, but not a sufficient condition for an enantiomer separation, $R' = 0$ means that there are no interactions between the selectand and selector and enantiomer separation cannot be expected. In analogy to complexation gas chromatography [53], the retention increase R' is given by the product of the association constant K and the molality of the cyclodextrin in the solvent (OV-1701) for low concentrations of the cyclodextrin (activity coefficient ≈ 1):

$$K_{(ass)} \times m_{cyclodextrin} = \frac{r - r_0}{r_0} = R'$$

If the cyclodextrin is capable of discriminating between enantiomers, different complexation constants K_R and K_S will result and $\Delta_{R,S}(\Delta G)$ can be calculated according to:

$$\Delta_{R,S}(\Delta G) = -RT \ln(K_R/K_S) = -RT \ln(R'_R/R'_S).$$

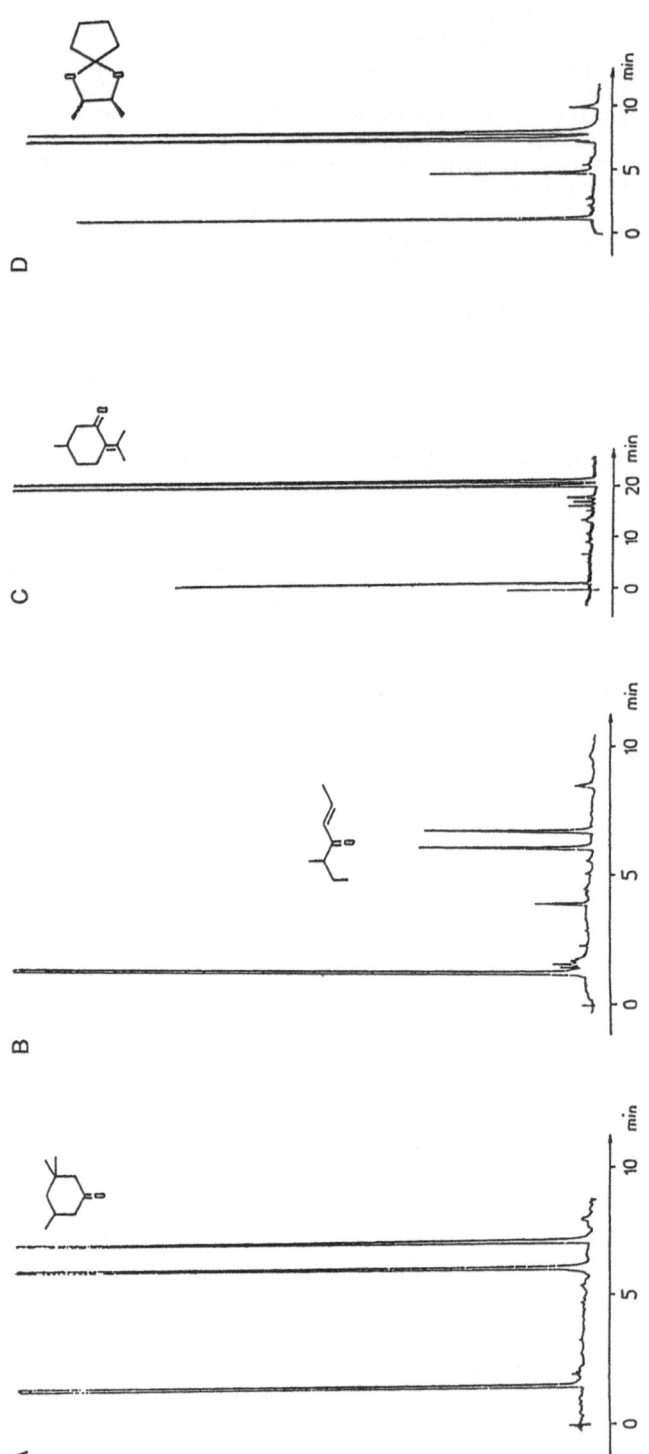

Fig. 2. (A) Enantiomer separation of racemic 3,3,5-trimethylcyclohexanone on a 40 m x 0.25 mm id glass open-tubular column, coated with 19% heptakis(2,6-O-dimethyl-3-O-heptafluoro-butanoyl)-β-cyclodextrin (Phase III) in OV-1701 (0.07 m). Oven temperature 120°C; inlet pressure 0.7 bar hydrogen. (B) Enantiomer separation of racemic E-5-methyl-hept-2-en-4-one, the hazelnut flavour compound Filbertone [55], on a 40 m x 0.25 mm id glass open-tubular column, coated with 19% heptakis(2,6-O-dimethyl-3-O-heptafluorobutanoyl)-β-cyclodextrin (Phase III) in OV-1701 (0.07 m). Oven temperature 100°C; inlet pressure 1.2 bar hydrogen. (C) Enantiomer separation of pulegone on a 40 m x 0.25 mm id glass open-tubular column, coated with 19% heptakis(2,6-O-dimethyl-3-O-heptafluorobutanoyl)-β-cyclodextrin (Phase III) in OV-1701 (0.07 m). Oven temperature 110°C; inlet pressure 1.2 bar hydrogen. (D) Enantiomer separation of racemic E-2,3-dimethyl-1,4-dioxaspiro[4.4]nonane on a 40 m x 0.25 mm id glass open-tubular column, coated with 19% heptakis(2,6-O-dimethyl-3-O-heptafluorobutanoyl)-β-cyclodextrin (Phase III) in OV-1701 (0.07 m). Oven temperature 90°C; inlet pressure 1.2 bar hydrogen.

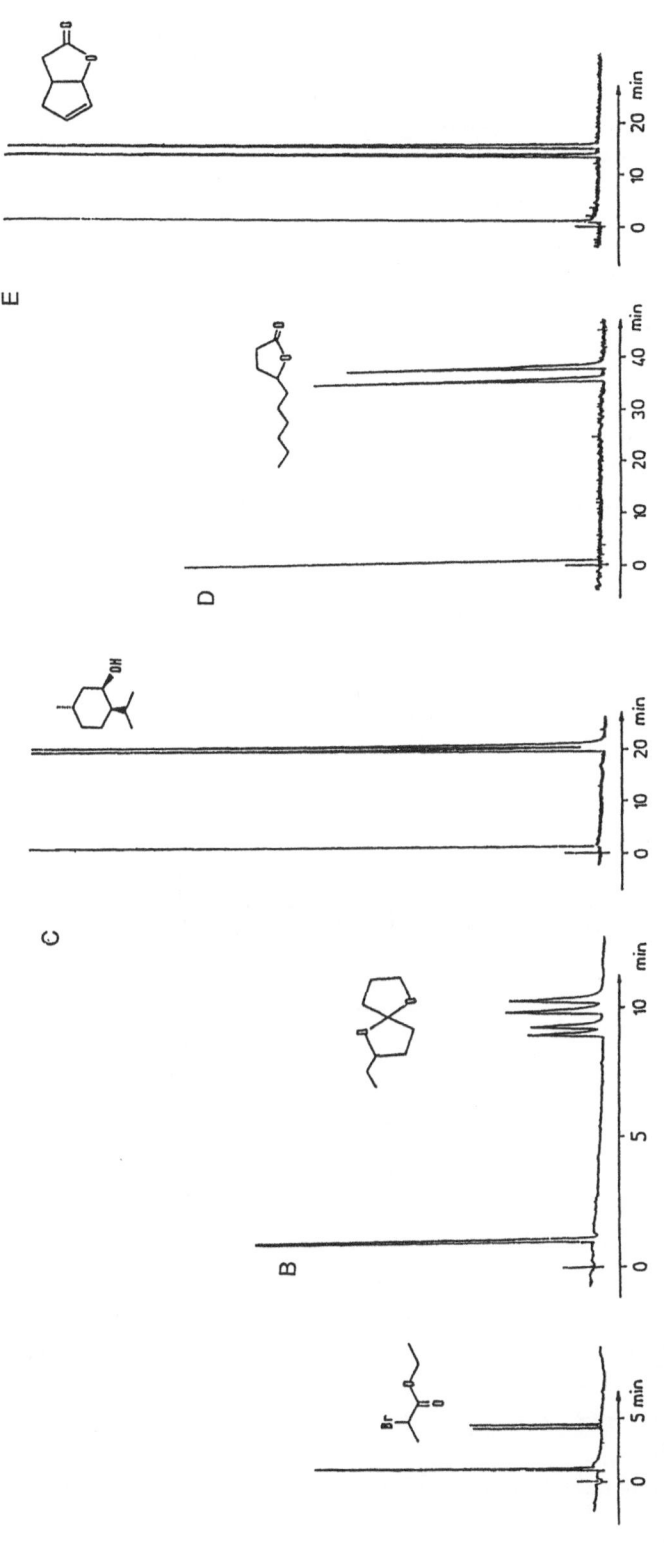

Fig. 3. (A) Enantiomer separation of racemic ethyl 2-bromopropionate on a 40 m x 0.25 mm id glass open-tubular column, coated with 19% heptakis(2,6-O-dimethyl-3-O-heptafluorobutanoyl)-β-cyclodextrin (Phase III) in OV-1701 (0.07 m). Oven temperature 100°C; inlet pressure 1.4 bar hydrogen. (B) Diastereomer and enantiomer separation of E- and Z-2-ethyl-1,6-dioxaspiro[4.4]nonane (Chalcogran) on a 40 m x 0.25 mm id glass open-tubular column, coated with 19% heptakis(2,6-O-dimethyl-3-O-heptafluorobutanoyl)-β-cyclodextrin (Phase III) in OV-1701 (0.07 m). Oven temperature 90°C; inlet pressure 1.0 bar nitrogen. (C) Enantiomer separation of racemic neomenthol on a 40 m x 0.25 mm id glass open-tubular column, coated with 19% heptakis(2,6-O-dimethyl-3-O-heptafluorobutanoyl)-β-cyclodextrin (Phase III) in OV-1701 (0.07 m). Oven temperature 100°C; inlet pressure 1.2 bar hydrogen. (D) Enantiomer separation of racemic γ-decalactone on a 40 m x 0.25 mm id glass open-tubular column, coated with 19% heptakis(2,6-O-dimethyl-3-O-heptafluorobutanoyl)-β-cyclodextrin (Phase III) in OV-1701 (0.07 m). Oven temperature 150°C; inlet pressure 1.5 bar hydrogen. (E) Enantiomer separation of racemic 2-oxa-bicyclo[3.3.0]oct-7-en-3-one on a 40 m x 0.25 mm id glass open-tubular column, coated with 19% heptakis(2,6-O-dimethyl-3-O-heptafluorobutanoyl)-β-cyclodextrin (Phase III) in OV-1701 (0.07 m). Oven temperature 150°C; inlet pressure 1.5 bar hydrogen.

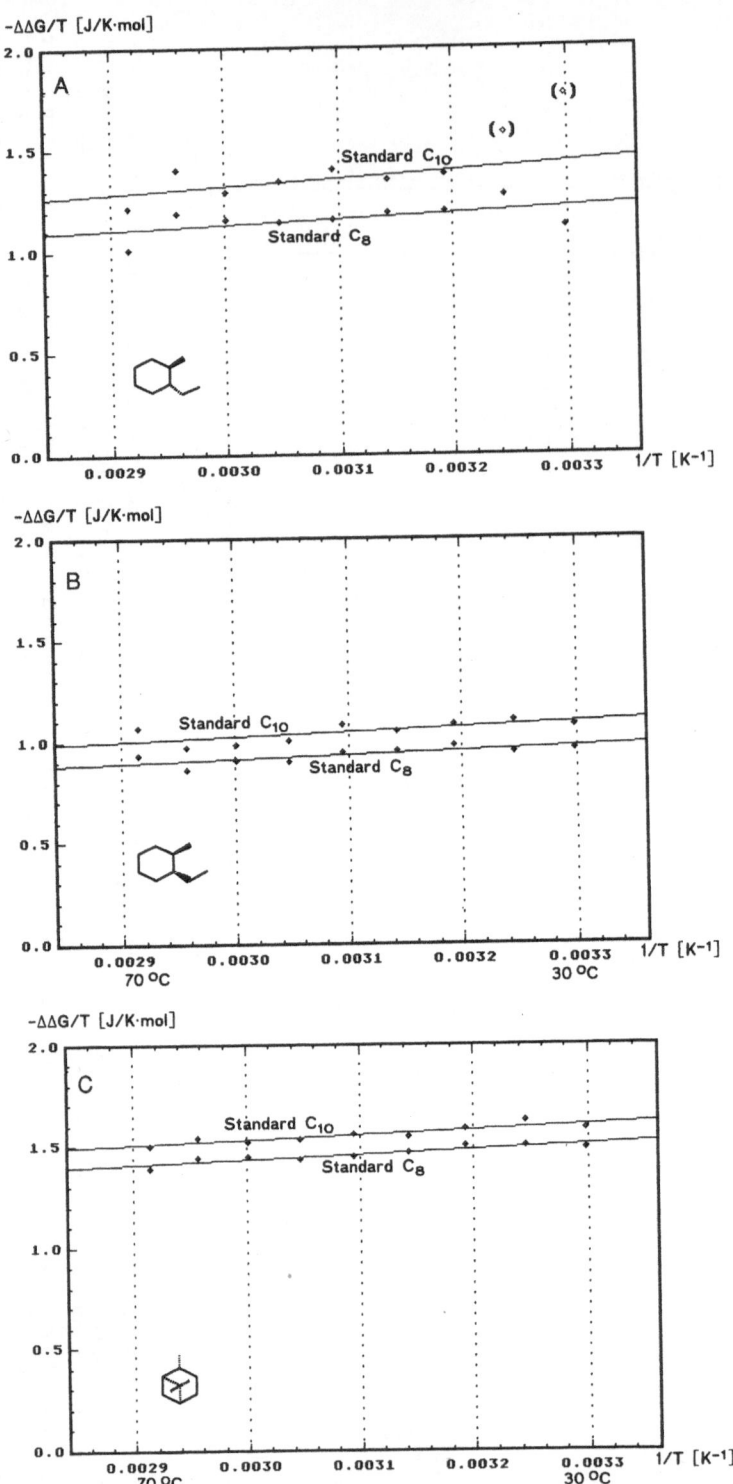

Fig. 4.   (A) Enantiomer separation of racemic 3,3,5-trimethylcyclohexanone on a 30 m x 0.25 mm id fused-silica-open-tubular column, coated with 10% heptakis(2,3,6-O-trimethyl-O-trimethyl)-β-cyclodextrin (Phase I) in OV-1701 (0.07 m). Oven temperature 90°C; inlet pressure 1.5 bar hydrogen. (B) Enantiomer separation of racemic 2,2-dimethyl-4-phenyl-1,3-dioxolane on a 30 m x 0.25 mm id fused-silica-open tubular column, coated with 10% heptakis(2,3,6-O-trimethyl-O-trimethyl)-β-cyclodextrin (Phase I) in OV-1701 (0.07 m). Oven

The subscript R arbitrarily refers to the enantiomer eluting as the second peak and S eluting as the first peak, respectively. Significantly, $\Delta_{R,S}(\Delta G)$ is independent of the concentration of the selector. Note that the expression, frequently used in HPLC, $\Delta_{R,S}(\Delta G) = -RT \ln\alpha$ will give erroneous results in the present treatment as it does not separate the partition equilibrium from the association equilibrium.

$\Delta_{R,S}(\Delta G)$ represents a quantitative thermodynamic measure for the ability of a cyclodextrin derivative to discriminate between enantiomers of a certain racemic compound at a given temperature. By measurements at different temperatures, the thermodynamic quantities $\Delta_{R,S}(\Delta H)$ and $\Delta_{R,S}(\Delta S)$ can be determined according to the Gibbs-Helmholtz equation, divided by the temperature T:

$$R \ln(R'_R/R'_S) = -\frac{\Delta_{R,S}(\Delta G)}{T} = -\frac{\Delta_{R,S}(\Delta H}{T} + \Delta_{R,S}(\Delta S).$$

The retention increase R' of five selectands and two selectors (Phase I and III) have been determined at different temperatures. The results are given in Tables 1-5. The data are interpreted as described in the following paragraphs.

As beforehand, n-octane and n-decane do not represent totally inert reference standards as n-alkanes show a small retention increase R' (up to ca. 0.1) in respect to each other. Thus, an error is introduced in the value for $R'_{selectand}$ as seen by the different values obtained by referring to different standards.

There is no direct correlation between $\Delta_{R,S}(\Delta G)$ and the separation factor α. For example, cis-pinane (selectand 5) exhibited a larger $\Delta_{R,S}(\Delta G)$ value on Phase III as compared to Phase I, although the separation factor and peak resolution was inferior for the former.

In most cases a large absolute value of R' is the prerequisite for efficient enantiomer separation. For example, the R' value of all five selectands on Phase I were large and separation was good, on the other hand, in no case has an enantiomer separation been observed for low R' values. However, there is no clear-cut relationship, as for instance an average R' value of 0.12 was sufficient for a quantitative enantiomer separation of selectand 5 on Phase III (due to a large difference in the R' values for the enantiomers) while the same R' value did not lead to a separation for selectand 4.

The absence of enantiomer discrimination for selectand 3 on Phase III was due to the absence of molecular interaction (R' ≈ 0).

The absence of enantiomer discrimination for selectand 4 on Phase III, for which R' was clearly different from 0, was due to poor chiral discrimination in spite of a favourable chemical interaction (the appearance of a broad peak may be explained by partial separation into enantiomers).

In interpreting the thermodynamic quantities $\Delta_{R,S}(\Delta H)$ and $\Delta_{R,S}(\Delta S)$ not only their quantities, but their relative sign is of prime importance. The following cases may be distinguished when $R \ln(R'_R/R'_S) = -\Delta_{R,S}(\Delta G)/T$ is plotted against 1/T.

1. $\Delta_{R,S}(\Delta H)$ and $\Delta_{R,S}(\Delta S)$ have the same sign. In this case the linear Gibbs-Helmholtz plot intersects the abscissa in the field of a positive absolute temperature, i.e., a so-called isoenantioselective temperature $T_{iso} = \Delta_{R,S}(\Delta H)/\Delta_{R,S}(\Delta S)$ will occur when enthalpy and entropy contributions to $\Delta_{R,S}(\Delta G)$ cancel each other and enantiomer separation is impossible [10,54].

For this case, precedents have been observed in complexation gas chromatography with the following representative values [55] (cf. ref 54):

Fig. 4 (continued)

temperature 130°C; inlet pressure 1.5 bar hydrogen. (C) Enantiomer separation of racemic 1,2-dialkylcyclohexanes [24] on a 30 m x 0.25 mm id fused-silica-open-tubular column, coated with 10% heptakis(2,3,6-O-trimethyl-O-trimethyl)-β-cyclodextrin (Phase I) in OV-1701 (0.07 m). Oven temperature 50°C; inlet pressure 1.5 bar hydrogen.

Table 1. Chromatographic and Thermodynamic Data of Representative Selectands on Phase I Heptakis(2,3,6-O-trimethyl)-β-cyclodextrin (0.7 m in OV-1701)*

Selectand	Standard	r_0	r	R'	$-\Delta\Delta G$ [J mol^{-1}]
1 Ph 125°C $\alpha = 1.09$	C_{10}	8.71(2)	10.47(0) 11.37(7)	0.201(7) 0.305(8)	1378
	C_{12}	2.70(9)	3.18(7) 3.46(3)	0.176(3) 0.278(2)	1510
2 125°C $\alpha = 1.09$	C_{10}	12.72(2)	19.64(8) 21.39(2)	0.544(4) 0.681(5)	683
	C_{12}	3.95(6)	5.98(1) 6.51(2)	0.511(8) 0.645(9)	770
3 35°C $\alpha = 1.06$	C_8	2.75(9)	4.35(2) 4.61(6)	0.577(6) 0.673(4)	393
	C_{10}	0.43(0)	0.60(5) 0.64(2)	0.408(3) 0.493(7)	487
4 35°C $\alpha = 1.06$	C_8	3.41(4)	6.23(7) 6.57(8)	0.827(0) 0.926(7)	292
	C_{10}	0.53(2)	0.86(8) 0.91(5)	0.630(6) 0.719(7)	339
5. 35°C $\alpha = 1.12$	C_8	5.26(9)	13.00(0) 14.50(5)	1.421(2) 1.701(7)	461
	C_{10}	0.83(7)	1.80(8) 2.01(8)	1.161(2) 1.411(6)	500

*All measurements were performed at least three times. The numbers in brackets are uncertain. Inlet pressure 1.2 bar hydrogen. Column: 40 m x 0.25 mm. Reference column: 25 m x 0.25 mm.

Isopropyl oxirane on nickel(II)-bis(3-heptafluorobutanoyl-(1R)-8-methyl-camphor): $\Delta_{R,S}(\Delta H) = 635$ J mol^{-1}, $\Delta_{R,S}(\Delta S) = 2.2$ J K.mol^{-1}, $T_{iso} = 18.4$°C.

Isopropyl oxirane on nickel(II)-bis(3-heptafluorobutanoyl-(1R)-8-methylene-camphor): $\Delta_{R,S}(\Delta H) = 1243$ J mol^{-1}, $\Delta_{R,S}(\Delta S) = 3.4$ J K.mol^{-1}, $T_{iso} = 88.7$°C.

Z-2-Ethyl-1,6-dioxaspiro[4.4]nonane (Chalcogran) on nickel(II)-bis(3-heptafluoro-butanoyl-(1R)-8-isobutylene-camphor): $\Delta_{R,S}(\Delta H) = 3591$ J mol^{-1}, $\Delta_{R,S}(\Delta S) = 10.1$ J K.mol^{-1}, $T_{iso} = 82.8$°C.

2. $\Delta_{R,S}(\Delta H)$ and $\Delta_{R,S}(\Delta S)$ have opposite signs. In this case the linear Gibbs-Helmholtz plot intersects the abscissa in the field of a negative absolute temperature and no isoenantioselective temperature occurs. Most important, the quantities $\Delta_{R,S}(\Delta H)$ and $\Delta_{R,S}(\Delta S)$ do not oppose, but favourably complement each other in determining the quantity of $\Delta_{R,S}(\Delta G)$.

3. $\Delta_{R,S}(\Delta H) = 0$. The Gibbs-Helmholtz plot will be parallel to the abscissa and the enantiomer discrimination is temperature independent.

Table 2. Chromatographic and Thermodynamic Data of Representative Selectands on Phase III Heptakis(2,6-O-dimethyl-5-O-heptafluorobutanoyl)-β-cyclodextrin (0.7 m in OV-1701) *

Selectand	Standard	r_o	r	R'	$-\Delta\Delta G$ [J mol^{-1}]
1 Ph 125°C	C$_{10}$	8.7(2)	9.79(4)	0.124(1)	—
	C$_{12}$	2.70(0)	3.15(5)	0.164(7)	—
2 125°C α = 1.09	C$_{10}$	12.72(2)	25.88(9) 28.14(6)	1.035(0) 1.212(4)	523
	C$_{12}$	3.95(6)	8.34(1) 9.06(8)	1.108(3) 1.292(1)	508
3 35°C	C$_8$	2.75(9)	2.76(7)	0.003(0)	—
	C$_{10}$	0.43(0)	0.42(1)	-0.019(8)	
4 35°C	C$_8$	3.41(4)	3.82(3)	0.119(7)	—
	C$_{10}$	0.53(2)	0.58(2)	0.094(0)	—
5 35°C α = 1.04	C$_8$	5.36(9)	5.99(1) 6.25(0)	0.115(9) 0.164(0)	889
	C$_{10}$	0.83(7)	0.91(2) 0.95(2)	0.090(4) 0.137(3)	1072

* Conditions as Table 1.

4. $\Delta_{R,S}(\Delta S) = 0$. The Gibbs-Helmholtz plot passes through the origin and the enantiomer discrimination improves with decreasing temperature.

5. $\Delta_{R,S}(\Delta H) = \Delta_{R,S}(\Delta S) = 0$. The Gibbs-Helmholtz plot coincides with the abscissa and enantiomer discrimination is impossible at any temperature.

The thermodynamic data of three saturated hydrocarbons on permethylated β-cyclodextrin in OV-1701 (Phase 1) at nine different temperatures with intervals of 5°C (Tables 3-5 and Figure 5) show the following results:

1. A large systematic error is introduced in $\Delta_{R,S}(\Delta H)$ and $\Delta_{R,S}(\Delta S)$ by the inadequacies of the reference standards used (n-octane versus n-decane) when the retention increase R' is small.

2. For Phase I the quantities of $\Delta_{R,S}(\Delta H)$ and $\Delta_{R,S}(\Delta S)$ were smaller than in complexation gas chromatography.

3. For Phase I the *ratio* of $\Delta_{R,S}(\Delta H)$ and $\Delta_{R,S}(\Delta S)$ was similar to that in complexation gas chromatography, and neither the enthalpy contribution nor the entropy contribution to enantiomer discrimination can be neglected.

127

Table 3. Temperature Dependence of the Enantiomer Discrimination of *trans*-1-Ethyl-2-methylcyclohexane on Phase I Heptakis(2,3,6-O-trimethyl)-β-cyclodextrin in OV-1701[*]

Temperature [°C]	Standard	r_0	r	$-\Delta\Delta G / R$	$-\Delta\Delta G$ [J mol⁻¹]	$-\Delta\Delta G/T$ [J K mol⁻¹]
30	C_8	2.80(0)	4.54(8)	0.624(7)	339	1.118
			4.80(0)	0.714(6)		
	C_{10}	0.41(6)	0.60(3)	0.451(3)	534	1.762
			0.64(8)	0.557(9)		
35	C_8	2.75(9)	4.35(2)	0.577(6)	393	1.275
			4.61(6)	0.673(4)		
	C_{10}	0.43(0)	0.60(5)	0.408(3)	487	1.579
			0.64(2)	0.493(7)		
40	C_8	2.73(1)	4.13(4)	0.513(5)	375	1.196
			4.35(1)	0.593(0)		
	C_{10}	0.44(6)	0.62(9)	0.411(0)	431	1.378
			0.66(2)	0.485(1)		
45	C_8	2.69(1)	3.96(8)	0.474(6)	379	1.191
			4.16(5)	0.547(6)		
	C_{10}	0.46(1)	0.64(1)	0.389(4)	446	1.351
			0.67(3)	0.458(2)		
50	C_8	2.65(7)	3.80(9)	0.433(8)	323	1.160
			3.98(2)	0.498(7)		
	C_{10}	0.48(1)	0.63(9)	0.328(0)	453	1.403
			0.66(8)	0.388(3)		
55	C_8	2.63(2)	3.72(3)	0.414(4)	377	1.148
			3.88(5)	0.475(8)		
	C_{10}	0.49(7)	0.66(0)	0.328(2)	441	1.345
			0.68(9)	0.385(8)		
60	C_8	2.60(7)	3.56(2)	0.366(4)	386	1.158
			3.70(5)	0.421(1)		
	C_{10}	0.51(2)	0.67(2)	0.311(5)	432	1:296
			0.69(9)	0.364(0)		
65	C_8	2.56(3)	3.48(1)	0.358(2)	402	1.189
			3.62(2)	0.413(3)		
	C_{10}	0.53(0)	0.68(1)	0.283(7)	474	1.400
			0.70(8)	0.355(7)		
70	C_8	2.54(1)	3.37(5)	0.328(3)	347	1.011
			3.48(3)	0.370(7)		
	C_{10}	0.55(1)	0.69(1)	0.254(4)	417	1.126
			0.71(4)	0.294(5)		

[*] Conditions as Table 1.

4.   For cyclodextrins, in contrast to complexation gas chromatography, $\Delta_{R,S}(\Delta H)$ and $\Delta_{R,S}(\Delta S)$ may in certain cases assume the opposite sign (Table 6).

The last striking observation implies that the enantiomer eluted as the second peak, as compared to that eluted as the first peak, forms the stronger bond (enthalpy being more negative) and the interaction is accompanied by greater disorder (entropy being more positive). The latter finding is unusual for an association process and might be interpreted as arising from the participation of the solvent (OV-1701) - being expelled from the cyclodextrin cavity and thereby gaining conformational freedom - in the association

Table 4. Temperature Dependence of the Enantiomer Discrimination of *cis*-1-Ethyl-2-methylcyclohexane on Phase I Heptakis(2,3,6-O-trimethyl)-β-cyclodextrin in OV-1701*

Temperature [°C]	Standard	r_0	r	$-\Delta\Delta G$ R'	$-\Delta\Delta G$ [J mol^{-1}]	$-\Delta\Delta G/T$ [J K mol^{-1}]
30	C$_8$	3.47(3)	6.55(2)	0.886(4)	292	0.961
			6.92(9)	0.995(1)		
	C$_{10}$	0.51(6)	0.88(4)	0.714(0)	479	1.077
			0.93(5)	0.812(7)		
35	C$_8$	3.41(4)	6.23(7)	0.827(0)	292	0.947
			6.57(8)	0.926(7)		
	C$_{10}$	0.53(2)	0.86(8)	0.630(6)	339	1.099
		0.91(5)	0.719(7)			
40	C$_8$	3.35(4)	5.81(5)	0.733(6)	307	0.979
			6.12(3)	0.825(3)		
	C$_{10}$	0.54(7)	0.88(4)	0.616(2)	339	1.081
45	C$_8$	3.30(8)	5.49(0)	0.661(9)	301	0.955
			5.75(6)	0.742(4)		
	C$_{10}$	0.56(6)	0.88(7)	0.565(7)	333	1.047
			0.93(0)	0.641(7)		
50	C$_8$	3.23(6)	5.20(1)	0.607(2)	321	0.946
			5.43(8)	0.680(4)		
	C$_{10}$	0.58(6)	0.87(2)	0.488(8)	350	1.082
			0.91(2)	0.556(8)		
55	C$_8$	3.19(3)	5.04(0)	0.578(3)	295	0.898
			5.25(0)	0.644(3)		
	C$_{10}$	0.60(3)	0.89(4)	0.482(1)	330	1.005
			0.93(1)	0.544(0)		
60	C$_8$	3.15(4)	4.75(7)	0.508(3)	303	0.908
			4.94(2)	0.566(9)		
	C$_{10}$	0.62(0)	0.89(7)	0.447(7)	328	0.984
			0.93(2)	0.503(9)		
65	C$_8$	3.08(3)	4.60(4)	0.493(5)	292	0.864
			4.77(1)	0.547(5)		
	C$_{10}$	0.63(8)	0.90(0)	0.411(5)	329	0.973
			0.93(3)	0.462(6)		
70	C$_8$	3.04(0)	4.39(1)	0.444(6)	321	0.934
			4.55(2)	0.497(4)		
	C$_{10}$	0.65(9)	0.90(0)	0.364(2)	367	1.068
			0.93(2)	0.414(1)		

* Conditions as Table 1.

process of the selectand and selector. Thus, both $\Delta_{R,S}(\Delta H)$ and $\Delta_{R,S}(\Delta S)$ - although their absolute values are quite low - favourably complement each other in determining enantiomer discrimination as expressed by $\Delta_{R,S}(\Delta G)$. The rather low $\Delta_{R,S}(\Delta H)$ value renders enantiomer separation quite insensitive to the temperature which bears an important merit in GLC since many involatile compounds can only be chromatographed at elevated temperatures. Moreover, an inversion of the elution order [10,54] on raising the temperature will not occur since a positive isoenantioselective temperature is absent.

Table 5. Temperature Dependence of the Enantiomer Discrimination of *cis*-Pinane on Phase I Heptakis(2,3,6-O-trimethyl)-β-cyclodextrin in OV-1701*

Temperature [°C]	Standard	r_0	r	$-\Delta\Delta G / R$	$-\Delta\Delta G$ [J mol^{-1}]	$-\Delta\Delta G/T$ [J K mol^{-1}]
30	C_8	5.50(4)	14.06(7)	1.555(6)	451	1.488
			15.74(5)	1.860(5)		
	C_{10}	0.81(7)	1.89(8)	1.321(9)	479	1.582
			2.12(4)	1.598(8)		
35	C_8	5.36(9)	13.00(0)	1.421(3)	461	1.497
			14.50(5)	1.701(7)		
	C_{10}	0.83(7)	1.80(8)	1.161(3)	500	1.623
			2.01(8)	1.411(6)		
40	C_8	5.24(8)	11.85(1)	1.258(4)	470	1.502
			13.15(8)	1.507(4)		
	C_{10}	0.85(6)	1.80(2)	1.105(1)	496	1.585
			2.00(1)	1.337(3)		
45	C_8	5.10(0)	10.89(5)	1.136(1)	467	1.468
			12.01(4)	1.355(5)		
	C_{10}	0.87(4)	1.73(9)	1.012(5)	491	1.545
			1.94(0)	1.219(2)		
50	C_8	4.96(7)	10.04(1)	1.021(5)	468	1.447
			11.00(5)	1.215(7)		
	C_{10}	0.89(9)	1.68(4)	0.872(7)	504	1.559
			1.84(6)	1.052(6)		
55	C_8	4.86(9)	9.55(6)	0.962(4)	472	1.437
			10.44(0)	1.144(0)		
	C_{10}	0.92(0)	1.69(6)	0.842(8)	503	1.532
			1.85(2)	1.013(3)		
60	C_8	4.78(3)	8.82(9)	0.846(2)	483	1.449
			9.60(0)	1.007(3)		
	C_{10}	0.94(0)	1.66(5)	0.772(0)	506	1.518
			1.81(1)	0.926(6)		
65	C_8	4.63(9)	8.35(3)	0.800(5)	487	1.439
			9.05(5)	0.951(8)		
	C_{10}	0.96(0)	1.63(4)	0.701(7)	522	1.542
			1.77(1)	0.844(7)		
70	C_8	4.54(1)	7.82(8)	0.723(7)	477	1.391
			8.42(6)	0.855(5)		
	C_{10}	0.98(5)	1.60(4)	0.627(9)	516	1.503
			1.72(6)	0.752(3)		

* Conditions as Table 1.

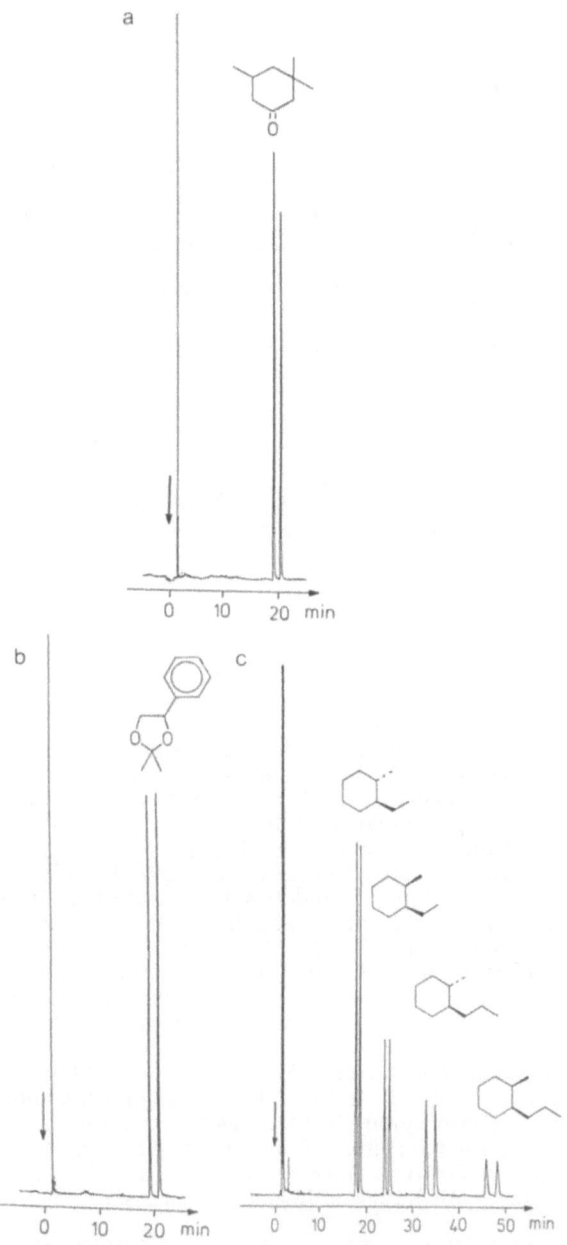

Fig. 5.   Gibbs-Helmholtz plots of enantiomer discrimination of three saturated hydrocarbons by Phase I heptakis(2,3,6-O-trimethyl)-β-cyclodextrin) (data from Tables 3-5).

Table 6. $\Delta_{R,S}(\Delta H)$ and $\Delta_{R,S}(\Delta S)$ of Enantiomer Discrimination of Saturated Hydrocarbons on Phase I Heptakis(2,3,6-O-trimethyl)-β-cyclodextrin

Selectand	Standard	$\Delta_{R,S}(\Delta H)$ [J mol^{-1}]	$\Delta_{R,S}(\Delta S)$ [J K mol^{-1}]
	C$_8$	-72	0.318
	C$_{10}$	-373	0.204
	C$_8$	-193	0.334
	C$_{10}$	-226	0.345
	C$_8$	-241	0.710
	C$_{10}$	-243	0.800

ACKNOWLEDGEMENTS

Generous support of this work by Deutsche Forschungsgemeinschaft and Fonds der chemischen Industrie is gratefully acknowledged.

REFERENCES

1. J. D. Morrison, "Asymmetric Synthesis", Vol. 1, Academic Press, Orlando (1985).
2. V. Schurig, "Asymmetric Synthesis", J. D. Morrison, ed, Vol. 1, Academic Press, Orlando, 77 (1985).
3. V. Schurig, Kontakte (Darmstadt), 1:3-22 (1986).
4. E. Gil-Av, B. Feibush and R. Charles-Sigler, Tetrahedr. Lett, 1009-1015 (1966).
5. H. Frank, G. J. Nicholson and E. Bayer, Angew. Chem., Int. Ed. Engl., 17:363 (1978).
6. W. A. König, "The Practice of Enantiomer Separation by Capillary Gas Chromatography", Hüthig, Heidelberg (1987).
7. V., Schurig, Chromatographia, 13:263-270 (1980).
8. V, Schurig and W. Bürkle, J. Am. Chem. Soc., 104:7573-7580 (1982).
9. V. Schurig, A. Ossig and R. Link, J. High Resolut. Chromatogr., 11:89 (1988).
10. V. Schurig and R. Link, in "Chiral Separations", D. Stevenson and I.D. Wilson, eds, Plenum Press, New York, 91-114 (1988).
11. V. Schurig, J. Chromatogr., 441:135-153 (1988).
12. V. Schurig, W. Bürkle, K. Hintzer and R. Weber, J. Chromatogr., 475:23-44 (1989).
13. T. Koscielski, D. Sybilska, S. Belniak and J. Jurczak, Chromatographia, 21:413-416 (1986).
14. T. Koscielski, D. Sybilska and J. Jurczak, J. Chromatogr., 364:297-303 (1986).
15. Z. Juvancz, G. Alexander and J. Szejtli, J. High Resolut. Chromatogr., 10:105 (1987).
16. V. Schurig and H.-P. Nowotny, in "Proceedings of Advances in Chromatography 1987", West Berlin, September 8-10, 1987; J. Chromatogr., 441:155-163 (1988).
17. W. A. König, S. Lutz and G. Wenz, Angew. Chem., Int. Ed. Engl., 27:747-748 (1988).
18. W. A. König, S. Lutz, P. Mischnick-Lübbecke, B. Brasset and G. Wenz, J. Chromatogr., 447:193-196 (1988).
19. W. A. König, S. Lutz, G. Wenz and E. von der Bey, J. High Resolut. Chromatogr., 11:506-509 (1988).
20. W. A. König, S. Lutz, W. Hagen, R. Krebber, G. Wenz, K. Baldenius, J. Ehlers and H. tom Dieck, J. High Resolut. Chromatogr., 12:35-39 (1989).
21. W. A. König, Nachr. Chem. Techn. Lab., 37:471-476 (1989).
22. H.-P. Nowotny, D. Schmalzing, D. Wistuba and V. Schurig, J. High Resolut. Chromatogr., 12:383-393 (1989).
23. V. Schurig, H.-P. Nowotny, M. Schleimer and D. Schmalzing, J. High Resolut. Chromatogr., 12:549-551 (1989).
24. V. Schurig, H.-P. Nowotny and D. Schmalzing, Angew. Chem., Int. Ed. Engl., 28:736-737 (1989).

25. D. Armstrong, in "Recent Advances in Chiral Separations", D. Stevenson and I. D. Wilson, eds, Plenum Press, New York, 169 (1990).
26. A. Villiers, C. R. Acad. Sci., 112:536 (1891).
27. F. Schardinger, Wien. Klin Wochenschr., 17:207 (1904); Zentralbl. Bakteriol. Parasitenkd. Infektionskr. Hyg. II, 29:188 (1911).
28. F. Cramer, "Einschlußverbindungen", Springer Verlag, Berlin (1954).
29. F. Cramer and H. Hettler, Naturwiss, 54:625-632 (1967).
30. M. L. Bender and M. Komiyama, "Cyclodextrin Chemistry", Springer Verlag, Berlin (1978).
31. J. Szejtli, Die Stärke - Starch, 26:26-33 (1977).
32. J. J. Stezowski, K. H. Jogun, E. Eckle and K. Bartles, Nature, 274:617-619 (1978).
33. K. H. Jogun and J. J. Stezowski, Nature, 278:667-668 (1979).
34. K. H. Jogun, J. M. Maclennan and J. J. Stezowski, Fifth European Crystallographic Meeting, Abstracts, 34 (1979).
35. M. M. Harding, J. M. Maclennan and R. M. Paton, Nature, 274:621-623 (1978).
36. J. A. Hamilton, M. N. Sebesan, L. K. Steinrauf and A. Geddes, Biochem. Biophys. Res. Commun., 73:659-664 (1976).
37. R. Tokuoka, T. Fufiwara and K. Tomita, Acta Cryst., B37:1158-1160 (1981).
38. J. J. Stezowski, M. Czugler and E. Eckle, I. Int. Symposium on Cyclodextrins, Budapest, 151-161 (1981).
39. A. P. Croft and R. A. Bartsch, Tetrahedron, 39:1417-1474 (1982).
40. J. Boger, R. J. Corcoran and J.-M. Lehn, Helv. Chim. Acta, 61:2190-2218 (1978).
41. B. Casu, M. Reggiani, G. G. Gallo and A. Vigevani, Tetrahedron, 25:803-821 (1968).
42. F. Cramer, G. Mackensen and K. Sensse, Chem. Ber., 102:494-508 (1969).
43. F. Cramer and W. Dietsche, Chem. Ber., 92:378-384 (1959).
44. D. W. Armstrong and W. de Mond, J. Chromatogr. Sci., 22:411 (1984).
45. D. W. Armstrong, T. J. Ward, J. D. Armstrong and T. E. Besley, Science, 232:1132 (1986).
46. V. Schurig, Angew. Chem., Int. Ed. Engl., 23:747-766 (1984).
47. K. Grob, "Making and Manipulating Capillary Columns for Gas Chromatography", Hüthig, Heidelberg, 1986, p. 124.
48. A. Venema and P. J. A. Tolsma, J. High Resolut. Chromatogr., 12:32 (1989).
49. J. Jauch, D. Schmalzing, V. Schurig, R. Emberger, R. Hopp, M. Köpsel, W. Silberzahn and P. Werkhoff, Angew. Chem., Int. Ed. Engl., 28:1022-1023 (1989).
50. E. Gil-Av and J. Herling, J. Phys. Chem., 66:1208-1209 (1962).
51. M. A. Muhs and F. T. Weiss, J. Am. Chem. Soc., 84:4697-4705 (1962).
52. J. H. Purnell, in "Gas Chromatography 1966", A. B. Littlewood, ed, Institute of Petroleum, London, 3 (1987).
53. V. Schurig, R. C. Chang, A. Zlatkis and B. Feibush, J. Chromatogr., 99:147-171 (1974).
54. V. Schurig, J. Ossig and R. Link, Angew. Chem., Int. Ed. Engl., 28:194-196 (1989).
55. H. Laderer, University of Tübingen, Thesis (1990).

CHIRAL CHROMATOGRAPHY ON THE PROCESS SCALE

Roger M. Smith, G. M. Hall* and G. Subramanian*

Departments of Chemistry and *Chemical Engineering
Loughborough University of Technology
Leicestershire LE11 3TU, UK

SUMMARY

Although in many cases it is possible to prepare enantiomerically pure agrochemicals and pharmaceuticals by stereospecific synthesis, it may often be more practical to use a less complicated synthesis followed by a chromatographic separation of the components of a racemic mixture. This will be true particularly when the unwanted isomers can be recycled back into the synthetic process. However, the relatively low capacity of preparative chromatographic systems makes this approach impractical for large volume chemicals, such as pesticides. However, larger samples can be handled on a process scale by overloading the column and carrying out the separation in the displacement mode.

Here the criteria for process scale separations are examined and compared with the criteria for analytical and preparative separations. Results from a series of studies to model high-capacity chiral separations on cyclodextrin and cellulose-based stationary phases are reported.

INTRODUCTION

In the last few years there has been an increased awareness of the importance of the stereochemistry of drugs and other physiologically active compounds and the need for testing and monitoring of the individual isomers [1,2]. Differences in the biological activity of enantiomers have also called into question much published pharmacological data [3,4]. These interests have fuelled the need for samples of optically pure enantiomers for use as standards and for toxicology. For many years it has been the goal of the synthetic organic chemist to devise routes to yield products of known and closely defined absolute and relative stereochemistry; however, despite the skills of the chemist, the routes needed to achieve chiral purity are often long, complex, and require the use of exotic reagents or conditions.

As part of this work, it is necessary to have analytical methods which can determine the enantiomeric purity of the product or intermediates. These have included the traditional measurements [5] of optical activity, including polarimetry, optical rotatory dispersion and circular dichroism spectroscopy, and the formation of diastereoisomers and their analysis by spectroscopic methods, including nuclear magnetic resonance (NMR) spectroscopy, and achiral chromatography [6]. These techniques are often now being displaced by direct chromatographic methods using chiral mobile or stationary phases in gas-liquid chromatography (GLC), high-performance liquid chromatography (HPLC), thin-layer chromatography (TLC), or supercritical fluid chromatography (SFC) [4,5,7,8]. Most of these methods, however, are not absolute and still require the individual isomers for calibration purposes.

Frequently, it may not be appropriate to invest the time required for the development of a chiral synthetic method until the enantiomers have been demonstrated to possess significant biological activity. The chemist must then rely on the separation of the individual isomers from a racemic synthetic mixture.

Recent Advances in Chiral Separations, Edited by D. Stevenson and
I. D. Wilson, Plenum Press, New York, 1991

SEPARATION OF ENANTIOMERS

Traditionally the resolution of enantiomers has been difficult. The methods have either involved careful crystallisation or the formation of the more easily separated diastereoisomers by reaction with a pure enantiomeric reagent followed by crystallisation and finally removal of the substituent groups [9]. Preparative chromatographic separation was clearly a possibility but by 1973 despite some 400 papers on the topic only a handful of successful separations had been reported [10]. With the advent, in the last 10 years, of successful chiral chromatography it is now possible to directly separate and collect enantiomers in high purity. Even relatively small analytical columns can be used to yield sufficient quantities for standards or for identification. However, if larger quantities are needed for further synthetic steps or for toxicological testing, larger preparative scale columns capable of separating 0.1 to 10 g at each injection and specialised equipment are needed.

PREPARATIVE CHIRAL CHROMATOGRAPHY

The technique of instrumental high-efficiency preparative chromatography has become the subject of considerable study in recent years both for achiral and chiral separations [11-14] with particular interest in liquid chromatography. The primary step is to increase the diameter of the column and hence the sample loading capacity. The high value of most fine chemicals can justify the time and expense in manpower and equipment required to develop and achieve a separation. Because of their low sample capacity gas chromatographic methods are rarely used for preparative separations and although removal of the mobile phase is easy with SFC so far it has been little used. The theoretical basis of HPLC methods have been examined in detail by Knox [15], Guiochon [16,17] and Snyder and Cox [18]. However, many of the separations have been carried out in industrial laboratories and most have not been reported.

The role of preparative chromatography for the separation of mixtures of enantiomers has been reviewed [8,19] and Allenmark has identified the following benefits compared to manual methods [19].

1. No loss of optically active material should occur.
2. The optical purity of the stationary phase is not crucial as long as separation can be achieved.
3. If the peaks are fully resolved then both enantiomers can be obtained in 100% purity and yield.
4. The recovery of the purified material can be readily automated.

A wide range of different combinations of mobile and stationary phases can be used in chiral chromatography and most have been employed for preparative separations but with differing degrees of success.

Chiral Mobile Phases

Chiral mobile phases have found little use, even in analytical chromatography, because of their cost and the limited range of readily available chiral compounds and have not been used preparatively.

Chiral Additives to the Mobile Phase

The separation of suitable enantiomers can be achieved by the inclusion in the mobile phase of cyclodextrins, ternary copper derivatives of amino acids, or chiral ion-pair reagents, such as quinine and camphor sulphonic acid [20-22]. Although this approach has been used for preparative chiral chromatography, there are problems in removing the analytes from the collected fractions.

Chiral Stationary Phases (CSP)

More than 40 CSPs are now available, many commercially, and these are generally the preferred route to achieve separations [23-25]. They have been grouped into four main classes.

Acceptor/donor (Pirkle)-type columns. These columns have very clearly defined sites of interaction, which can be specific to individual analytes. This usually results in very high separation of the enantiomers of polyfunctional compounds with the appropriate configurations. Their use for preparative chromatography has been described [26]. Samples up to 8 g have been resolved but each chromatographic run was slow taking 4 hr.

Because of the need for the presence of a specific interaction these columns are not generally applicable for a wide range of analytes. However, it will often be possible to introduce suitable interacting groups into an analyte by derivatisation, it a suitable functional site is near the chiral centre. However, this derivative group will need to be removed after the separation has taken place. Alternatively, if the analyte molecule contains appropriate functional sites it may also be feasible to synthesise a corresponding specifically matching stationary phases, but the need for the separation of the analyte will have to be sufficiently high for the development costs to be justified.

Protein-based chiral columns. Columns in which a protein is bound to a silica matrix have been found to resolve a wide range of analytes, however they have very low sample capacities. Their use for preparative separations has been described as "impossible" [27]. Although preparative separations have been claimed, the loading was very low (0.25 mg on a 22 x 500 mm column [28]). It has also been reported that overloading the column can cause reversal of the order of elution of enantiomers [29].

Cellulose esters. The ready availability and relatively low cost of natural polymers suggest that they should have wide applicability as stationary phases. With these columns the presence of functional groups in the analyte has not been found to be necessary for resolution [25]. The most widely used material is cellulose triacetate but it is mechanically soft and can be sensitive to compression. The loading can be quite good with up to 450 mg being separated at a time on a column (Figure 1) [30]. The mechanical properties can be improved by coating the cellulose esters onto a silica base [31].

Host-guest or inclusion materials. Another group of natural polymers which can be used to achieve chiral separations are the cyclodextrins whose interaction is based on the selective inclusion of the analyte into a chiral cavity. However, the resolution of enantiomers is often low [32]. Cyclodextrins have been used successfully for a number of separations on the preparative scale [33] but the kinetics of the inclusion step are slow and rapid elution is not possible so that each run may take several hours [25].

Other column materials. These include the helical polymers of triphenylmethyl methylacrylate in which the overall structure of the polymer provides the chirality of the stationary phase. These matrices have been used for the preparative separation of *trans*-2,3-diphenyloxirane and 200 mg has been separated successfully on a 2.2 x 30 cm column [34]. Ligand exchange phases in which a ternary metal complex is bound to the silica surface can also be used preparatively but are limited to the separation of amino acids and related compounds [27].

A general problem with many of these CSP columns is that the separation process is kinetically slow, each separation taking from 1 to 15 hr to complete, so that although the resolution of enantiomers is essentially complete the mass throughput is low. This is not a serious problem during the investigative phases of drug developments as cost is less important than quality.

PROCESS SCALE SEPARATIONS

Once the biological activity, efficacy and safety of an enantiomer have been demonstrated on a small scale there will be a need for large scale manufacture for clinical trials and toxicology and eventually for sale. At this stage either a specific synthetic route must be developed or separation of the enantiomers on a process scale system, with 10 g to 1 kg injections, will be needed. The costs of such process scale separations to prepare a specified purity now become a major criterion. These depend on the throughput of the separation system and its capital and running costs. These include the cost of eluents, recovery or disposal of waste eluents and of separated impurities (which could include the "wrong" isomer). Care must also be taken to avoid contamination of the product by dissolved silica or impurities from the eluent. The cost will also need to be compared with the cost of developing a synthetic route.

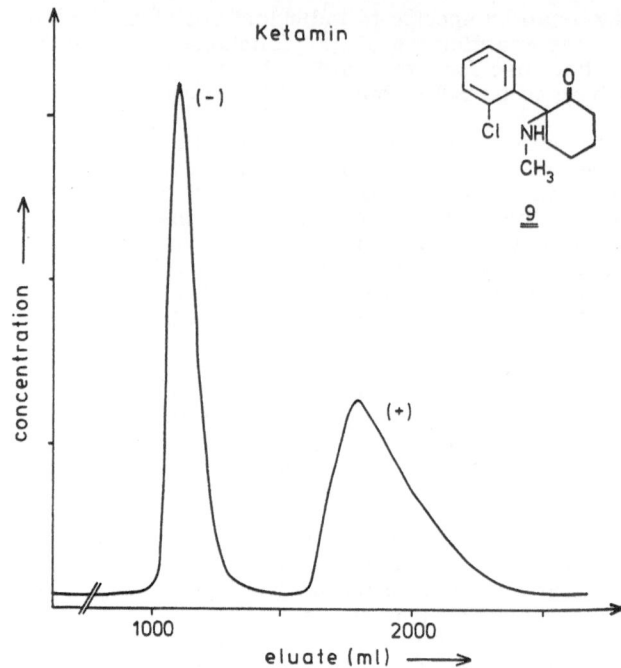

Fig. 1. Separation of 450 mg ketamin on cellulose triacetate [30].

The separation work recently reported by Pirkle (this symposium) using chiral membranes is of considerable significance in this large scale approach. However, very specific interactions with the membrane are required and an active group may need to be introduced into the compound to be resolved, which may subsequently have to be removed. An alternative is to increase the throughput of a chiral chromatographic system. Both these approaches suffer from the problem that they will yield both isomers. One of these will often be essentially a waste product whose disposal must be included in the costing. These routes are therefore unlikely to be of major importance unless the unwanted isomer can be recycled back into the synthetic route or can be racemised to yield additional desired isomer.

An analysis of the factors to be considered in scaling up to process chromatography has recently been reported with particular note of the need to satisfy Current Good Manufacturing Practice [35]. The throughput of a chromatographic system can be increased by deliberately grossly overloading the column and operating in the displacement rather than the normal elution mode [36]. In this case the components of the mixture, rather than being in equilibrium between the mobile and stationary phases, compete with one another for the active sites on the surface of the stationary phase. As the sample passes down the column this sets up a displacement train of bands in which components which are more strongly attracted to the stationary phase displace the weaker components in turn (Figure 2). A very strongly interacting displacer compound is added to the eluent after the sample is injected to force the analytes through the column without tailing. Unlike preparative and analytical separations, in displacement chromatography increasing the length of the column does not increase separation power as long as sufficient column length is available for the displacement train to become established. The other advantage of displacement chromatography is that because the analytes are displaced from the column within one column volume, the amount of mobile phase needed is limited and the samples are not greatly diluted, thus simplifying their recovery. Even though there are overlap regions between the bands these fractions can be recycled.

Although this technique has been known for many years it is not widely applied although the theoretical background has been recently investigated by Guiochon [16] and Horvath [36]. One of the reasons for the relative obscurity of the technique is that the separation conditions cannot be directly transferred from an analytical chromatogram.

138

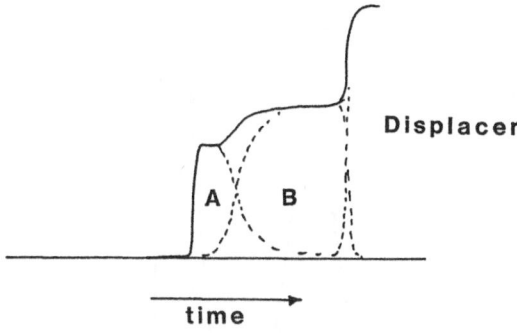

Fig. 2. Displacement train established on a column.

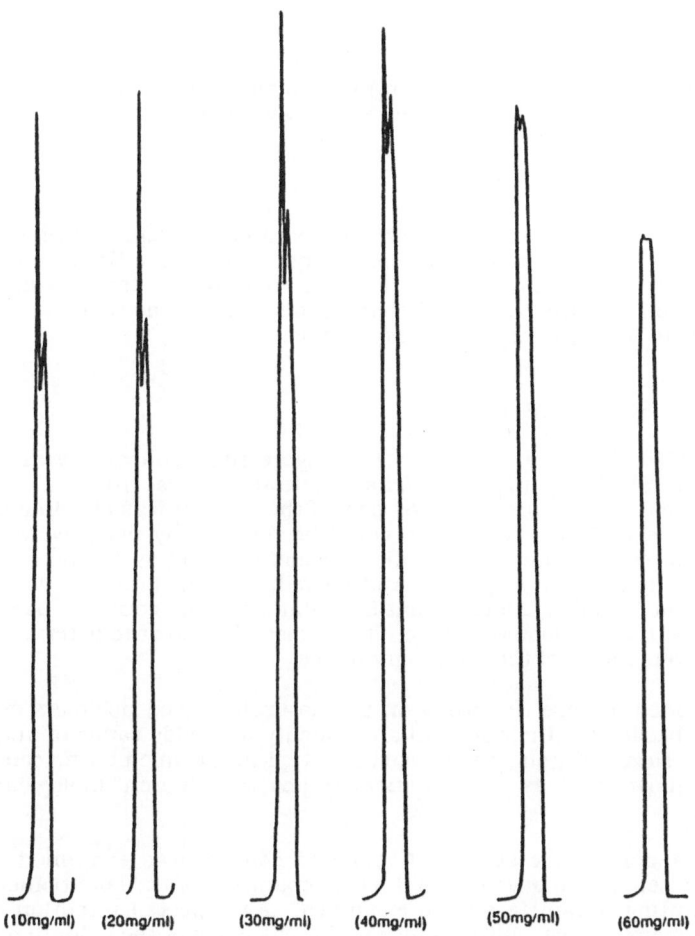

(10mg/ml)   (20mg/ml)   (30mg/ml)   (40mg/ml)   (50mg/ml)   (60mg/ml)

Fig. 3.   Separation of hexobarbitone on cyclodextrin column (25 cm x 0.46 cm) with methanol as eluent at a flow rate of 1 ml min^{-1}. Detection system UV spectroscopy at 254 nm.

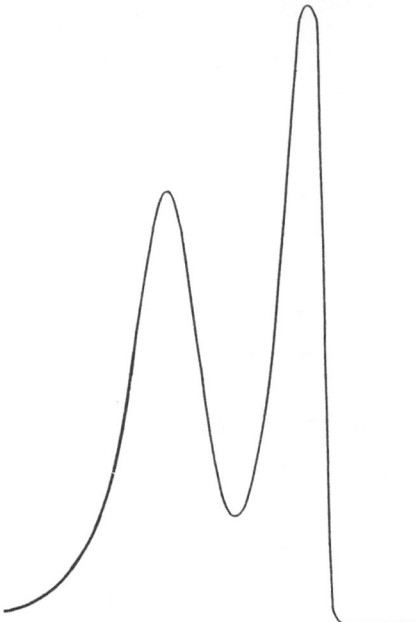

Fig. 4.   Separation of hexobarbitone (250 mg) on Chiralcel CA column (50 cm x 2 cm id) with propan-2-ol-n-hexane (1:9 v/v) as eluent at a flow rate of 8 ml min^{-1}. Detection system UV spectroscopy at 254 nm.

The technique has been used for a chiral separation by Vigh [37] who has generated a series of specifically designed displacer compounds. These have the useful properties that after the separation they can be flushed from the column by a temperature change which alters the permeability of the stationary phase. This reduces the time needed to regenerate the column prior to the injection of the next sample.

MODEL PROCESS SCALE SYSTEMS

In our studies we have set out to investigate the applicability and limitations of process scale chiral chromatography. The eventual aim was to examine displacement chromatography for chiral separations. As part of the study it was also intended to examine whether the permeability of the column could be reduced by using wide-pore stationary phases so that lower back pressures would be generated and as a consequence the design requirements of the separation plant could be reduced. In practice, wide-bore columns would be used but it was considered that 5-10 mm columns could be used as reasonable model systems for the development stage. It was also of interest to determine the feasibility of on-column regeneration of the stationary phase.

It was decided to base the study on the naturally based polymers cyclodextrin and cellulose esters because of their general applicability to a wide range of analytes and their moderate costs. Hexobarbitone was chosen as a typical racemic pharmaceutical compound as it has been shown that its two enantiomers possess different biological activity as an anaesthetic [38].

In the first part of the study a Cyclobond column was examined but after some development the separation gave only a limited separation of the enantiomers (Figure 3). A number of operating parameters were examined. As expected retention decreased with increases in temperature with little change in the efficiency of the column and only a slight decline in the resolution. However, attempts to significantly increase the sample concentration were limited by the solubility in the mobile phase used as the injection solvent and overload conditions could not be reached.

To find an alternative material that was inherently chiral we examined cellulose triacetate (Chiralcel CA). Methanol containing an increasing proportion of water was examined as the eluent. Separation of hexobarbitone was only achieved at 30% water but the column material rapidly deteriorated and the back pressure increased.

On a second Chiralcel CA1 column the alternative eluent propanol-hexane (1:9) was examined. This gave a fair separation of hexobarbitone (Figure 4) but again it was difficult to achieve high sample concentrations.

A major problem is that non-polar mobile phases are needed for these separations but these cannot solublise the polar pharmaceuticals of principal interest in this study.

ACKNOWLEDGEMENTS

The authors thank SERC Biotechnology Directorate for a research grant and J. T. Baker and Crosfields for the donation of stationary phases and for technical assistance.

REFERENCES

1. E. J. Ariens, J. J. S. van Rensen and W. Welling, eds, "Stereoselectivity of Pesticides", Elsevier, Amsterdam (1988).
2. W. H. De Camp, *Chirality*, 1:2-6 (1989).
3. E. J. Ariens, *Eur. J. Drug Metab. Pharmacokinetics*, 13:307-308 (1988).
4. E. J. Ariens in "Chiral Separations by HPLC", A.M. Krstulovic, ed, Ellis Horwood, Chichester, 31 (1989).
5. S. G. Allenmark, "Chromatographic Enantioseparation: Methods and Applications", Ellis Horwood, Chichester, 27 (1988).
6. S. G. Allenmark, "Chromatographic Enantioseparation: Methods and Applications", Ellis Horwood, Chichester, 42 (1988).
7. W. J. Lough, ed, "Chiral Liquid Chromatography", Blackie, Glasgow (1989).
8. M. Zief and L. J. Crane, eds, "Chromatographic Chiral Separations", Chromatographic Science Series, Vol. 40, Marcel Dekker, New York (1988).
9. A, Collet in "Chiral Separations by HPLC", A. M. Krstulovic, ed, Ellis Horwood, Chichester, 81-104 (1989).
10. G. Blaschke, *Angew. Chem. Internat. Ed.*, 19:13-24 (1980).
11. K. Hostettman, M. Hostettman and A. Marson, "Preparative Liquid Chromatography," Springer Verlag, Berlin (1986).
12. B. A. Bidlingmeyer, "Preparative Liquid Chromatography," J. Chromatography Library, Vol. 38, Elsevier, Amsterdam (1987).
13. K. Jones, *Chromatographia*, 25:547-559 (1988).
14. E. Grushka, ed, "Preparative-Scale Chromatography", Chromatographic Science Series, Vol. 46, Marcel Dekker, New York (1988).
15. J. H. Knox and H. M. Pyper, *J. Chromatogr.*, 363:1-30 (1986).
16. G. Guiochon and A. Katti, *Chromatographia*, 24:165-189 (1987).
17. A. Katti and G. Guiochon, *Anal. Chem.*, 61:982-990 (1989).
18. L. R. Snyder, C. A. Cox and P. E. Antle, *Chromatographia*, 24:82-96 (1987).
19. S. G. Allenmark, "Chromatographic Enantioseparation: Methods and Applications," Ellis Horwood, Chichester, 192 (1988).
20. N. Nimura, in "Chiral Separations by HPLC", A.M. Krstulovic, ed, Ellis Horwood, Chichester, 107-123 (1989).
21. C. Pettersson, in "Chiral Separations by HPLC", A. M. Krstulovic, ed, Ellis Horwood, Chichester, 124-146 (1989).
22. D. Sybilska and J. Zukowski, in "Chiral Separations by HPLC", A. M. Krstulovic, ed, Ellis Horwood, Chichester, 147-172 (1989).
23. I. W. Wainer, *Trends Anal. Chem.*, 6:125-134 (1987).
24. A. C. Mehta, *J. Chromatogr.*, 426:1-13 (1988).
25. W. H. Pirkle and T. C. Pochapsky, *Chem. Rev.*, 89:347-362 (1989).
26. W. H. Pirkle and J. M. Finn, *J. Org. Chem.*, 47:4037-4040 (1982).
27. R. Dappen, H. Arm and V. R. Meyer, *J. Chromatogr.*, 373:1-20 (1986).
28. P. Erlandson, L. Hansson and R. Isaksson, *J. Chromatogr.*, 370:475-483 (1986).
29. S. G. Allenmark, *Chem. Sci.*, 20:5-10 (1982).
30. G. Blaschke, *J. Liq. Chromatogr.*, 9:341-368 (1986).
31. A. Ichida, T. Shibata, I. Okamoto, Y. Yuki, H. Namikoshi and Y. Toya, *Chromatographia*, 19:280-284 (1984).
32. S. M. Han and D. W. Armstrong, in "Chiral Separations by HPLC", A.M. Krstulovic, ed, Ellis Horwood, Chichester, 208-284 (1989).
33. K. H. Rimbock, F. Kastner and A. Mannschreck, *J. Chromatogr.*, 329:307-310 (1985).

34. Y. Okamoto, S. Honda, K. Hatada and H. Yuki, *J. Chromatogr.*, 350:127-134 (1985).
35. S. Ostrove, *LC GC Magazine*, 7:550-554 (1989).
36. Cs. Horvath, J. Frenz and Z. El Rassi, *J. Chromatogr.*, 255:273-293 (1983).
37. G. Vigh, G. Quintero and G. Farkas, *J. Chromatogr.*, 506:481-493 (1990).
38. G. Wahlstrom, *Life Science*, 5:1781-1790 (1966).

A DIODE LASER POLARIMETRIC HIGH-PERFORMANCE LIQUID CHROMATOGRAPHY DETECTOR AND APPLICATIONS TO ENANTIOMERIC PURITY DETERMINATIONS AND ENANTIOSELECTIVE REACTIONS

David M. Goodall*, David K. Lloyd† and Zecai Wu

Department of Chemistry, University of York
Heslington, York YO1 5DD, UK

SUMMARY

The principles of operation of a polarimetric high-performance liquid chromatography (HPLC) detector based on an 820 nm diode laser are discussed. Root-mean-square noise is 4 $\mu°$ (1 sec time constant), and detection limits are found in the range 0.10-2 μg dependent on specific rotation and chromatographic peak width. Determinations of enantiomeric purity have been carried out using achiral chromatography with dual polarimetric/absorbance detection. Applications to the pharmaceuticals ephedrine hydrochloride, pseudoephedrine hydrochloride (in a cough linctus) and the agrochemicals paclobutrazol and fluazifop-P-butyl demonstrate the accuracy and precision of the technique and its potential in quality control. Determinations of enantiospecificity in enzyme and free radical reactions are reviewed. Polarimetric detection with chiral chromatography of complex reaction mixtures has allowed identification of enantiomers, and with achiral chromatography of near racemates has given enantiomeric purities to 0.1%.

INTRODUCTION

The development of laser-based polarimetric detectors specially designed for HPLC [1-4] has added a new dimension to the fields of chiral synthesis, separations and analysis. Here the principles of polarimetric methods for analysis of chiral molecules will be outlined, together with recent applications of a diode laser polarimeter in enantiomeric purity determinations and enantioselective reactions. A full review of polarimetric detection in HPLC is given elsewhere [5].

Polarimetry measures optical rotation, which is a dispersive manifestation of optical activity. The relationship between circular dichroism (CD) and optical rotation (OR) [6,7] is illustrated in Figure 1. Differential absorption of left and right circularly polarised light, A_L-A_R, may be linked via the Kronig-Kramers transform to the wavelength-dependent optical rotation, α, of the plane of plane-polarised light. It is important to note that an optically-active transition gives rise to optical rotatory dispersion (ORD) extending from the vacuum UV to the near IR, whereas CD can only be studied within the optically-active absorption band. Polarimetric measurements at the sodium D line wavelength (589 nm) have for many years been used to characterise the purity of chiral molecules, and single-wavelength polarimetric HPLC detectors can be similarly used as universal detectors for chiral molecules. The coupling of polarimetric detection with HPLC provides a powerful combination of techniques for determining enantiomeric purity of compounds present in impure form, e.g., in a reaction mixture or a pharmaceutical preparation, since HPLC allows the compound of interest to be separated from interfering compounds before its

---

* Author to whom correspondence should be addressed.
† Current address: Department of Oncology, McGill University, 3655 Drummond, Montreal, P.Q., H3G 1Y6, Canada.

Recent Advances in Chiral Separations, Edited by D. Stevenson and
I. D. Wilson, Plenum Press, New York, 1991

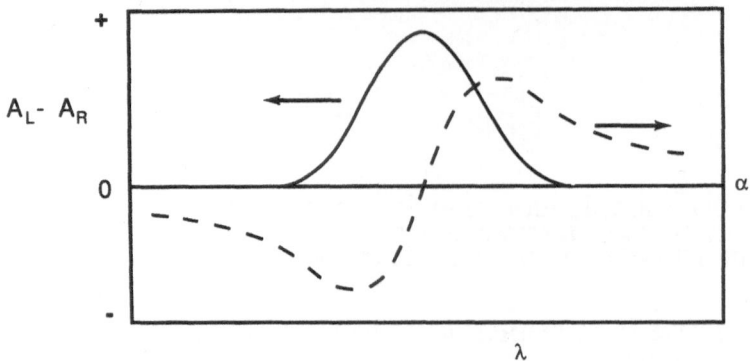

Fig. 1.   Relationship between differential absorption of left and right circularly polarised light, A_L - A_R, optical rotation, α, and wavelength of the polarised light beam, λ.

optical activity is measured. The coupling of CD with HPLC [8] is less routine but can offer advantages in characterising eluted species and determining stereochemistry from their CD spectra [9].

Early applications of polarimetry in HPLC involved the separation of sugars [10]. In this case a commercial polarimeter with a mercury arc source was modified with a special 1 dm pathlength flow cell only 0.65 mm in diameter. This gave a cell volume of 33 µl, no significant band broadening in the separation, and detection limits on column of 5-20 µg. Shortly afterwards Yeung et al. [1] described the first laser-based polarimeter designed specifically for use in HPLC. With three orders of magnitude increase in power of the source in moving from an arc lamp to an argon-ion laser, together with an optical design optimised to take account of the laser noise characteristics, the detection limit for fructose was improved to 500 ng. The best reported sensitivity with an argon-ion laser polarimeter is 11 ng for fructose, using microbore HPLC and a 1 µl flowcell [11].

The use of absorbance and polarimetric detectors coupled in series to monitor peaks eluting from an achiral column [12-14] has been shown to provide a simple method for determination of the enantiomeric purity of compounds for which a standard of known purity is available. All applications involve ratioing the response of the two detectors; absorbance is proportional to the analyte concentration, whilst optical rotation depends on the product of the concentration and the difference between mole fractions of each enantiomer in the analyte. This method has the potential for routine use in quality control of compounds which are used in enantiomerically-pure form. First applications were to the clinical chelating agent penicillamine [15] and the insecticide *trans*-permethrinic acid pentafluorobenzyl ester [12]. Subsequently the method was used in the analysis of ophthalmic epinephrine solutions [13], important since (R)-(-)-epinephrine is active and (S)-(+)-epinephrine inactive in the treatment of glaucoma. In this analysis results were obtained with accuracy comparable to but in a much shorter time than the USP procedure using a multistep sample treatment followed by conventional polarimetry.

Applications to free [14] and dansylated [16] amino acids have been described using an argon-ion laser polarimeter. Mannschreck et al. have devised a method using dual detectors which allows enantiomeric purities to be determined for compounds which are only partially resolved on a CSP [17], and have shown a range of applications in organic [18,19] and pharmaceutical chemistry [20]. Enantiomeric purity measurements for ephedrine hydrochloride and pseudoephedrine hydrochloride [21] and the agrochemicals paclobutrazol and fluazifop-P-butyl [22] which we have recently obtained with a diode laser polarimeter will be described here. Novel applications of achiral chromatography with coupled detectors to enantioselective enzyme reactions [23] and to enantiospecific equithermal oxidations [24] will also be reviewed.

Diode Laser Polarimeter and Limit of Detection

The polarimetric HPLC detector (ACS ChiraMonitor) used to obtain all the results reported in this paper is based on the design of a diode laser polarimeter described by Lloyd et al. [4]. A block diagram of the instrument is given in Figure 2. The plane of polarisation of light at 820 nm from a continuous wave diode laser is modulated via the Faraday effect by passing an AC current at frequency f through a coil surrounding a double extra-dense flint

144

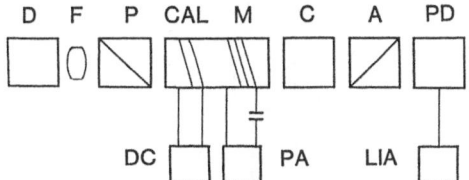

Fig. 2.   Polarisation-modulated polarimeter block diagram: D, light source; F, focussing lens; P, polariser; CAL, calibrator; M, modulator; C, cell; A, analyser; PD, photodetector; DC, calibrator DC power supply; PA, power amplifier; LIA, lock-in amplifier.

glass rod. With the polarisation axis of the analyser set at 90° to that of the polariser, the light flux at the detector is modulated at a frequency $2f$. Introduction of a sample which rotates the plane of polarisation by an angle α gives rise to a signal at frequency f proportional to α. For signal recovery the photodiode output is sent to a lock-in amplifier which is referenced to the modulator drive current.

The principal advantage of using a diode laser is the low flicker noise of such a device. A signal-to-noise analysis of the laser polarimeter has been presented, and a typical root-mean-square (RMS) noise level of 4 μ° observed with a 1 sec time constant is in accord with theoretical predictions [4]. From this the limit of detection (LOD) for an optical rotation signal is calculated to be 12 μ° using the equation [22,25] with E_{RMS} the RMS error

$$LOD = 3 \times E_{RMS}. \tag{1}$$

This may be converted to LOD in concentration of a chiral analyte, since from Biot's law

$$\alpha = [\alpha]cl, \tag{2}$$

where c is the mass concentration and l the cell pathlength. For fructose, $[\alpha]_{820} = 45°$ ml g^{-1} dm^{-1}. The ChiraMonitor has $l = 0.16$ dm, and substituting values into Eq. 2 gives LOD for concentration = 1.7 μg ml^{-1} in the OR cell.

Chromatograms with OR and UV detection of a mixture of fructose and sucrose (2 μg of each injected on column) are shown in Figure 3. From the measured peak width for fructose a dilution factor of 6.6 between injection and OR peak maximum was determined. Multiplying the concentration LOD in the OR cell by this factor and by the injected volume of 20 μl gives LOD = 0.22 μg for mass of fructose injected.

In a number of studies [5,21,22] we have found LOD values with the ChiraMonitor to be between 0.2 and 1 μg on column (1 sec time constant), with low LOD values found for compounds with high specific rotation and/or low chromatographic peak widths. Because most chiral compounds have $|\ [\alpha]_{589}\ |$ values in the range 10 to 100° ml g^{-1} dm^{-1}, we can make the generalisation that detection limits will generally be found in the range 0.1-2 μg for most compounds for which an efficient achiral separation is available. If required, estimates of $[\alpha]_{820}$ can readily be made from specific rotation at the sodium D line tabulated in standard compilations [26,27]. For a compound with one dominant chiral chromophore, centred at wavelength λ_0, $[\alpha]$ scales with λ according to the Drude relationship [28]:

$$[\alpha]_\lambda \propto (\lambda^2 - \lambda_0^2)^{-1}. \tag{3}$$

From Eq. 3, $[\alpha]_{820}/[\alpha]_{589} \approx 0.5$ for $\lambda_0 \leq 280$ nm , which is the case for most compounds of interest to analysts. Polarimetry, with relatively low variations of $[\alpha]$ both for change of wavelength in the visible - near IR region and for change of compound, may be contrasted with CD, where there is no response outside an absorption band and $\varepsilon_L - \varepsilon_R$ values can vary substantially within absorption bands. Thus polarimetry rather than CD should normally prove to be the preferred optical activity method for use in routine HPLC analysis of chiral molecules. With unknown materials CD spectra may give more information [9], but at the expense of LOD.

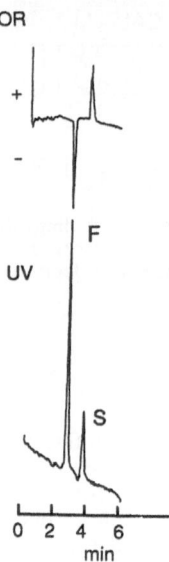

Fig. 3.   Achiral chromatograms of fructose (F) and sucrose (S) with dual optical rotation and UV absorbance detection. Column: Spherisorb C_{18} (25 cm x 4.6 mm id). Mobile phase: water. Flow rate: 1.0 ml min^{-1}; 2 μg of each compound injected.

The linear range of the technique is illustrated by Figure 4, a plot of log (optical rotation response) versus log (concentration) for paclobutrazol, which has a slope of 1.005 ± 0.005 and correlation coefficient 0.99998 over the range 0.02 to 10 mg ml^{-1} (20 μl injection) [22]. The lower limit is set by the LOD (0.19 μg in this case), whilst the upper limit is set by column capacity. It should be noted that the linear range of polarimetric measurements extends into highly concentrated solutions, and that 820 nm is an ideal wavelength to avoid any attenuation of the light flux due to sample absorption [4].

Enantiomeric Purity Determination

The observed rotation, α, of a compound containing mole fractions of (+)- and (-)-enantiomers x and $(1 - x)$ respectively is given from Biot's law (Eq. 2) as

$$\alpha = [\alpha] \, (2x - 1)cl. \tag{4}$$

By using achiral chromatography with polarimetric and spectrophotometric detectors in series, enantiomeric purity can be calculated from the ratio of the optical rotation and absorbance response, α/A, for unknown and standard samples [21]

$$\frac{(\alpha/A)_U}{(\alpha/A)_S} = \frac{2x_U - 1}{2x_S - 1}. \tag{5}$$

This follows from Eq. 4 since absorbance, A, is directly proportional to c. The (+)-enantiomer mole fraction of the standard, x_S, is normally obtained by chiral chromatography.

This method has been applied to two pharmaceutical compounds, ephedrine hydrochloride and pseudoephedrine hydrochloride [21] and two agrochemicals, paclobutrazol and fluazifop-P-butyl [22]. For all these compounds the desired biological activity resides in a single enantiomer, so it is of considerable interest to establish the accuracy and precision with which enantiomeric purity measurements can be carried out. The results of blind trials in which unknown mixtures of single enantiomer and racemate were analysed are given in Table 1. Mean values are presented with their 95% confidence limits [25] and the number of replicate injections, n. All values are seen to be in agreement within the 95% confidence limits, and the achiral chromatographic method gives enantiomer mole fractions to 1% for fluazifop-P-butyl and to 0.4% for ephedrine

146

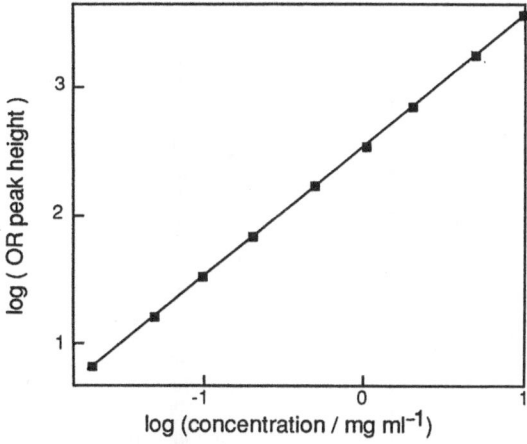

Fig. 4. Linear response of the polarimetric detector. Plot of log (OR peak height) versus log (concentration/mg ml^{-1}) for (2R,3R)-paclobutrazol.

hydrochloride. Both compounds have similar specific rotation ($[\alpha]_{820}$ = 16 and 18° ml g^{-1} dm^{-1}, respectively) and peak widths were similar (~15 sec fwhm, flow rate 1 ml min^{-1}). The difference in 95% confidence limits correlates with sample loading, 50 µg for ephedrine hydrochloride and 20 µg for fluazifop-P-butyl.

The achiral method carries a little more uncertainty than does chiral chromatography, in part because it is a relative method and errors in x_S must be taken into account as well as those in $(\alpha/A)_u$ and $(\alpha/A)_S$ [21]. For routine quality control measurements where a well-defined standard is available the α/A method offers advantages of using a normal achiral column instead of a less robust and more expensive chiral column. In contrast to polarimetric measurements on bulk samples, the achiral chromatographic technique with dual polarimetric/absorbance detection can be used with impure samples. This is illustrated by our measurements of the enantiomeric purity of (+)-pseudoephedrine hydrochloride in Sudafed cough linctus (Figure 5), which with a 7.5 µg loading of the active constituent in the presence of a large excess of sucrose was found to have x_u = 0.99 ± 0.01 with seven replicate injections [21].

Table 1. Enantiomer Mole Fraction of (+)-Enantiomer in Blind Trials

Compound	Sample	Achiral Loading (µg)	x_u* Actual	$x_u \pm t_{95}s_{x_u}/\sqrt{n}$ (n) Chiral Separation	Achiral Chromatography
Ephedrine hydrochloride	Standard			0.9989 ± 0.0010 (4)	
	1	50	0.974		0.972 ± 0.004 (5)
	2	50	0.950		0.950 ± 0.003 (6)
Fluazifop-P-butyl	Standard			0.948 ± 0.005 (9)	
	1	~20	0.940		0.942 ± 0.016 (5)
	2	~20	0.931		0.938 ± 0.012 (5)
	3	~20	0.919		0.923 ± 0.008 (7)
	4	~20	0.910		0.901 ± 0.010 (7)
	5	~20	0.926	0.928 ± 0.004 (4)	0.916 ± 0.010 (8)
	6	~20	0.902	0.904 ± 0.008 (6)	0.902 ± 0.012 (8)
	7	~20	0.886	0.884 ± 0.007 (7)	0.880 ± 0.011 (6)

* Uncertainties in x_u are calculated from errors in pipetting and enantiomeric purity of the standards to be ± 0.002 (ephedrine hydrochloride) and ± 0.005 (fluazifop-P-butyl).

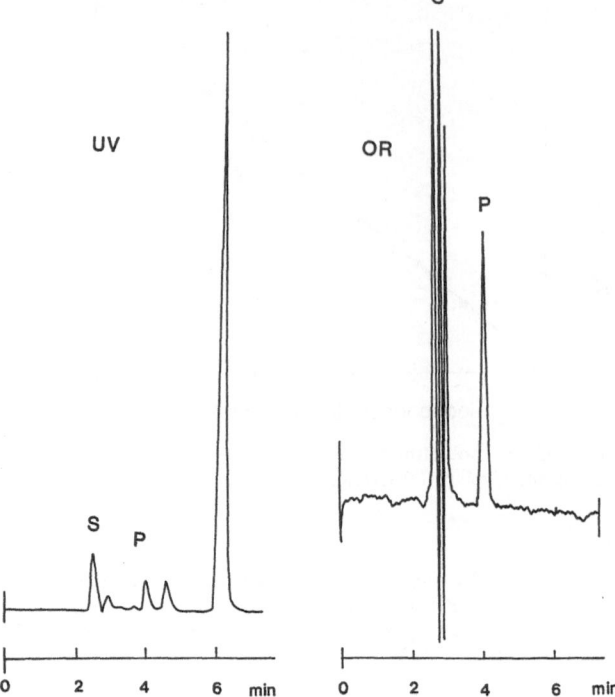

Fig. 5.   Chromatograms of Sudafed cough linctus with dual OR/UV detection. Column: Spherisorb 5 C_8. Mobile phase: methanol-water (0.8% ammonium nitrate by weight) (55:45). Flow rate: 1.0 ml min^{-1}. Pharmaceutical preparation diluted 20 times and filtered before injection. Sucrose (S) and pseudoephedrine hydrochloride (P) peaks labelled, 7.5 µg P injected.

ENANTIOSELECTIVE REACTIONS

Enantioselectivity in enzyme reactions has been characterised using the diode laser polarimeter [23].   Racemic mixtures of bicyclic ketones were reduced with various dehydrogenase enzymes, and the time-dependent growth of ketone and alcohol peaks in the OR chromatograms was compared to the decay of ketone and growth of alcohol peaks in the UV chromatograms. $3\alpha20\beta$-Hydroxysteroid dehydrogenase was found to selectively reduce (+)-dichlorobicycloheptenone but to show no enantioselectivity towards dihydrobicyclo-heptenone, whilst alcohol dehydrogenase from *Thermoanaerobium brockii* selectively reduced (-)-dihydrobicycloheptenone.   The invariance of α/A with time for the product alcohols from reductions at 22°C showed that enantioselectivity is retained as the reactant is depleted.   However, at 62°C the decrease in α/A at high conversions suggested that temperature may be affecting the specificity of the enzyme.

We have used HPLC with coupled polarimetric and absorbance detection to study the enantioselective oxidation of alcohols by chiral fenchelyl nitroxide radicals [24].   These have been termed "equithermal" reactions, in which enantioselectivity is thought to be associated with chiral discrimination in the transition state [29], which can adopt geometry similar to that of a Pirkle complex.   Single enantiomers of the radicals R1 and R2 (Figure 6) were reacted with *rac*-2-methyl-1-phenyl-1-propanol and *rac*-2,2-dimethyl-1-phenyl-1-propanol.   Excess alcohol was present over that required for the stoichiometric reaction

$$
\begin{array}{ccccc}
\backslash & \backslash & heat & \backslash & \backslash \\
NO + 2\ CHOH & \rule{1.5cm}{0.4pt} & & 2\ NOH + C{=}O\,. \\
/ & / & & / & / \\[6pt]
R & A & & Ac & K
\end{array}
$$

(R = radical, A = alcohol, Ac = hydroxamic acid, K = ketone)

148

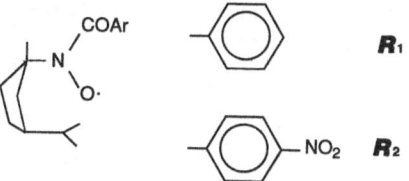

Fig. 6. R1, Benzoyl fenchelyl nitroxide; R2, paranitrobenzoyl fenchelyl nitroxide.

A chiral separation of one of the reaction mixtures is shown in Figure 7. The OR chromatogram gives fewer peaks than the UV chromatogram, allowing the enantiomer peaks to be identified and α/A values for single enantiomers to be determined. From achiral chromatography, $(\alpha/A)_U$ for the alcohol peak in the mixture and $(\alpha/A)_S$ for a pure single enantiomer standard gave the enantiomeric purity of the residual alcohol (Eq. 5). The precision with which enantiomer mole fractions were obtained is shown in Table 2. Enantioselectivities were between 0.1 and 0.3%, which although small were outside the 95% confidence limits on the mean values, typically ±0.06% for quintuple injections. The α/A method for small enantiomeric excesses is more accurate than taking the difference between the areas of two large UV peaks obtained by separating enantiomers on a CSP. It is also simpler and far more accurate than previously used methods of polarimetry on the bulk reaction mixture or chiral derivatisation and subsequent separation of diastereoisomers [30].

CONCLUSIONS

The development of laser polarimeters has improved detection limits and increased the precision and accuracy of the coupled detector technique using polarimetric and absorbance detectors in series. Experience with our 820 nm diode laser polarimeter has shown that enantiomeric purities can be determined within 1% or better for many chiral compounds with sample loading of 8 to 50 µg on an achiral column. With applications demonstrated on chemically pure materials and for a pharmaceutical preparation, this

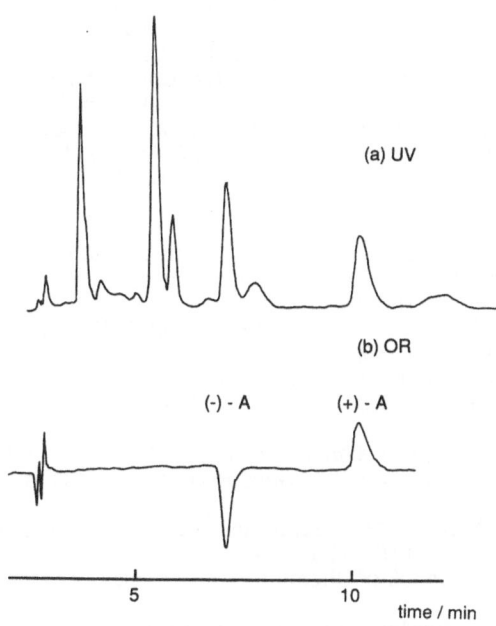

Fig. 7.   Enantiomeric separation of 2,2-dimethyl-1-phenyl-1-propanol (A) in a complex reaction mixture on a Daicel OD chiral column (25 cm x 4.6 mm id). Mobile phase: hexane: isopropanol (95:5 v/v). Flow rate: 1.0 ml min^{-1}, 15 µg of each enantiomer injected.

Table 2. Oxidation of 2-Methyl-1-phenyl-1-propanol by Nitroxide R1 at 60°C

Radical Enantiomer	Percentages of Enantiomers of A Remaining	
	(+)	(-)
(+)-R1	50.17 ± 0.06	49.83 ± 0.06
(-)-R1	49.91 ± 0.05	50.09 ± 0.05

promises to be an important method for quality control of compounds which are used in single enantiomer form. Further developments in data acquisition and processing of outputs from the coupled detectors should result in further improvement in sensitivity, and in the ability to determine specific rotations and enantiomeric purities of materials without ever having to fully resolve enantiomers.

ACKNOWLEDGEMENTS

We wish to thank Shandong Teachers' University and the Chinese Government for a visiting scholarship to Zecai Wu, and ICI Agrochemicals and the Laboratory of the Government Chemist for support of some of this work.

REFERENCES

1. E. S. Yeung, L. E. Steenhoek, S. D. Woodruff and J. C. Kuo, *Anal. Chem.*, 52:1399-1402 (1980).
2. D. R. Bobbitt and E. S. Yeung, *Appl. Spectrosc.*, 40:407-410 (1986).
3. D. M. Goodall and D. K. Lloyd, *in* "Chiral Separations", D. Stevenson and I. D. Wilson, eds, Plenum, New York, 131-134 (1988).
4. D. K. Lloyd, D. M. Goodall and H. Scrivener, *Anal. Chem.*, 61:1238 -1243 (1989).
5. D. K. Lloyd and D. M. Goodall, *Chirality*, 1:251-264 (1989).
6. S. F. Mason, "Molecular Optical Activity and the Chiral Discriminations," Cambridge University Press, Cambridge (1982).
7. N. Purdie and K. A. Swallows, *Anal. Chem.*, 61:77A-89A (1989).
8. A. F. Drake, J. M. Gould and S. F. Mason, *J. Chromatogr.*, 202:239-245 (1980).
9. A. F. Drake and G. D. Jonas, *Chromatography and Analysis*, 3:11-14 (1989).
10. W. A. Boehme, *Chromatogr. Newsletter*, 8:38-41 (1980).
11. D. R. Bobbitt and E. S. Yeung, *Anal. Chem.*, 59:1577-1581 (1984).
12. W. A. Boehme, G. Wagner and U. Oehme, *Anal. Chem.*, 54:709-711 (1982).
13. B. D. Scott and D. L. Dunn, *J. Chromatogr.*, 319:419-426 (1985).
14. B. H. Reitsma and E. S. Yeung, *J. Chromatogr.*, 362:353-362 (1986).
15. J. L. DiCesare and L. S. Ettre, *J. Chromatogr.*, 251:1-16 (1982).
16. B. H. Reitsma and E. S. Yeung, *Anal. Chem.*, 59:1059-1061 (1987).
17. A. Mannschreck, M. Mintas, G. Becher and G. Stuhler, *Angew. Chem. Int. Ed.*, 19:469-470 (1980).
18. A. Mannschreck, A. Eigelsperger and G. Stuhler, *Chem. Ber.*, 115:1568-1575(1982).
19. A. Mannschreck, D. Andert, A. Eigelsperger, E. Gmahl and H. Buchner, *Chromatographia*, 25:182-188 (1988).
20. R. W. Hartmann, C. Batzl, A. Mannschreck and J. K. Seydel, *in* "Trends in Medicinal Chemistry," 821-838 (1988).
21. Z. Wu, D. M. Goodall and D. K. Lloyd, *J. Pharm. Biomed. Anal.*, 8:357-364 (1990).
22. Z. Wu, D. M. Goodall, D. K. Lloyd, P. R. Massey and K. C. Sandy, *J. Chromatogr.*, 513:209-218 (1990).
23. J. Leaver and G. Foster, *Biotechnol. Techniques*, 3:179-184 (1989).
24. D. J. Brooks, M. J. Perkins, D. M. Goodall and D. K. Lloyd, *J. Org. Chem.*, submitted (1990).
25. J. C. Miller and J. N. Miller, "Statistics for Analytical Chemistry," 2nd edn, Ellis Horwood, Chichester, 115-117 (1988).
26. R. C. Weast, ed, "Handbook of Chemistry and Physics," 66th edn, CRC Press, Inc., Boca Raton (1985).
27. J. Buckingham, ed, "Dictionary of Organic Compounds," 5th edn, Chapman and Hall, London (1982).
28. C. Djerassi, "Optical Rotatory Dispersion," McGraw-Hill, New York (1960).
29. M. J. Perkins, *Rev. Chem. Intermed.*, 7:133-141 (1986).
30. C. Berti and M. J. Perkins, *Angew. Chem. Int. Ed.*, 18:864-865 (1979).

HIGH-PERFORMANCE THIN-LAYER CHROMATOGRAPHY:

A QUANTITATIVE TECHNIQUE FOR ENANTIOMER SEPARATION

Peter E. Wall

BDH Limited, Poole
Dorset BH12 4NN, UK

SUMMARY

Chiral stationary phases for thin-layer chromatography (TLC)/high-performance thin-layer chromatography (HPTLC) are a recent development compared to high-performance liquid chromatography (HPLC). Here the advantages offered by enantiomer resolution in TLC are described and the progress made in quantitative separations reviewed. Using test compounds, sensitivity levels and the possibility of quantitative determinations are investigated for N-(3,5-dinitrobenzoyl)-R-(-)-α-phenylglycine derivatised propylamino silica gel layers.

INTRODUCTION

Although chiral stationary phases for HPLC are now well established with a wide variety of column packings now commercially available, sorbents for enantiomer separations by TLC and HPTLC are still very much in their early development. The types of phases so far prepared mimic those already available in the column format. Ligand exchange layers are presently available commercially as "Chiral plates" from Macherey-Nagel (Duren, FRG) and as a HPTLC precoated plate, "CHIR", with a concentration zone from E. Merck (Darmstadt, FRG). A modified procedure according to Davankov et al. is used in the formation of these layers [1,2]. After impregnation of a C_{18}-bonded silica gel plate with a copper salt and a chiral selector (2-hydroxydodecyl-4-hydroxyproline) and air drying, the plate is ready for use. These plates have so far proved successful for the separation of individual amino acids, peptides, α-halogenated carboxylic acids, α-hydroxy carboxylic acids, penicillamine, butyrolactone and thiazolidine derivatives [2-4].

Most of the other success with chiral stationary phases for TLC/HPTLC has revolved around the use of those reagents originally used by Pirkle et al. [5,6] for the preparation of column packing materials. N-(3,5-Dinitrobenzoyl)-R-(-)-α-phenylglycine, N-(3,5-dinitrobenzoyl)-L-leucine and R-(-)-1-(1-naphthyl) ethyl urea bonded to silica gel have been shown to be suitable phases for the separation of enantiomers by TLC [7-9] including, in the case of the first two, both ionic and covalent forms. The plates are prepared very simply by chemical reaction on the precoated amino-bonded layers. Aminopropyl-bonded silica gel plates are impregnated with the chiral reagent which reacts readily with the primary amine function of the bonded silica gel. After air drying the plates are ready for use. Racemates of hexobarbital, benzodiazepines, β-amino alcohols, 1,1'-bi-2-naphthol, 2,2,2-trifluoro-1-(9-anthryl)ethanol, and α-methylarylacetic acids have been resolved with these phases [7-9].

Advantages and Disadvantages of TLC/HPTLC

Undoubtedly the many advantages and benefits of the thin-layer technique outweigh its apparent problems. This is particularly noticeable with enantiomer separations where some of the strict criteria essential for resolution in HPLC are not so necessary in TLC, thus:

Recent Advances in Chiral Separations, Edited by D. Stevenson and
I. D. Wilson, Plenum Press, New York, 1991

1.   A wide range of eluant and detection reagent compositions can be used including those that are aqueous based or involve "aggressive" reagents, e.g., acetonitrile, water, alcohols, acetic acid and phosphoric acid.

2.   Plates are quickly and easily prepared and for the most part have good stability [8]. In all instances the silica gel layers can be dipped in the chiral reagents for surface reaction and are then almost immediately ready for use.

3.   Little or no sample pretreatment is required before chromatography. As the major part of the impurities remains at the origin during development there is no need for complicated purification of the sample.

4.   As modern spectrodensitometers are capable of *in situ* scanning of developed TLC/HPTLC plates by UV absorption, fluorescence and fluorescence quenching, the detection range is extensive. This can be further widened by reaction of the sample spots or bands with derivatisation reagents.

5.   TLC and HPTLC chromatograms often provide semi-permanent records of the separation achieved (steps will need to be taken to prevent oxidation of some components). Further analytical methods can then be applied to determine the nature of the sample (UV spectra, NMR, etc).

6.   HPTLC offers further benefits: (i) All standards can be run *in situ* with the analytes. (ii) Multiple standards can be applied in a single chromatogram (usually four or five leading to the determination of sample components with a high degree of accuracy (%RSD of the order of 1%). (iii) Many samples (up to ~75) can be analysed in a single chromatographic run. (iv) Solvent usage is low. Even ordinary TLC tanks (n-chambers) only require 20 ml of eluant. Twin-trough and sandwich type tanks may only use a few millilitres.

The major problems found with enantiomer separations by TLC/HPTLC are either variable detection levels or background interference leading to high "noise" levels during scanning with the spectrodensitometer. Unfortunately many racemates and hence the pure enantiomers require reaction with suitable derivatising reagents in order for detection at nanogram or even microgram levels to occur. These reagents will often also react with the background chiral selector of the stationary phase usually causing a darkening of the layer. As the spectrodensitometer relies on reflectance measurement, based on subtraction of the light received for the background taken from that for the sample spot, the sensitivity can be greatly reduced. Often the minimum detectable amount is at the microgram level.

Quantification

There is no doubt that for analytically reliable and reproducible results in TLC, HPTLC (high performance layers) are the plates of choice [10]. Enantiomer separations prove to be no different in this respect. Mack et al. have previously shown that the "CHIR" HPTLC plate based on a chiral selector which functions via a ligand exchange mechanism can be used to determine very low concentrations (nanograms) of one enantiomer in the presence of micrograms of the other, e.g., D- and L-tryptophan. They further demonstrated that the reflectance curve for the isomers gave the expected polynomial fit for a series of standards of known concentration [3,4].

It has been possible also to demonstrate that the same is true of HPTLC plates derivatised with N-(3,5-dinitrobenzoyl)-R-(-)-α–phenylglycine using the test compound (±)-2,2,2-trifluoro-1-(9-anthryl)ethanol. The quantitative separation of (±)-1,1'-bi-2-naphthol was also investigated and some preliminary conclusions drawn.

EXPERIMENTAL

Materials

HPTLC Silica gel 60 F254 NH_2, 10 x 10 cm plates and N-(3,5-dinitrobenzoyl)-R-(-)- α-phenylglycine (E. Merck, Darmstadt, FRG), (±)-2,2,2-Trifluoro-1-(9-anthryl)ethanol and (±)-1,1'bi-2-naphthol (Aldrich Chemical Co., Gillingham, Dorset, UK), HiPerSolv tetrahydrofuran, hexane and propan-2-ol (BDH Limited, Poole, Dorset, UK) were used. Spotting of standards was achieved with a Nano-applicator (CAMAG, Muttenz, Switzerland)

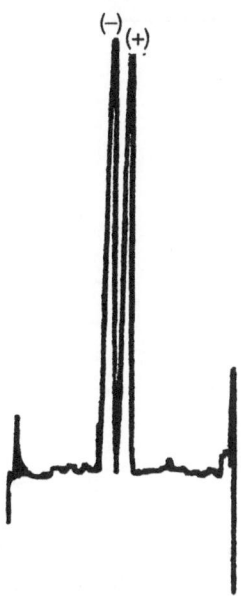

Fig. 1.   Separation of enantiomers (±)-2,2,2-trifluoro-1-(9-anthryl)ethanol on CSP1 using hexane-propan-2-ol (85:15 v/v) as mobile phase.

and the developing chambers were of the twin-trough type (CAMAG). The spectro-densitometer was a Scanner I also from CAMAG.

Ionically bonded N-(3,5-dinitrobenzoyl)-R-(-)-α-phenylglycine aminopropyl silica gel, 10 x 10 cm HPTLC plates (CSP1 plates) were prepared as previously described [8]. A set of standards of (±)-2,2,2-trifluoro-1-(9-anthryl)-ethanol in ethanol were prepared at concentrations of 0.5 g, 0.75 g, 1.0 g, 1.5 g and 2 g per 100 ml. Using a Nano-applicator, 100 nl aliquots were spotted on to the CSP1 plate at 1 cm intervals, 1 cm from a plate edge. After air drying the sample applications were identified as yellow spots.

The plate was developed in a twin-trough chamber with a mixture of hexane-propan-2-ol (85:15 v/v) without presaturation. Migration of the mobile phase was allowed to continue for 70 mm, after which the plate was removed and air dried.

Maximum wavelength absorption was determined at 380 nm using a mercury source. The plate was scanned in the direction of migration at a scan width of 3 mm for a length of 90 mm. The tracks for each optical isomer were then scanned across the plate based on the points of maximum density, and hence absorption, of the components. The peak height was measured and a calibration curve according to a polynomial fit prepared.

Table 1.   Chromatographic Separation of the Enantiomers of 2,2,2-Trifluoro-1-(9-anthryl) ethanol and 1,1'-Bi-2-napthol using HPTLC on an N-(3,5-Dinitrobenzoyl)-R-(-)-α-phenylglycine Derivatised Aminopropyl-bonded Silica Gel HPTLC Plate

Compound	Isomer	R_f	R_s	α	% Isomer
2,2,2-Trifluoro-1-	(-)	0.50			48.4
(9-anthryl)ethanol	(+)	0.57	1.5	1.02	51.6
1,1'-Bi-2-naphthol	(+)	0.28			59.4
	(-)	0.36	2.0	1.14	40.6

153

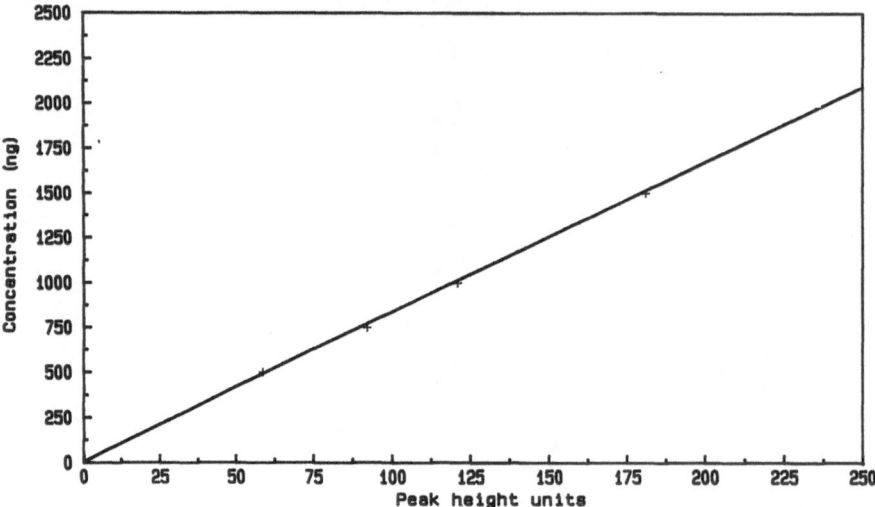

Fig. 2.   Calibration curve for the (-)-isomer of 2,2,2-trifluoro-1-(9-anthryl)ethanol obtained following separation on an N-(3,5-dinitrobenzoyl)-R-(-)-α-phenylglycine derivatised aminopropyl-bonded silica gel HPTLC plate (CSP1) with hexane-propan-2-ol (85:15 v/v) as solvent.

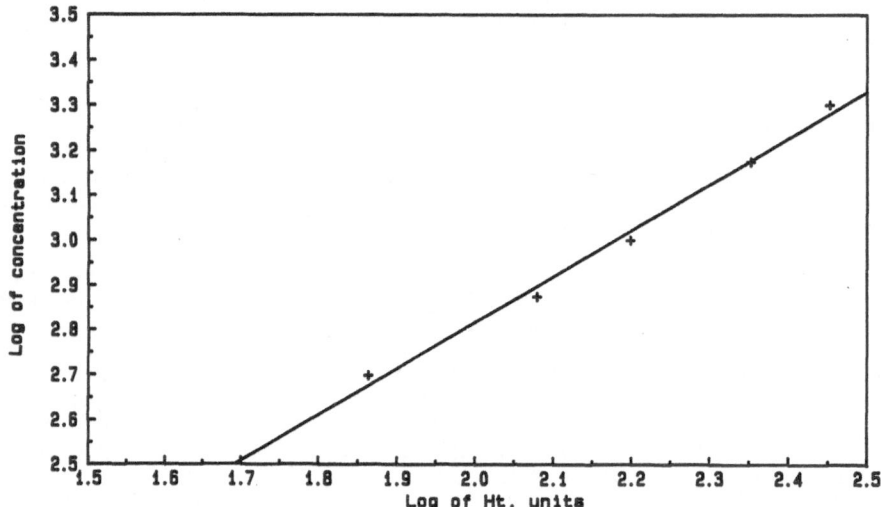

Fig. 3.   A linear calibration curve for the (+)-isomer of 2,2,2-trifluoro-1-(9-anthryl)ethanol obtained following separation on an N-(3,5-dinitrobenzoyl)-R-(-)-α-phenylglycine derivatised aminopropyl-bonded silica gel HPTLC plate (CSP1) with hexane-propan-2-ol (85:15 v/v) as solvent. Data has been linearised using the Kubelka-Munk equation.

The same procedure was adopted with the (±)-1,1'-bi-2-naphthol using standards with concentrations of 1.0 g, 2.0 g, 3.0 g and 8.0 g per 100 ml ethanol. The mobile phase was hexane-propan-2-ol (75:25). Migration of the solvent front was allowed to continue to 80 mm.

The wavelength for maximum absorption of the isomers on the HPTLC plate was determined at 410 nm using a mercury source. The plate was scanned as before at a slit

154

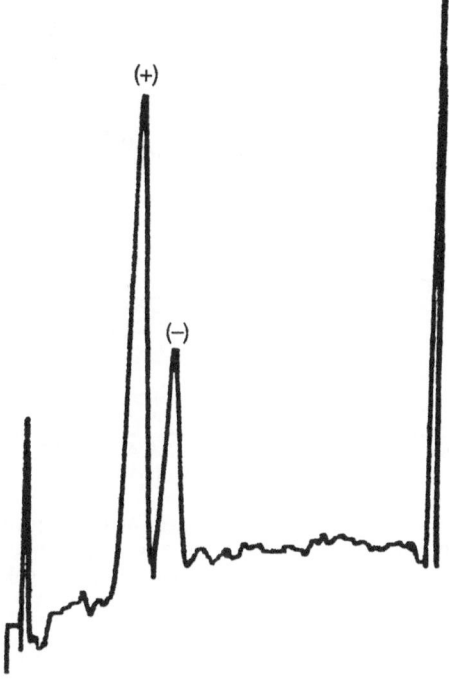

(+)

(−)

Fig. 4.   Separation of 1,1'-bi-2-naphthol on CSP1 using hexane-propan-2-ol (75:25 v/v) as mobile phase.

width of 3 mm for a length of 90 mm. Calibration curves were plotted according to peak height data and concentration of the isomers.

RESULTS

For the test compound (±)-2,2,2-trifluoro-1-(9-anthryl)ethanol, good resolution with base line separation of enantiomers was obtained with an acceptable "noise" level from the plate background. A typical chromatogram is shown in Figure 1. The limit of quantitative detection without post-chromatographic derivatisation was 100 ng. The relative percentages of each isomer (Table 1) indicated the compound was a true racemate. The order of migration of the enantiomers was verified by running a chromatogram of a pure optical isomer alongside the racemate.

The x-power curves for each enantiomer were essentially superimposable (that for the (−)-enantiomer is shown in Figure 2) and showed the expected results indicating that reliable quantification could be attained. Although this is more than adequate for thin-layer chromatographers working with reflectance measurements from a scanner, for complete verification the results were then linearised according to the Kubelka-Munk expression [11]

$$Log\ R_e = a_0 + a_1 \log C,$$

where R_e is the reflectance signal, represented here as the peak height units, and C is the concentration of sample and a straight line fit applied to the plot of the logarithm of peak height units against the logarithm of concentration as shown in Figure 3 for the (+)-enantiomer.

The enantiomers of 1,1'-bi-2-naphthol also appeared well resolved (Figure 4) but the (+)-isomer was of much higher concentration than expected, comprising some 59.4% of the mixture compared to 40.6% for the (−)-isomer. The reflectance measurements were linearised as before using the Kubelka-Munk expression. The results are graphically represented for the (+)- and (−)-isomer of 1,1'-bi-2-naphthol in Figure 5. With this test racemate the limit of detection without derivatisation was 200 ng.

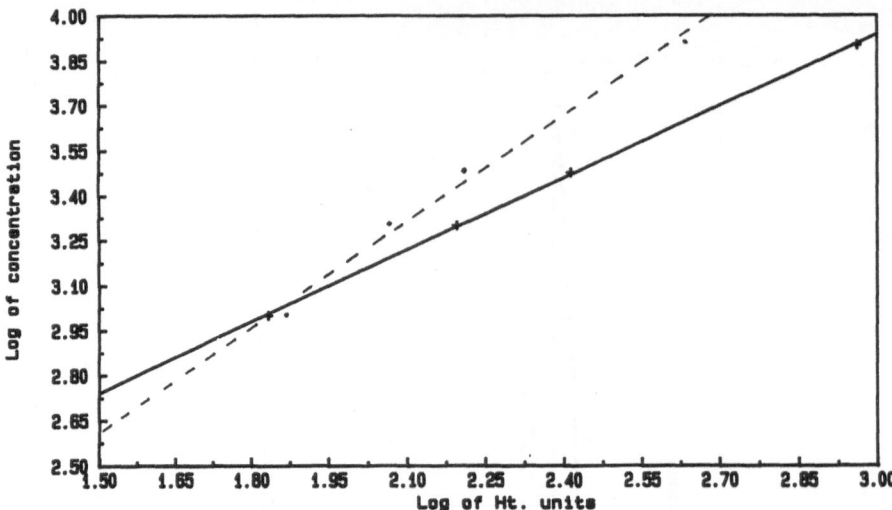

Fig. 5.   Linear calibration curves for the (+) and (-)-isomers of 1,1-bi-2-naphthol obtained following linearisation using the Kubelka-Munk equation after separation on N-(3,5-dinitrobenzoyl)-R-(-)-α-phenylglycine derivatised aminopropyl-bonded silica gel HPTLC plate (CSP1) with hexane-propan-2-ol (75:25 v/v) as solvent. Key: —— (+)-isomer; ---- (-)-isomer.

CONCLUSIONS

There is no doubt that quantitative TLC of enantiomer separations is a reliable analytical technique. For good reproducibility and low relative standard deviation HPTLC layers are recommended where the average particle size of the modified silica gel is 5-6 μm. It has already been demonstrated that ligand exchange plates are effective for quantitative work down to the low nanogram level (order of 10 ng) with amino acids. For the test compounds tried, N-(3,5-dinitrobenzoyl)-R-(-)-α-phenylglycine derivatised plates gave reliable results to the limit of 100 ng.

The α value obtained (Table 1) for the enantiomer separation of 2,2,2-trifluoro-1-(9-anthryl)ethanol is slightly lower than that for the comparable HPLC results [12]. There may be a number of factors involved here. TLC relies upon separation on the layer with migration of the mobile phase through a dry sorbent whereas in HPLC the sorbent is pre-wetted. Excess silanol groups in the TLC layer may also cause unwanted interactions as the stationary phase will not have been end-capped. The number of theoretical plates will also vary due to length of column, shape of sorbent particle and particle size.

The results obtained with 1,1'-bi-2-naphthol do raise a few presently unanswered questions. Unexpectedly the concentrations of the two enantiomers were not similar. Either the compound provided was not a true racemate or, more likely, there was a difference in detector response for each enantiomer resulting from differences in their interaction with the stationary phase itself. Investigations into this phenomenon are continuing.

REFERENCES

1.   V. A. Davankov, A. S. Bochkov and A. A. Kurganov, *Chromatographia*, 13:677 (1980).
2.   K. Günther, *J. Chromatogr.*, 448:11-30 (1988).
3.   M. Mack and H. E. Hauck, *Chromatographia*, 26:197-205 (1988)
4.   M. Mack, H. E. Hauck and H. Herbert, *J. Planar Chromatogr.*, 1:304-308 (1988).
5.   W. H. Pirkle, D. W. House and J. M. Finn, *J. Chromatogr.*, 192:143-158 (1980).

6.   W. H. Pirkle, J. M. Finn, J. L. Schreiner and B. C. Hamper, *J. Am. Chem. Soc.*, 103:3964-3966 (1981).
7.   I. W. Wainer, C. A. Brunner and T. D. Doyle, *J. Chromatogr.*, 264:154 (1983).
8.   P. E. Wall, *J. Planar Chromatogr.*, 2:228-232 (1989).
9.   C. A. Brunner and I. Wainer, *J. Chromatogr.*, 472:277-283 (1989).
10. U. B. Hezel *in* "HPTLC High Performance Thin-layer Chromatography", A. Zlatkis and R.E. Kaiser, eds, J. Chromatography Library Vol. 9, Elsevier, Amsterdam, Ch. 8, 147-199 (1977).
11. C. F. Poole and S. Katib *in* "Quantitative Analysis Using Chromatographic Techniques", E. Katz, ed, John Wiley, Chichester, Ch. 6, 240-243 (1987).
12. J. M. Finn *in* "Chromatographic Chiral Separations", M. Zief and L.J. Crane, eds, Chromatographic Science Series Vol. 40, Marcel Dekker, New York, Ch. 3, 63 (1988).

CHIRAL SEPARATIONS BY THIN-LAYER CHROMATOGRAPHY

I. D. Wilson, T. D Spurway, L. Witherow,
R. J. Ruane and K. Longden

ICI Pharmaceuticals, Mereside, Alderley Park
Macclesfield, Cheshire SK10 4TG, UK, and
*V.A. Howe Ltd, Beaumont Close, Banbury
Oxfordshire OX16 7RG, UK

SUMMARY

Recent progress in thin-layer chromatography (TLC) as a means for the separation of enantiomers is reviewed. The areas discussed include separations using ligand exchange, ion-pair, cyclodextrin and "Pirkle"-based chiral TLC systems. The use of these systems for the chromatographic separation of a variety of enantiomers is described and the robustness, reliability and problems associated with each of these methods considered.

INTRODUCTION

There can be no doubt that the ready availability of reliable, robust and efficient methods for the resolution of enantiomers by thin-layer chromatography (TLC) would represent a useful addition to the available techniques of chiral separations. As a technique TLC has many advantages over column-based methods, especially for qualitative analysis. In addition to its use as a qualitative, or semi-quantitative technique, the major advances made over the last decade in the instrumentalisation of TLC also mean that it can be applied to problems involving a requirement for quantitative analysis (see Wall, this volume pp 151-157). TLC also has a role as a "learning" technique and in principle could be used for "scouting" for suitable chiral separation systems for subsequent column chromatography (although this is by no means always easy). Thus, many TLC systems can be evaluated rapidly and simultaneously in a way which is impractical for other techniques.

Another useful feature of TLC is, of course, that unlike columns the TLC plate is designed to be used once and then discarded. The analyst need therefore have no qualms about attempting separations by TLC on samples which he would not contemplate injecting onto an expensive, and often fragile, chiral high-performance liquid chromatography (HPLC) column.

However, as we noted in our previous review of the area of chiral TLC separations [1] despite these apparent advantages relatively little progress has been made. To date the only commercially available chiral TLC plates are those based on ligand exchange [2, 3] which, with a few exceptions, are of most use for the separation of amino acid enantiomers.

Here we review recent developments in chiral TLC and describe our experiences with various different approaches.

Recent Advances in Chiral Separations, Edited by D. Stevenson and
I. D. Wilson, Plenum Press, New York, 1991

MATERIALS AND METHODS

Test Compounds

Racemic amino acids were obtained from the Sigma Chemical Co. Ltd (Poole, Dorset, UK). 2,2,2-Trifluoro-(9-anthryl)ethanol was obtained from Aldrich Chemical Co. Ltd (Gillingham, UK). Samples were applied to the plates as solutions in methanol.

Reagents

N-(3,5-Dinitrobenzoyl)-L-leucine, N-benzoxycarbonyl-glycyl-L-proline (ZGP) and β-cyclodextrin were obtained from Sigma. Ninhydrin, urea and triethylamine were purchased from BDH Ltd (Poole, UK) and Rathburn Chemicals (Walkerburn, UK) respectively. Solvents were of HPLC grade and were obtained from BDH Ltd.

TLC Plates

Ligand exchange plates were obtained from Camlab Ltd (Cambridge, UK) (Machery-Nagel "Chiral plate", 10 x 20 cm TLC plate, glass backed) or BDH Ltd (E. Merck, "CHIR" plate, 10 x 10 cm HPTLC plate, glass backed). Diol- and amino-bonded HPTLC plates (E. Merck, 10 x 10 cm glass backed with fluorescent indicator), silica gel and C_{18}-bonded silica gel TLC plates (E. Merck, 10 x 20 cm glass-backed with fluorescent indicator) were also purchased from BDH Ltd.

The various solvent, reagent and plate pretreatment systems employed, together with visualisation techniques used, are discussed at the appropriate point in the text.

Sample Application and Chromatography

Samples were applied to the origin of the TLC or HPTLC plates either manually using disposable 1, 2 or 5 µl capillaries or instrumentally as 1 cm bands using a Camag Linomat IV sample applicator (Camag, Switzerland). Plates were then developed using ascending chromatography in glass TLC tanks. Scanning densitometry was performed as described in the text using a Shimadzu CS9000 scanning densitometer (V.A. Howe Ltd).

ENANTIOMER RESOLUTION BY LIGAND EXCHANGE

The use of ligand exchange for TLC, particularly for the resolution of amino acids, is well developed. A number of systems have been described involving various different types of chiral selector, few of which are commercially available [2-5]. In the case of the commercial products the ligand exchange chiral selector employed is (2S, 4R, 2'RS)-4-hydroxy-1-(2'-hydroxydodecyl)- proline/copper[II] acetate as described by Gunther et al. [2]. Alternative systems have been demonstrated by Weinstein with N,N-di-n- propyl-L-alanine [4] and Sinibaldi et al. with poly-L-phenylalaninamide [5] copper complexes. In both cases the TLC plates impregnated with the selector were of C_{18}-bonded silica gel.

Machery Nagel's "Chiral plate" has been available for some time and many applications have been described for its use with amino acids and related compounds such as dipeptides, α-methylamino acids, N-alkylamino acids, halogenated amino acids, α-hydroxycarboxylic acids and a variety of heterocyclic compounds [2]. More recently Merck have introduced the "CHIR" plate, which differs from the "Chiral plate" in being prepared from high performance (HP)TLC grade material and in possessing a preconcentration zone [3,6,7]. Like the "Chiral plate" the "CHIR" plate has also been applied to various classes of compound including amino acids and their derivatives, α-hydroxy and α-halogenated carboxylic acids, mandelic acid and derivatives and di- and tripeptides [3]. Results for the separation of the enantiomers of mandelic acid, α-hydroxyacids and related compounds on the "Chiral plate", taken from Gunther [2], are given in Table 1. Interestingly, the solvent systems used for the "Chiral TLC plate" and the "CHIR" HPTLC plate included not only the methanol-water-acetonitrile mixtures previously reported for this type of separation, but also chloroform-methanol, dichloromethane-methanol and dichloromethane-ethanol mixtures [2,3]. We have briefly examined the properties of both types of plate. Like Brinkman and Kamminga [8] we have found that wide variations in the composition of the recommended methanol-water-acetonitrile solvent systems used were possible with surprisingly small effects on separations. Thus changing the methanol-water-acetonitrile ratio from 0:100:30 to 75:25:30 or 100:0:30 (v/v) gave R_f values for the enantiomers of valine of 0.50, 0.46; 0.60, 0.54 and 0.58, 0.54, respectively, on the "CHIR" HPTLC plate. Brinkman

Table 1. Separation of α-Hydroxycarboxylic Acids [2]

Compound	R_f (Configuration where known)	
3,4-Dihydroxymandelic acid	0.32	0.38
Hydroxyisoleucine	0.56	0.63 (L)
Hydroxyleucine (sodium salt)	0.56	0.60 (L)
3-Hydroxymandelic acid	0.34	0.39
4-Hydroxymandelic acid	0.32	0.36
Hydroxymethionine	0.52	0.58 (L)
Hydroxyphenylalanine	0.56	0.62 (L)
Hydroxyvaline	0.52	0.60 (L)
Mandelic acid	0.46	0.53 (L)
Vanillylmandelic acid	0.42	0.55

Solvent: dichloromethane-methanol (45:5 v/v).

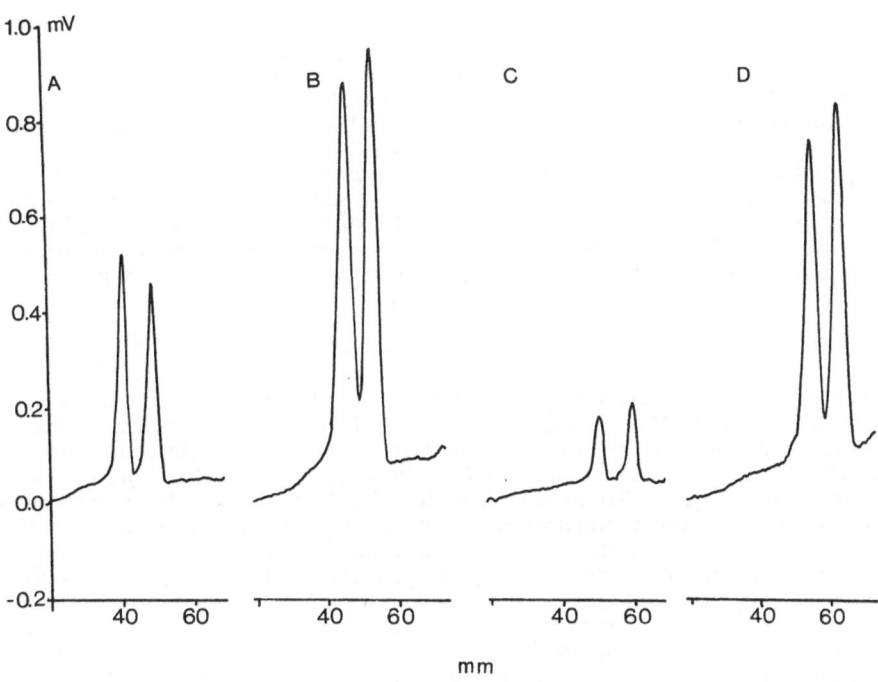

Fig. 1. Separation of the enantiomers of isoleucine (A, B) and leucine (C, D) on the "CHIR" HPTLC plate (A, C) or "Chiral plate" (B, D) using methanol-water-acetonitrile (50:50:200 v/v) as solvent with detection using ninhydrin. Scanning densitometry was performed at 535 nm.

and Kamminga in fact recommended the use of an acetonitrile-water 80:20 (v/v) solvent system [8].

We observed with both types of plate that loadings of 1 µg/spot gave good results but that increasing this to 2 µg/spot resulted in very poor "V" shaped spots (with methanol-water-acetonitrile 50:50:200 v/v as solvent). Typical chromatograms (535 nm) for the separation of the enantiomers of isoleucine and leucine on both types of plate following visualisation with ninhydrin are shown in Figure 1 and R_f values for a number of amino acids are given in Table 2. Both types of plate proved to be reliable and robust. One limitation which was apparent, however, was the similarity in R_f values for all the amino acids that we examined. The plates appear therefore to be unsuited to the analysis of enantiomer composition of complex mixtures of amino acids. As Grinberg and Weinstein have previously shown [9], this can be achieved by two-dimensional chromatography with the amino acids separated from each other in the first dimension and into enantiomers in the second. This possibility is not available with the commercially available plates. Another limitation of this type of plate is that for certain amino acids (e.g., lysine) the ligand exchange approach seems unsuitable.

The use of these ligand exchange plates for the control of optical purity does seem possible with as little as 0.1% of an enantiomer of tyrosine detectable in the presence of the other reported for the "Chiral plate" [10] and similar results for L-tryptophan (10 µg) in the presence of D-tryptophan (10 ng) reported for the "CHIR" plate [7].

N-BENZOXYCARBONYL-GLYCYL-L-PROLINE (ZGP) FOR ENANTIOMER SEPARATION OF β-BLOCKERS

Ion-pair systems have been used by a number of workers for the HPLC resolution of enantiomers [11]. Recently the use of N-benzoxycarbonyl-glycyl-L-proline (ZGP) for the resolution of the enantiomers of propranolol and alprenolol on TLC has been reported [12]. Chromatography was performed on "diol"-bonded silica gel HPTLC plates which were prewashed in dichloromethane and then dipped in the mobile phase (a 5 mM solution of ZGP and 0.4 mM ethanolamine in dichloromethane). Following air drying samples of propranolol and alprenolol were applied and the plate developed, using continuous development, to approximately 4 cm up the plate for 10 to 20 min. The plates were then removed, dried and the process repeated to obtain the spot reconcentrating effect of multiple development. In our hands this technique has proved to be much less robust and reliable than, for example, ligand exchange with poor day-to-day reproducibility. Unlike Tivert and Backman [12] we have been unable to use dichloromethane alone as solvent but have had to add some ethanol (2% v/v) to the mobile phase in order to cause propranolol to migrate from the origin. The concentration of ethanol in the solvent appears to be rather critical, with higher concentrations resulting in high R_f values and a loss of resolution. The separation was also sensitive to the amount of material applied to the plate with best resolution obtained with low loadings of propranolol (5 µg applied as a 1 cm band).

ENANTIOMER SEPARATION ON "PIRKLE" TYPE PHASES

First example of a separation of enantiomers on a "Pirkle"- type stationary phase was reported by Wainer et al. [13]. This group separated the enantiomers of 2,2,2-trifluoro-1-(9-anthryl)ethanol on (R)-N-(3,5-dinitrobenzoyl)phenylglycine ionically bonded to amino-propyl bonded silica gel. Since then there have been several reports of the use of "Pirkle"-type phases of various types. These have included both ionic and covalently bonded types of "Pirkle" phase [1,14]. In our studies we have employed commercially available aminopropyl bonded silica gel HPTLC plates (E. Merck) ionically bonded to (L)-N-(3,5-dinitro-benzoyl)leucine [1]. In our hands this reagent gave slightly better separations of (±)2,2,2-trifluoro-1-(9-anthryl)ethanol than the dinitrobenzoylphenylglycine phase employed by Wainer et al. [13]. A similar conclusion was arrived at by Wall [14] who separated the enantiomers of a range of compounds including hexobarbitol, oxazepam, lorazepam and the isocyanatonaphthalene derivatives of propranol, atenolol and metoprolol (Table 3). In this study both ionically and covalently bonded "Pirkle" plates were examined with the latter apparently proving to be more stable. However, the chromatographic performance of both ionic and covalent "Pirkle" phases was very similar and, given the extra work involved in preparing covalently bonded material, the former seems better suited to general use.

In Figure 2 a typical separation of (±)-2,2,2-trifluoro-1-(9-anthryl)ethanol on an ionically bonded (L)-N-(3,5-dinitrobenzoyl)leucine phase is shown. Scanning densitometry

162

Table 2 Separation of Amino Acid Enantiomers by Ligand Exchange on TLC* and HPTLC† Plates

Amino Acid	R_f	
	HPTLC	TLC
Isoleucine	0.53	0.60
	0.40	0.48
Leucine	0.56	0.59
	0.47	0.50
Methionine	0.53	0.59
	0.47	0.51
Norleucine	0.56	0.58
	0.44	0.46
Phenylalanine	0.60	0.61
	0.47	0.48
Proline	0.44	0.45
	0.37	0.39
Serine	0.47	0.50
	0.41	0.46
Tryptophan	0.63	0.65
	0.49	0.52
Tyrosine	0.69	0.70
	0.57	0.60
Valine	0.53	0.56
	0.43	0.46

Solvent: methanol-water-acetonitrile (50:50:200 v/v).
* "Chiral plate" from Macherey Nagel.
† "CHIR" plate from E. Merck.

was performed at 380 nm. The plate was prepared very simply by dipping an aminopropyl-bonded HPTLC plate into a 0.05 M solution of the chiral selector in tetrahydrofuran (THF). The plate was then washed in THF to remove unbound chiral selector and allowed to air-dry in a fume cupboard. Chromatography was performed using propan-2-ol-hexane (20-80 v/v) as solvent. Scanning densitometry gave the relative proportions of each enantiomer as 47 and 53%, respectively, a result which is in excellent agreement with that of Wall ([14], see also this volume Wall pp 151-157) and clearly illustrates the potential of TLC linked to scanning densitometry as a quantitative technique. An improved separation of the two enantiomers of this analyte was obtained using two-dimensional TLC with propan-2-ol-hexane (20:80 v/v) for both dimensions.

The results illustrated in Figure 2 were possible because 2,2,2-trifluoro-1-(9-anthryl)ethanol appeared as yellow spots or bands when present on the "Pirkle" phase. These spots were readily detectable visually at ca. 2 µg mm⁻¹ and just detectable at ca. 0.5 µg mm⁻¹. This fortunate effect enabled us to perform many exploratory studies and allowed us to optimise the system. However, for many of the compounds which one might wish to determine detection is somewhat more problematic. This is because many of these compounds do not provide a coloured, readily visible spot and can no longer be detected by fluorescence quenching due to the presence of the chiral selector. In his studies Wall employed a variety of different spray reagents to achieve detection but reported the difficulty of finding reagents which were both sensitive and sufficiently selective [14]. Indeed problems were encountered because often the CSP itself gave rise to a colour reaction with the reagent!

An alternative and readily implemented approach is to coat only a portion of the plate with the chiral selector. By careful selection of solvent and chromatographic conditions the enantiomers can be separated on the CSP portion of the plate and then detected on the

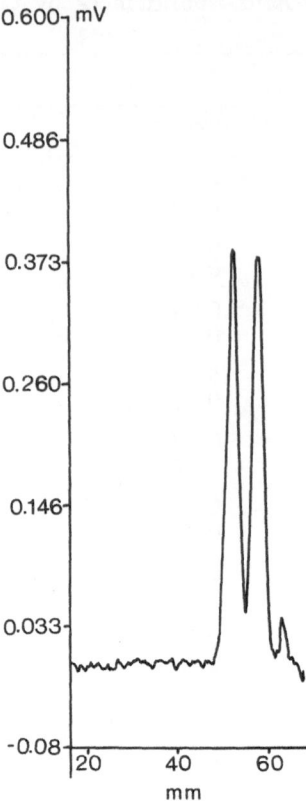

Fig. 2.   Separation of the enantiomers of 2,2,2-trifluoro-(9-anthryl) ethanol on (L)-N-(3,5-dinitobenzoyl)leucine ionically bonded to aminopropyl-bonded silica gel HPTLC plates using n-hexane-propan-2-ol (80:20 v/v) as solvent. Scanning densitometry was performed at 380 nm. Spots appeared yellow on a pink background.

non-CSP-coated part of the plate using fluorescence detection. Chromatography of this sort can be performed in a single dimension with the lower half of the plate coated with the chiral selector and the resolved enantiomers detected on the underivatised upper half, as shown in Figure 3, or by two-dimensional chromatography. This methodology has also been implemented by other groups with considerable success (Pirkle, personal communication).

This approach can be likened to column switching in HPLC where the combination of a chiral and achiral separation has been shown to be particularly powerful [15].

The "Pirkle" phase described above was robust and reliable as a means for the separation of suitable enantiomers by TLC. No doubt many of the other phases of this type which have been developed will prove to be equally well suited to TLC.

ENANTIOMER RESOLUTION USING CYCLODEXTRINS AS MOBILE PHASE ADDITIVES

The use of bonded cyclodextrin phases in TLC was reported in 1986 [16], however as yet such plates are not commercially available. More recently Armstrong et al. have described the resolution of enantiomers and diastereoisomers on TLC with β-cyclodextrin [17], or hydroxypropyl- and hydroxyethyl-derivatised β-cyclodextrin as mobile phase additives [18].

Separations with β-cyclodextrin were only possible with solutions of this material in saturated urea solution to increase its water solubility (normally only 1.67×10^{-2} M at 25°C [17]). The authors also added 0.6 M sodium chloride to the mobile phase to stabilise the

164

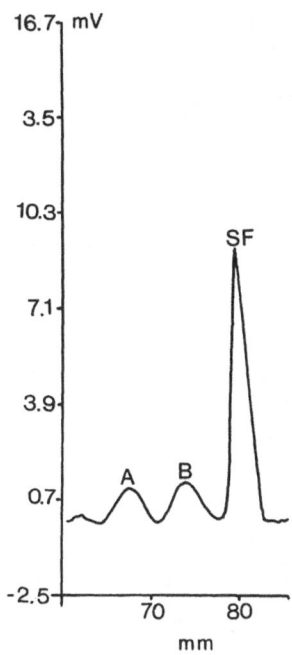

Fig. 3.   Separation of the enantiomers of 2,2,2-trifluoro-(9-anthryl) ethanol (A, B) on (L)-N-(3,5-dinitrobenzoyl)leucine ionically bonded to aminopropyl-bonded silica gel HPTLC plates (see Figure 2 for solvents). The plate was only partially coated with the chiral selector and the resolved enantiomers were allowed to migrate onto the uncoated portion of the plate and were detected by UV quenching at 254 nm. Scanning densitometry was performed at 230 nm. Only the uncoated portion of the plate is shown. SF = Solvent front.

Table 3. Racemic Substances Separated on "Pirkle" Phases [14]

Compound	R_f	
	Isomer 1	Isomer 2
Atenolol*	0.11	0.14
Hexobarbital	0.65	0.70
Lorazepam	0.20	0.23
Metoprolol*	0.15	0.33
1,1'-Bi-2-naphthol*	0.29	0.39
Oxazepam	0.20	0.23
Propranolol*	0.40	0.43

Eluent: hexane-2-propanol (75:25 v/v).
* Isocyanatonapthalene derivatives.

binder used to prepare the C_{18} reversed-phase TLC plates (Whatman KC18F, 5 x 20 and 20 x 20 cm). Significant resolution of racemic analytes such as dansyl DL-glutamic acid was obtained with mobile phases containing more than 0.04 M β-cyclodextrin with optimum resolution attained between 0.08 and 0.12 M. This range of optimum concentrations varied slightly with the compound under study and more so with the amount of organic modifier (acetonitrile or methanol). With high concentrations of β-cyclodextrin resolution was lost

Table 4. Separation Data for Enantiomers including Dansyl-Amino Acids by TLC using β-Cyclodextrin in the Mobile Phase [17]

Compounds	R_{f1}	R_{f2}	α	R_S	Mobile Phase†
DL-Alanine-2-naphthylamide hydrochloride	0.59	0.66	1.12	1.2	Methanol-0.163 M B-CD (35:65)
(±)2,2'-Binaphthyldiyl-N-benzylmonoaza-16-crown-5	0.05	0.08	1.60	0.6	Methanol-0.265 M B-CD (60:40)
N'-Benzylnornicotine	0.29	0.34	1.17	1.7	Methanol-0.200 M B-CD (60:40) ‡
(±)2-Chloro-2-phenylacetyl chloride	0.02	0.07	3.50	0.55	Acetonitrile-0.151 M B-CD (30:70)
Dansyl-DL-x-amino-n- butyric acid*	0.42	0.47	1.12	1.5	Acetonitrile-0.151 M B-CD (30:70)
Dansyl-DL-aspartic acid*	0.64	0.70	1.09	1.8	Acetonitrile-0.133 M B-CD (25:75)
Dansyl-DL-glutamic acid*	0.65	0.72	1.11	2.0	Methanol-0.163 M B-CD (35:65)
Dansyl-DL-leucine*	0.30	0.35	1.17	2.0	Acetonitrile-0.151 M B-CD (30:70)
Dansyl-DL-methionine*	0.34	0.38	1.12	2.1	Acetonitrile-0.151 M B-CD (30:70)
Dansyl-DL-norleucine*	0.24	0.28	1.17	1.6	Acetonitrile-0.151 M B-CD (30:70)
Dansyl-DL-norvaline*	0.32	0.34	1.06	1.4	Acetonitrile-0.151 M B-CD (30:70)
Dansyl-DL-phenylalanine*	0.35	0.39	1.11	1.4	Acetonitrile-0.151 M B-CD (30:70)
Dansyl-DL-serine*	0.41	0.47	1.15	1.5	Acetonitrile-0.133 M B-CD (20:80)
Dansyl-DL-tryptophan*	0.43	0.45	1.05	0.8	Acetonitrile-0.231 M B-CD (35:65)
Dansyl-DL-threonine*	0.42	0.51	1.24	2.0	Methanol-0.151 M B-CD (30:70)
Dansyl-DL-valine	0.36	0.43	1.19	2.5	Acetonitrile-0.151 M B-CD (30:70)
(±)S-(1-Ferrocenylethyl) thiophenol	0.38	0.42	1.07	0.5	Acetonitrile-0.151 M B-CD (30:70)
(±)S-(1-Ferrocenyl-2-methylpropyl)thioethanol	0.42	0.51	1.21	2.0	Acetonitrile-0.125 M B-CD (15:85)
(1R,2S,5R)-(-)-Menthyl-(S)- and (1S,2R,5S)-(+)-Menthyl-(R)-p-toluenesulfinate	0.06	0.08	1.33	0.6	Acetonitrile-0.151 M B-CD (30:70)
Mephenytoin	0.32	0.38	1.19	1.5	Methanol-0.308 M B-CD (35:65)
N'-(2-Naphthylmethyl) nornicotine	0.19	0.24	1.26	1.7	Methanol-0.200 M B-CD (60:40) ‡

* D-Isomer was eluted first.
† Solutions also contained urea and sodium chloride.
‡ 1% Aqueous triethyl ammonium acetate (pH 7.1).

as the spots merged near the solvent front. Development times were 6 to 8 hr at 20°C. The results obtained for the resolution of some 12 dansylated amino acids together with an additional nine racemic compounds are given in Table 4. Interestingly the authors noted that firstly, certain compounds could be resolved on TLC with β-cyclodextrin in the mobile phase but not on β-cyclodextrin-bonded HPLC columns (and vice versa) and that secondly, for those racemates which could be resolved by both TLC and HPLC the elution order was reversed.

More recently Armstrong et al. [18] have reported the use of partially substituted hydroxypropyl- and hydroxyethyl-β-cyclodextrins for the resolution of a variety of racemates on TLC. These materials have the particular advantage over unmodified β-cyclodextrin of enhanced water solubility as well as showing a different chromatographic selectivity. Indeed, depending upon the degree of substitution (average molar substitution, AMS), the solubility of these substituted β-cyclodextrins can be in excess of 1 g ml^{-1} at 25°C [18]. However, weighing against this greater solubility is the disadvantage that at high concentrations the viscosity of the mobile phase is such that TLC development times become very long (greater than 40 hr for a 5 x 20 cm plate with acetonitrile-water (30:70 v/v) containing 0.4 M, 0.6 AMS hydroxypropyl-β-cyclodextrin) [18]. Armstrong et al. also noted that as the AMS increased the amount of derivatised cyclodextrin needed to achieve enantiomer separation was also increased, concluding that, on balance, the 0.6 AMS hydroxypropyl β-cyclodextrin gave the best results. Acetonitrile (between 20 and 40% v/v) appeared to be the organic modifier of choice with streaking less common with this solvent than methanol. The results obtained for the separation of a variety of racemates including dansylated amino acids are given in Table 5. A range of diastereoisomers were also separated.

Table 5 Separation Data for Enantiomers by TLC with Derivatised β-Cyclodextrin in the Mobile Phase [18]

Compound	R_f*	R_S	Mobile Phase (HP-B-CD†-ACN-H$_2$O)	
DL-Alanine-B-naphthylamide	0.66	1.0	0.3 M	25:75
R-(+)Benzyl-2-oxazolidinone S-(-)Benzyl-2-oxazolidinone	0.54	0.9	0.3 M	35:65
Dansyl-DL-leucine‡	0.44	3.6	0.4 M	30:70
Dansyl-DL-methionine‡	0.37	1.6	0.4 M	30:70
Dansyl-DL-norleucine‡	0.20	1.8	0.4 M	30:70
Dansyl-DL-phenylalanine‡	0.36	1.3	0.4 M	35:65
Dansyl-DL-threonine‡	0.58	0.9	0.4 M	30:70
Dansyl-DL-valine‡	0.37	1.3	0.4 M	30:70
Mephenytoin	0.39	2.0	0.4 M	30:70
DL-Methionine-B-napthylamide	0.60	1.8	0.3 M	35:65
5-(4-Methylphenyl)-5-phenylhydantoin	0.24	0.8	0.3 M	35:65

*This is the R_f of the fastest moving enantiomer.
†HP-B-CD Stands for the 0.6 AMS hydroxypropyl-B-CD.
‡ D-Isomer was eluted first.

As we have reported elsewhere our own studies with β-cyclodextrin coated onto the TLC plate provided somewhat equivocal results [19]. In the absence of a suitable β-cyclodextrin bonded phase for TLC the use of these compounds as mobile phase additives is an attractive alternative. The application of this approach is, however, not always straightforward and our attempts to use β-cyclodextrin-containing mobile phases on TLC to separate enantiomers readily resolved on β-cyclodextrin HPLC columns have as yet met with little success.

CONCLUSIONS

Since we last described the state of the art of chiral separations on TLC [1] there have been a number of promising developments. These have included the introduction of a second commercially available ligand exchange plate, the use of cyclodextrins as mobile phase additives and more detailed studies demonstrating the utility of "Pirkle"-type phases for TLC. The important conclusion is that the separation of enantiomers using TLC is possible. Whilst much remains to be done to achieve the degree of understanding that is already available for HPLC, this is clearly achievable and will lead to further improvements in chiral TLC.

REFERENCES

1. I. D. Wilson and R. J. Ruane in "Chiral Separations", D. Stevenson and I.D. Wilson, eds, Plenum, New York, 135-144 (1988).
2. K. Gunther, *J. Chromatogr.*, 448: 11-30 (1988).
3. M. Mack, H.E. Hauck and H. Herbert, *J. Planar Chromatogr.*, 1: 304-308 (1988).
4. S. Weinstein, *Tetra. Lett.*, 25: 985-987 (1984).
5. M. Sinibaldi, A. Messina and A.M. Girell, *Analyst*, 113: 1245-1247 (1988).
6. M. Mack and H. E. Hauck, *J. Planar Chromatogr.*, 2: 190-193 (1989).
7. M. Mack and H. E. Hauck, *Chromatographia*, 26: 197-205 (1988).
8. U. A. Th. Brinkman and D. Kamminga, *J. Chromatogr.*, 330: 375-378 (1985).
9. N. Grinberg and S. Weinstein, *J. Chromatogr.*, 303: 251-258 (1984).
10. R. Raush in "Recent Advances In Thin-Layer Chromatography", F.A.A. Dallas, H. Read, R.J. Ruane and I.D. Wilson, eds, Plenum, New York, 151-161 (1988).

11. C. Petterson and G. Schill, *J. Chromatogr. Sci.*, 40: 283-313 (1988).
12. A-M Tivert and A. Backman, *J. Planar Chromatogr.*, 2: 472-473 (1989).
13. I. W. Wainer, L. A. Brunner and T.D. Doyle, *J. Chromatogr.*, 264: 154 (1983).
14. P. Wall, *J. Planar Chromatogr.*, 3: 228-232 (1989).
15. I. W. Wainer and R. M. Stiffin, *J. Chromatogr.*, 424: 158-162 (1988).
16. A. Alak and D. W. Armstrong, *Anal. Chem.*, 58: 582-584 (1986).
17. D. W. Armstrong and S.M. Han, *J. Chromatogr.*, 448: 345-354 (1988).
18. D. W. Armstrong, J. R. Faulkner and S. M. Han, *J. Chromatogr.*, 452: 323-330 (1990).
19. I. D. Wilson *in* "Bioactive Analytes Including CNS Drugs, Peptides and Enantiomers", E. Reid, B. Scales and I.D. Wilson, eds, Plenum, New York, 227-281 (1986).

APPENDIX 1 — ABSTRACTS

The abstracts provided here are for those papers and posters for which no manuscript was received for publication. They have been subjected to minimal editing before being retyped but are otherwise printed essentially as received.

NEW CHIRAL STATIONARY PHASES FOR CAPILLARY GAS CHROMATOGRAPHY AND LIQUID CHROMATOGRAPHY

Daniel W. Armstrong

Department of Chemistry, University of Missouri-Rolla
Rolla, MO 65401, USA

The use of cyclodextrin (CD)-bonded phases in the reversed-phase liquid chromatographic (LC) separation of enantiomers and other isomers is well known. Recently, new classes of derivatised CDs have been developed which may be more useful and widely applicable than the original native CDs. We have developed several liquid CD derivatives that are effective stationary phases in capillary gas chromatography (GC). The polar permethylhydroxypropyl-CDs and trifluoroacetyl derivatives seem to be the most useful. With these phases racemic molecules can be resolved that cannot be separated by any existing LC method. Because of the lack of solvent in GC, enantiomers with very little functionality can be resolved. The factors that control the physical state of CD as well as the GC separation mechanism will be discussed.

Several aromatic derivatives of CD were made and used as an LC-bonded stationary phase in the normal phase mode. Nearly 100 enantiomers could be resolved by these new normal-phase columns that cannot be resolved by the original reversed-phase CD columns. The structure of these phases and their separation properties will be reviewed. In addition to having unique selectivities, these columns seem to be more durable and stable than other normal-phase columns in that they can be exposed to water and polar solvents without deleterious effects.

CHIRAL ANALYSIS OF PHENOXYPROPIONIC HERBICIDES (POSTER)

B. Blessington and N. Crabb

Department of Pharmaceutical Chemistry, School of Pharmacy
University of Bradford, Bradford BD7 1DP, UK

Recent years have seen an increased awareness of the role of chirality in producing biological response. This is true of pharmaceuticals, pesticides, insecticides, herbicides and flavouring compounds. One isomer (eutomer) may display much greater potency than its enantiomer (distomer). More worrying however are the well documented cases (e.g., Thalidomide) where one enantiomer dominates the adverse effects. Regulatory authorities will no longer permit the marketing of racemic material if efficacy or toxicological responses are sensitive to chirality.

This presentation will illustrate development in analytical methodology for the important group of phenoxypropionate herbicides. These are currently being produced worldwide at an annual rate of hundreds of thousands of tons, in their racemic (±) form, even though only the (+)-isomer has herbicidal activity.

Legislation has been introduced into several European countries to restrict sales to the optically and herbicidally active (+)-isomer and to ban racemic material on environmental grounds. BASF have recently introduced a commercially-viable, biotechnology synthesis of one such active isomer Duplosan®.

At the heart of legislative control, as well as production QA control, are analytical methods for determining chiral purity. Examples will be presented to illustrate "direct chiral analysis" using α-glycoprotein columns (AGP) as well as Pirkle π interaction columns. As examples of "indirect chiral chromatography" separations on Pirkle columns and capillary-GC methods using diastereomer derivatives will be discussed.

Depending upon the specific problem (e.g., the analysis of reference samples, raw material, formulations, mixture formulations or environmental residue studies) the analyst must select from a range of methods. This presentation will give some guidance on selection criteria.

ENANTIOSELECTIVITY OF DRUG DISPOSITION AND ITS BIOLOGICAL SIGNIFICANCE

John Caldwell

Department of Pharmacology and Toxicology
St. Mary's Hospital Medical School, London W2 1PG, UK

Recent developments in stereospecific analysis and synthesis have led to a resurgence of interest in the biological significance of the chirality of drugs and other chemicals which may enter the body, deliberately or accidentally. Since the animal body is a chiral environment, composed of specifically-handed macromolecules, it is not surprising that there occur differential interactions of chiral drugs with proteins, carbohydrates and nucleic acids. These macromolecules are determinants of the actions of drugs and other chemicals as targets, such as receptors, membrane structural elements, enzymes and ion channels, and of their disposition in terms of active transport processes both in to and out of the body, protein binding and enzymic metabolism. The enantioselectivity of these various processes of drug disposition influences the enantiomeric composition of a drug in the body, notably when a racemic drug has been administered, which will in turn have a marked bearing upon their biological activity.

The increased attention currently being directed to the stereochemical aspects of drug disposition is giving enhanced insights into a number of aspects of pharmacology and toxicology, including better discernment of concentration-effect relationships, insight into toxic mechanisms and human interindividual variation in drug metabolism and response, and provides new criteria for the design and selection of more selective drugs and impetus for the use of single enantiomers of chiral drugs.

CHIRAL DRUG SEPARATIONS WITH MOBILE PHASE ADDITIVES ON
POROUS GRAPHITIC CARBON

Brian J. Clark

Department of Pharmaceutical Chemistry, School of Pharmacy
University of Bradford, Bradford BD7 1DP, UK

In order to discriminate between drug enantiomers, chemical derivatisation and other more specific assay methods such as isotope dilution and enzymatic methods have been employed for some years. Often, however, these methods suffer from operational difficulties and there has recently been considerable progress in high-performance liquid chromatography (HPLC) [1]. These methods mainly resolve around the use of chiral stationary phases (CSPs) which, although regularly producing elegant separations, can suffer from limited lifetimes in addition to the high initial purchase cost.

Alternatively, a range of mobile phase additives can be exploited as chiral selectors with the possibilities of substantial cost benefits and operational flexibility [2]. In conjunction with the chiral additives ODS and diol-bonded silica, achiral stationary phases have been regularly used. But the operation with these materials is often characterised by extended retention times and poor chromatographic peak shapes. One possible cause of this is the inability to adjust the operating pH to the ion-suppression point due to the limited operating range for bonded phase silica. However, the introduction of porous graphitic carbon (PGC) as stationary phase can alleviate these problems [3,4].

In this presentation the use of PGC as stationary phase will be demonstrated with a number of mobile phase additives to include: inclusion complexation with cyclodextrins and derivatised cyclodextrins and bioaffinity with proteins for the chiral resolution of a range of drugs and related compounds.

REFERENCES

1.   I. W. Wainer, *Trends Anal. Chem.*, 6:125-134 (1987).
2.   C. Pettersson, *Trends Anal. Chem.*, 7:209-217 (1988).
3.   J. E. Mama, A. F. Fell and B. J. Clark, *Anal. Proc.*, 26:71-73 (1989).
4.   B. J. Clark, *J. Chromatogr. Anal.*, 1(6): in press (1989).

RESOLUTION OF DRUGS AND RELATED COMPOUNDS BY INCLUSION COMPLEXATION WITH A CROWN ETHER CHIRAL STATIONARY PHASE (POSTER)

Brian J. Clark, Alan L. Holmes and Ian Johnstone*

Department of Pharmaceutical Chemistry, School of Pharmacy
University of Bradford, Bradford BD7 1DP, UK, and
* J.T. Baker UK, PO Box 9, Hayes Gate House
Hayes, Middlesex UB4 0JD, UK

Following the synthesis of optically active "crown ethers" in 1973, Cram and coworkers [1] quickly realised the potential for inclusion complexation in the resolution of chiral compounds. This was based on ammonium ions; the size of the crown ether cavity and the potential for multiple hydrogen bonds with the ether groups.

Although chiral recognition was reported with amino acids and their derivatives, this initial interest was not sustained and only recently has a commercial chiral stationary phase (CSP) column material become available.

This work reports the applications of the column material to the resolution of chiral drugs and their related compounds of origin other than the amino acids.

REFERENCE

1.   G. Dotsevi, Y. Sogah, D. H. Hoffman and D. J. Cram, *J. Am. Chem. Soc.*, 100:4569 (1978).

COMPUTER-AIDED METHOD DEVELOPMENT IN CHIRAL LIQUID CHROMATOGRAPHY OF DRUGS, IMPURITIES AND METABOLITES

A. F. Fell, T. A. G. Noctor, G. Ley and B. Kaye*

Department of Pharmaceutical Chemistry
University of Bradford, Bradford BD7 1DP, UK, and
*Metabolism Department, Pfizer Central Research
Sandwich, Kent CT13 9NJ, UK

In developing a suitable analytical method for the chiral analysis of a bulk drug and its impurities, or of a drug and its metabolites in the presence of endogenous compounds, it is important to assess the relative significance of the key parameters that define the method. For a chiral column these would include eluent composition (and pH and molarity for aqueous components), temperature and flow rate. In the case of an achiral column with addition of chiral recognition agent to the eluent, a key parameter would also be the type and concentration of the chiral additive.

The relative importance of these parameters can be assessed in a rational manner by full factorial design [1]. Thus for three variables examined at two levels a total of 2^3 experiments is carried out and the effect of each variable calculated. Having ranked the effect of the variable parameters, their interactions can also be calculated. Provided that the initial starting conditions are suitable, the full factorial design or its variant, the central composite design, yield valuable information on the most important parameters to be further subjected to systematic optimisation.

Sequential simplex optimisation [2,3] affords a valuable approach in developing chiral separations of drugs and related compounds on a variety of chiral systems, including chiral-AGP (Chrom Tech AB) and b-Cydodextrin (Astex Inc). In developing automated optimisation techniques, a variety of computer-aided methods are available, based on the multichannel detector, both to identify individual components in cases of peak crossover (recognising that enantiomers themselves give identical spectra, and that their elution order is unlikely to change on these columns), and to assess peak homogeneity. Of these methods, the second derivative (in the time domain) gives a valuable tool for enhancing the resolution of overlapping peaks during optimisation other methods, such as the absorbance ratio, spectral suppression and spectral normalisation, can be valuable for both achiral and achiral components in assessing peak homogeneity. Clearly a chiroptical detector is required to characterise chiral peaks and early experience in using a laser-based detector will be presented.

These techniques for method development in chiral LC will be exemplified with respect to the analysis of doxazosin enantiomers and the pharmacokinetic profiling of oxamniquine in drug metabolism studies.

REFERENCES

1. J. C. Berridge "Automated Optimisation in HPLC", John Wiley, New York (1985).
2. A. F. Fell and T. A. G. Noctor in "Chiral Separations", D. Stevenson and I. D. Wilson, eds, Plenum Press, New York, 121-125 (1989).
3. A. F. Fell, T. A. G. Noctor, J. E. Mama and B. J. Clark, *J. Chromatogr.*, 434:377-384 (1988).

RESOLUTION OF RACEMIC DRUGS ON A CHIRAL α_1-ACID GLYCOPROTEIN COLUMN

Jörgen Hermansson

ChromTech AB, Box 512
S-145 63 Norsborg
Sweden

α_1-Acid glycoprotein (AGP) is an important transport protein for drugs in human plasma. This protein has been used for the preparation of the Chiral-AGP column. The protein tolerates organic solvents and high temperatures, and can be used in a wide pH range without being irreversible denatured. The characteristics of both immobilised and nonimmobilised AGP will be discussed.

The AGP column demonstrates enantioselectivity for a very large number of chiral drugs and compounds from many different classes (amines, acids and nonprotolytes) have been resolved, without precolumn derivatisation [1].

Retention and enantioselectivity can be regulated by uncharged or charged organic modifiers, as well as by the pH of the mobile phase and the column temperature. Chiral selectivity can be induced by adding a certain modifier to the mobile phase, which is the result of a reversible change of the protein structure. It has been demonstrated that the hydrogen bonding properties of an uncharged modifier is very important for the chiral selectivity. For example, high enantioselectivity can be obtained for certain solutes using the proton acceptor acetonitrile as modifier, whereas no resolution is obtained using a modifier with both proton accepting and donating properties.

The possibility of creating a "new" chiral phase by a dynamic modification of the chiral phase is a very important property of the AGP column, which means that the already broad applicability can be extended.

Results will also be presented from the use of the AGP column for the determination of the enantiomers of drugs in human plasma.

REFERENCE

1. J. Hermansson, *Trends Anal. Chem.*, 8:251-259 (1989).

SELECTIVITY OF CELLULOSE BENZOATE AND 4-METHYLBENZOATE
AS CHIRAL STATIONARY PHASES (POSTER)

Tohru Shibata, Akito Ichida, Yoshitaka Fukui and Kyozo Mori

Daicel Chemical Industries Ltd, 1239 Shinzaike
Aboshi-ku, Himeji 671-12, Japan

We have discovered that some cellulose carboxylates exhibit good chiral recognition. Among these, cellulose tribenzoate and tris-4-methylbenzoate (recently commercialised under the trade name of Chiralcel OJ) are potent. In this presentation, the difference in the analyte selectivity of these compounds and the possible applications of the latter compound will be described.

The analyte selectivity of these cellulose carboxylates were compared in a series of structurally related analytes, a series of α-arylalcohols and α-substituted carbonyl compounds. In both series, 4-methylbenzoate gave larger α values than did benzoate to an analyte of a larger molecular size. For example, α values given by 4-methylbenzoate and benzoate, respectively, were 1.17 and 1.57 to 1-phenyl ethanol, 1.58 and 1.20 to 1-phenylpropyn-2-ol, and 1.54 and 1.37 to 1-(1-naphthyl)ethanol. Consequently, cellulose benzoate can resolve a number of simple aliphatic compounds whereas 4-methylbenzoate gives better separation to an aromatic racemate in most cases. It is not clear how the methyl group has such a large effect on analyte selectivity. It may expand the favoured adsorption site or change the conformation of the main chain. These points are currently being investigated.

Cellulose 4-methylbenzoate can be utilised to the chiral separation of many useful compounds. Arylpropionic acid and its esters, diarylmethane derivatives like clofedianol, and some dihydropyridine were also resolved in addition to agrochemicals such as aryloxypropionic acid derivatives and insecticidal thionophosphates. However, the selectivity of methylbenzoate towards a variety of central nervous system depressants is quite outstanding. Mephobarbital, secobarbital (whose asymmetric center locates on the side chain) methsuximide, its demethylated metabolite glutethimide, ethotoin, chlormezanone and methaqualone were resolved, in some cases with a large α value.

INVESTIGATION OF THE GRAFTING MODE APPLIED TO CHIRAL STATIONARY PHASES
DERIVING FROM (S)-3,5-DNB-TYROSINE:IMPROVEMENT OF THE EFFICIENCY (POSTER)

Laurent Siret, André Tambute*, Marcel Caude and Robert Rosset

Laboratoire de Chimie Analytique del'Ecole Supérieure
de Physique et Chimie Industrielles de Paris
10 rue Vauquelin, 75231 Paris Cedex 05, France, and
*Direction des Recherches et Etudes Techniques
Centre d'Etudes du Bouchet, BP No. 3, Le Bouchet
91718 Vert-le-Petit, France

Very recently, the enantiorecognition ability and the wide scope of application of two novel chiral stationary phases (CSPs) deriving from (S)-tyrosine, coined (S)-*thio*-DNBTyr-A (CSP 1a) and (S)-*thio*-DNBTyr-E (CSP 1b) (see Scheme) have been emphasised. Numerous racemates such as phosphine oxides, sulfoxides, ibuprofen derivatives, oxazepam (CNS depressant), albendazole sulfoxide (anthelmintic) or lactams and lactones of pharmacological interest were separated.

The main interest of these CSPs consists in their complementarity with regard to the early commercially available Pirkle (R)-DNBPG CSP. A comparative study between CSPs 1a and 1b and (R)-DNBPG applied to a new family of compounds (ter-butyl N-arylsulphinamoyl esters) is presented. Whereas no resolution occurs on (R)-DNBPG, selectivity values up to 2 are observed on CSP 1A.

R	$C-NH-(CH_2)_3-CH_3$ $\parallel$ O	$-C-O-CH_3$ $\parallel$ O
$-S-(CH_2)_3$	(S)-thio-DNBTyr-A CSP 1a	(S)-thio-DNBTyr-E CSP 1b
$NH-CH_2-\overset{*}{CH}-CH_2$ $\vert$ OH	(S)-azol-DNBTyr-A CSP 2a	(S)-azol-DNBTyr-E CSP 2b

With a view to extending the efficiency of CSPs 1a and 1b, an investigation of the grafting mode was performed. By grafting the chiral selector via a glycidyloxy group onto LiChrosorb-NH$_2$, two novel CSPs were obtained: CSPs 2a and 2b (see Scheme). A comparative study between CSPs 1a/2a and 1b/2b was carried out using various racemates. Whereas no significant change occurred concerning selectivity values (5% increase as an average), a 40% increase in resolution was observed. Several examples of improved separations are presented. The influence of the polar modifier used in the mobile phase is also discussed: the high selectivity values obtained with chlorinated solvents and the strong efficiency observed with ethanol may be maintained by using a ternary mobile phase, leading to very high resolution values per unit of time.

CHIRAL CHROMATOGRAPHY: A POWERFUL TOOL FOR RESOLUTION
OF ORGANIC COMPOUNDS CONTAINING HETEROATOMS SUCH
AS SULPHUR, PHOSPHORUS, NITROGEN AND OXYGEN (POSTER)

André Tambute, Frédéric Guir and Laurent Siret*

Direction des Recherches et Etudes Techniques
Centre d'Etudes du Bouchet, BP No. 3, Le Bouchet
91710 Vert-le-Petit, France, and
*Laboratoire de Chimie Analytique de l'Ecole Supérieure
de Physique et Chimie Industrielles de Paris
10 rue Vauquelin, 75231 Paris Cedex 05, France

Chiral chromatography applied to the resolution of organic compounds containing heteroatoms such as sulphur, phosphorus, nitrogen and oxygen is going to experience great development thanks to two novel chiral stationary phases (CSPs) recently developed in our laboratory.

Numerous examples of their effectiveness and their versatility are given. CSP 1a and CSP 1b were readily obtained from commercially available (S)-tert-Boc-tyrosine by binding chiral selectors CS 1a [(S)-N-(3,5-dinitrobenzoyl) tyrosine-O-(2-propen-1-yl) methyl ester] or CS 1b {(S)-N-(3,5-dinitrobenzoyl) tyrosine-O-2-(propen-1-yl)-n-butyl amide] onto γ-mercaptopropyl silanised silica gel (these CSPs still remain efficient after a long working period).

Examples are given on the field of analytical chromatography for determination of optical purity of an enantiomer issued from enantioselective synthesis or fractional recrystallisation of diastereomeric salts. This analytical technique is generally more efficient and accurate than the NMR method when using chemical shift reagents.

(S)-CSP 1a :　R$=-\overset{\overset{\displaystyle O}{\|}}{C}-OCH_3$　　　　　　　thio-DNBTyr-E

(S)-CSP 1b :　R$=-\overset{\overset{\displaystyle O}{\|}}{C}-NH-(CH_2)_3-CH_3$　　thio-DNBTyr-A

Nowadays chemists are increasingly concerned with the preparation of highly optically pure compounds in amounts ranging from a few milligrams to many grams. Preparative scale chromatography provides an alternative to the above mentioned methods and is particularly useful for compounds devoid of a reactive functional group(s).

APPENDIX 2 — CHIRAL CHROMATOGRAPHY LITERATURE 1988-1989

In the volume which contained the proceedings of the previous meeting in this series ("*Chiral Separations*", D. Stevenson and I.D. Wilson, eds, Plenum, 1988) we took the opportunity of including, in addition to the papers from the symposium, a literature survey for the year of the meeting (1987) and part of the subsequent year up to the time that the manuscripts were sent for printing. The main reason for the inclusion of that survey was to complement the symposium papers and ensure topicality in a rapidly expanding area of chromatography.

For essentially the same reason we have decided to include a second survey, covering the whole of 1988 and 1989, in this volume. As before the literature has been subdivided into subject areas (HPLC, GC, TLC, SFC, etc) and the references listed in alphabetical order, by first author, in the various categories. Some articles cover more than one area, and are therefore listed more than once, with a total of over 400 entries, providing a graphic illustration of the continued interest in this area.

In order to increase the usefulness of the survey it has been indexed on the basis of author (all authors) and subject. The subject index is further subdivided into compound, derivatisation reaction, stationary and mobile phase indexes.

Once again we hope that the survey will prove useful to all those engaged in the investigation and use of chiral chromatography and that it actively complements the content of the symposium papers.

I.D. Wilson and R.J. Ruane

GENERAL REVIEWS (GEN REV)

1. Ahnoff M., Einarsson S.
CHIRAL DERIVATIZATION.
In "Chiral Liquid Chromatography". Ed:
W.J. Lough, Blackie, 37-73, (1989).

2. Allenmark S.G.
CHROMATOGRAPHIC
ENANTIOSEPARATION: METHODS AND
APPLICATIONS.
Ellis Horwood, (1988).

3. Armstrong D.W., Han S.M.
ENANTIOMERIC SEPARATIONS IN
CHROMATOGRAPHY.
CRC Crit. Rev. Anal. Chem., 19, 175-224,
(1988).

4. Bhushan R.
METHODS OF TLC RESOLUTION OF
ENANTIOMERIC AMINO ACIDS AND
THEIR DERIVATIVES.
J. Liq. Chromatogr., 11, 3049-3065,
(1988).

5. Blaschke G.
SUBSTITUTED POLYACRYLAMIDES AS
CHIRAL PHASES FOR THE
RESOLUTION OF DRUGS.
J. Chromatogr. Sci, 40, 179-198, (1988).

6. Bopp R.J., Kennedy J.H.
PRACTICAL CONSIDERATIONS FOR
CHIRAL SEPARATIONS OF
PHARMACEUTICAL COMPOUNDS.
LC-GC, 6, 514-522, (1988).

7. Caldwell J., Darbyshire J.F., Winter S.M.,
Hutt A.J.
PITFALLS IN THE ENANTIOSELECTIVE
ANALYSIS OF CHIRAL DRUGS.
In "Bioanalysis of Drugs and Metabolites,
Especially Anti-Inflammatory and
Cardiovascular". Eds: E. Reid, J.D.
Robinson and I.D. Wilson, Plenum.,
257-261, (1988).

8. Coventry L.
CYCLODEXTRIN INCLUSION
COMPLEXATION.
In "Chiral Liquid Chromatography". Ed:
W.J. Lough, Blackie, 148-165, (1989).

9. Davankov V.A.
SEPARATION OF ENANTIOMERIC
COMPOUNDS USING CHIRAL HPLC
SYSTEMS. A BRIEF REVIEW OF
GENERAL PRINCIPLES, ADVANCES AND
DEVELOPMENT TRENDS.
Chromatographia, 27, 475-482, (1989).

10. Dolphin J.
ENANTIOMER SEPARATIONS.
Lab. Pract., 38, 71-73, (1989).

11. Doyle T.D.
SYNTHETIC MULTI-INTERACTION CHIRAL
BONDED PHASES.
In "Chiral Liquid Chromatography". Ed:
W.J. Lough, Blackie, 102-128, (1989).

12 Dyas A.M., Robinson M.L., Fell A.F.
NEW APPROACHES TO CHIRAL
RESOLUTION OF DRUG ENANTIOMERS
AND RELATED COMPOUNDS BY HPLC.
Anal. Proc., 26, 346-348, (1989).

13. Feitsma K.G., Drenth B.F.H.
CHROMATOGRAPHIC SEPARATION OF
ENANTIOMERS.
Pharm. Weekbl., Sci. Ed., 10, 1-11, (1988).

14. Hare P.E.
CHIRAL MOBILE PHASES FOR THE
ENANTIOMERIC RESOLUTION OF AMINO
ACIDS.
Chromatogr. Sci., 40, 165-177, (1988).

15. Hermansson J.
ENANTIOMERIC SEPARATION OF DRUGS
AND RELATED COMPOUNDS BASED ON
THEIR INTERACTION WITH α1-ACID
GLYCOPROTEIN.
Trends Anal. Chem., 8, 251-259, (1989).

16. Hutt A.J., Caldwell J.
STEREOSPECIFIC ANALYTICAL
METHODOLOGY.
Metab. Xenobiot. Eds: J.W. Gorrod, H.
Oelschlaeger and J. Caldwell, Taylor &
Francis, 335-344, (1988).

17. Ichida A.
CHROMATOGRAPHIC RESOLUTION ON
CHIRAL STATIONARY PHASES.
Am. Lab., 20, 100-103, (1988).

18. Ichida A., Shibata T.
CELLULOSE DERIVATIVES AS
STATIONARY CHIRAL PHASES.
J. Chromatogr. Sci., 40, 219-243, (1988).

19. Johns D.M.
BINDING TO CELLULOSE DERIVATIVES.
In "Chiral Liquid Chromatography". Ed:
W.J. Lough, Blackie, 166-76, (1989).

20. Johns D.M.
BINDING TO SYNTHETIC POLYMERS.
In "Chiral Liquid Chromatography". Ed:
W.J. Lough, Blackie, 177-184, (1989).

21. Krstulovic A.M.
CHIRAL STATIONARY PHASES FOR THE
LIQUID CHROMATOGRAPHIC
SEPARATION OF PHARMACEUTICALS.
J. Pharm. Biomed. Anal., 6, 641-656,
(1987).

22. Lam S.
CHIRAL LIGAND EXCHANGE
CHROMATOGRAPHY.
In "Chiral Liquid Chromatography"., Ed:
W.J. Lough, Blackie, 83-101, (1989).

23. Lough W.J.
OTHER DIRECT CHIRAL RESOLUTION
METHODS.
In "Chiral Liquid Chromatography". Ed:
W.J. Lough, Blackie, 203-208, (1989).

24. Lough W.J., Matlin S.A.
CONSIDERATION OF OTHER
TECHNIQUES.
In "Chiral Liquid Chromatography". Ed:
W.J. Lough, Blackie, 213, (1989).

25. Lough W.J., Wainer I.W.
CHOICE OF CHIRAL LC SYSTEMS.
In "Chiral Liquid Chromatography". Ed:
W.J. Lough, Blackie, 223-234, (1989).

26. Mack M., Hauck H.E.
SEPARATION OF ENANTIOMERS IN THIN-
LAYER CHROMATOGRAPHY.
Chromatographia, 26, 197-205, (1988).

27. Mehta A.C.
DIRECT SEPARATION OF DRUG
ENANTIOMERS BY HIGH-PERFORMANCE
LIQUID CHROMATOGRAPHY WITH
CHIRAL STATIONARY PHASES.
J. Chromatogr., 426, 1-13, (1988).

28. Mosandl A.
CHIRALITY IN FLAVOR CHEMISTRY-
RECENT DEVELOPMENTS IN SYNTHESIS
AND ANALYSIS.
Food Rev. Int., 4, 1-43, (1988).

29. Nation R.L.
ENANTIOSELECTIVE DRUG ANALYSIS:
PROBLEMS AND RESOLUTIONS.
Clin. Exp. Pharmacol. Physiol., 16,
471-477, (1989).

30. Noctor T.A.G., Fell A.F., Kaye B.
OPTIMIZATION.
In "Chiral Liquid Chromatography". Ed:
W.J. Lough, Blackie, 235-246, (1989).

31. Okamoto Y., Hatada K.
OPTICALLY ACTIVE POLY
(TRIPHENYLMETHYL METHACRYLATE) AS
A CHIRAL STATIONARY PHASE.
J. Chromatogr. Sci., 40, 199-218, (1988).

32. Pettersson C.
LIQUID CHROMATOGRAPHIC
SEPARATION OF ENANTIOMERS USING
CHIRAL ADDITIVES IN THE MOBILE
PHASE.
Trends Anal. Chem., 7, 209-217, (1988).

33. Pettersson C., Schill G.
ENANTIOMER SEPARATION IN ION-
PAIRING SYSTEMS.
J. Chromatogr. Sci., 40, 283-313, (1988).

34. Pirkle W.H., Pochapsky T.C.
CONSIDERATIONS OF CHIRAL
RECOGNITION RELEVANT TO THE LIQUID
CHROMATOGRAPHY SEPARATION OF
ENANTIOMERS.
Chem. Rev., 89, 347-362, (1989).

35. Purdie N., Swallows K.A.
ANALYTICAL APPLICATIONS OF
POLARIMETRY, OPTICAL ROTATORY
DISPERSION, AND CIRCULAR
DICHROISM.
Anal. Chem., 61, 77A-89A, (1989).

36. Pryde A.
CHIRAL LIQUID CHROMATOGRAPHY:
PAST AND PRESENT.
In "Chiral Liquid Chromatography". Ed:
W.J. Lough, Blackie, 23-33, (1989).

37. Stevenson D., Williams G.
THE BIOLOGICAL IMPORTANCE OF
CHIRALITY AND METHODS AVAILABLE
TO DETERMINE ENANTIOMERS.
In "Chiral Separations". Eds: D. Stevenson
and I.D. Wilson, Plenum, 1-10, (1988).

38. Szepesi G.
ION-PAIRING.
In "Chiral Liquid Chromatography". Ed:
W.J. Lough, Blackie, 182-202, (1989).

39. Taylor D.R.
FUTURE TRENDS AND REQUIREMENTS.
In "Chiral Liquid Chromatography". Ed:
W.J. Lough, Blackie, 247-266, (1989).

40. Wainer I.W.
IMMOBILIZED PROTEINS AS HPLC
CHIRAL STATIONARY PHASES.
In "Chiral Liquid Chromatography". Ed:
W.J. Lough, Blackie, 129-146, (1989).

41. Wainer I.W., Alembik M.C.
THE ENANTIOMERIC RESOLUTION OF
BIOLOGICALLY ACTIVE MOLECULES
ON COMMERCALLY AVAILABLE LIQUID
CHROMATOGRAPHIC CHIRAL
STATIONARY PHASES.
J. Chromatogr. Sci., 40, 355-384, (1988).

42. Wainer I.W., Stiffin R.M., Chu Y-Q.
DRUG ANALYSIS USING HIGH-
PERFORMANCE LIQUID
CHROMATOGRAPHIC (HPLC) CHIRAL
STATIONARY PHASES: WHERE TO BEGIN
AND WHICH ONE TO USE.
In "Chiral Separations". Eds: D. Stevenson
and I.D. Wilson, Plenum, 11-22, (1988).

43. Wilson I.D., Ruane R.J.
PROSPECTS FOR CHIRAL THIN-LAYER
CHROMATOGRAPHY.
In "Chiral Separations". Eds: D. Stevenson
and I.D. Wilson, Plenum, 135-144, (1988).

44. Zief M.
PREPARATIVE ENANTIOMERIC
SEPARATION.
J. Chromatogr. Sci., 40, 337-353, (1988).

HIGH-PERFORMANCE LIQUID CHROMATOGRAPHY/ LIQUID CHROMATOGRAPHY

1. Aboul-Enein H.Y., Islam M.R.
DIRECT LIQUID CHROMATOGRAPHIC
RESOLUTION OF RACEMIC AMINO-
GLUTETHIMIDE AND ITS ACETYLATED
METABOLITE USING A CHIRAL α_1-ACID
GLYCOPROTEIN COLUMN.
J. Chromatogr. Sci., 26, 616-619, (1988).

2. Aboul-Enein H.Y., Islam M.R., Bakr S.A.
DIRECT HPLC RESOLUTION OF RACEMIC
NOMIFENSINE HYDROGEN MALEATE
USING A CHIRAL β-CYCLODEXTRIN-
BONDED STATIONARY PHASE.
J. Liq. Chromatogr., 11, 1485-1493,
(1988).

3. Akanya J.N., Taylor D.R.
ATTEMPTS ON THE SEMI-PREPARATIVE
RESOLUTION OF RACEMIC ESTERS OF
AMINO ACIDS ON A CHIRAL LIQUID
CHROMATOGRAPHIC COLUMN DERIVED
FROM N-FORMYLISOLEUCINE.
Chromatographia, 25, 923-924, (1988).

4. Akanya J.N., Taylor D.R.
THE EFFECT OF TEMPERATURE ON THE
RESOLUTION OF RACEMATES ON A
CHIRAL BONDED PACKING
INCORPORATING N-FORMYL-
PHENYLALANINE.
Chromatographia, 25, 639-642, (1988).

5. Akanya J.N., Taylor D.R.
THE ROLE OF THE MOBILE PHASE IN
THE RESOLUTION OF RACEMIC ESTERS
OF AMINO ACIDS BY NORMAL-PHASE
HIGH-PERFORMANCE LIQUID
CHROMATOGRAPHY.
Chromatographia, 25, 636-638, (1988).

6. Alembik M.C., Wainer I.W.
RESOLUTION AND ANALYSIS OF
ENANTIOMERS OF AMPHETAMINES BY
LIQUID CHROMATOGRAPHY ON A CHIRAL
STATIONARY PHASE: COLLABORATIVE
STUDY.
J. Assoc. Off. Anal. Chem., 71, 530-533,
(1988).

7. Allenmark S., Andersson S.
CHIRAL LIQUID CHROMATOGRAPHIC
MONITORING OF ASYMMETRIC
CARBONYL REDUCTION BY SOME YEAST
ORGANISMS.
Enzyme Microb. Technol., 11, 177-179,
(1989).

8. Allenmark S., Andersson S.
OPTICAL RESOLUTION OF SOME
BIOLOGICALLY ACTIVE COMPOUNDS BY
CHIRAL LIQUID CHROMATOGRAPHY ON
BSA-SILICA (RESOLVOSIL) COLUMNS.
Chirality, 1, 154-160, (1989).

9. Allenmark S., Andersson S., Bojarski J.
DIRECT LIQUID CHROMATOGRAPHIC
SEPARATION OF ENANTIOMERS ON
IMMOBILIZED PROTEIN STATIONARY
PHASES. VI. OPTICAL RESOLUTION OF A
SERIES OF RACEMIC BARBITURATES:
STUDIES OF SUBSTITUENT AND MOBILE
PHASE EFFECTS.
J. Chromatogr., 436, 479-483, (1988).

10. Andersson L.
PREPARATION OF AMINO ACID ESTER-
SELECTIVE CAVITIES FORMED BY NON-
COVALENT IMPRINTING WITH A
SUBSTRATE IN HIGHLY CROSS- LINKED
POLYMERS.
React. Polym., Ion Exch., Sorbents, 9,
29-41, (1988).

11. Andersson S., Allenmark S.
INFLUENCE OF AMPHIPHILIC MOBILE
PHASE ADDITIVES UPON THE DIRECT
LIQUID CHROMATOGRAPHIC OPTICAL
RESOLUTION BY MEANS OF BSA-BASED
CHIRAL SORBENTS.
J. Chromatogr., 12, 345-357, (1989).

12. Arai T., Koike H., Hirata K., Oizumi H.
SEPARATION OF PYRIDONE CARBOXYLIC
ACID ENANTIOMERS BY HIGH-
PERFORMANCE LIQUID
CHROMATOGRAPHY USING COPPER (II)-
L-AMINO ACID AS THE ELUENT.
J. Chromatogr., 448, 439-444, (1988).

13. Arai T., Matsuda H., Oizumi H.
DETERMINATION OF OPTICAL PURITY BY
HIGH-PERFORMANCE LIQUID
CHROMATOGRAPHY ON CHIRAL
STATIONARY PHASES: PANTOTHENIC
ACID AND RELATED COMPOUNDS.
J. Chromatogr., 474, 405-410, (1989).

14. Armani E., Barazzoni L., Dossena A.,
Marchelli R.
BIS(L-AMINO ACID AMIDATO)COPPER(II)
COMPLEXES AS CHIRAL ELUENTS IN THE
ENANTIOMERIC SEPARATION OF D,L-
DANSYLAMINO ACIDS BY REVERSED-
PHASE HIGH-PERFORMANCE LIQUID
CHROMATOGRAPHY.
J. Chromatogr., 441, 287-298, (1988).

15. Armani E., Dossena A., Marchelli R., Virgili R.
COPPER(II) COMPLEXES OF DIAMINO-DIAMIDO-TYPE LIGANDS AS CHIRAL ELUENTS IN THE ENANTIOMERIC SEPARATION OF D,L-DANSYLAMINO ACIDS BY REVERSED-PHASE HIGH-PERFORMANCE LIQUID CHROMATOGRAPHY.
J. Chromatogr., 441, 275-286, (1988).

16. Armstrong D.W., Han Y.I., Han S.M.
LIQUID CHROMATOGRAPHIC RESOLUTION OF ENANTIOMERS CONTAINING SINGLE AROMATIC RINGS WITH β-CYCLODEXTRIN-BONDED PHASES.
Anal. Chim. Acta, 208, 275-281, (1988).

17. Armstrong D.W., Heng L.J.
LIQUID CHROMATOGRAPHIC SEPARATION OF ANOMERIC FORMS OF SACCHARIDES WITH CYCLODEXTRIN BONDED PHASES.
Chirality 1, 27-37, (1989).

18. Ballasteros P., Claramunt R.M., Elguero J., Gallego-Preciado M., Roussel C., Chemlal A.
ENANTIOMERIC RESOLUTION OF 3,5'-DIMETHYL-4,4'-DIBROMO-1,1'-BISPYRAZOLYLPHENYLMETHANE BY LIQUID CHROMATOGRAPHY ON TRIACETYLCELLULOSE.
Heterocycles, 27, 351-356, (1988).

19. Balmer K., Persson B.A., Schill G.
OPTIMIZATION OF DETECTION SENSITIVITY FOR ENANTIOMERS OF METOPROLOL ON SILICA-BONDED α_1-ACID GLYCOPROTEIN.
J. Chromatogr., 477, 107-118, (1989).

20. Berry B.W., Jamali F.
STEREOSPECIFIC HIGH-PERFORMANCE LIQUID CHROMATOGRAPHIC (HLPC) ASSAY OF FLUOROBIPROFEN IN BIOLOGICAL SPECIMENS.
Pharm. Res., 5, 123-125, (1988).

21. Blaschke G.
SUBSTITUTED POLYACRYLAMIDES AS CHIRAL PHASES FOR THE RESOLUTION OF DRUGS.
J. Chromatogr. Sci., 40, 179-198, (1988).

22. Blessington B., Crabb M.
CHIRAL HIGH-PERFORMANCE LIQUID CHROMATOGRAPHIC ANALYSIS OF PHENOXY HERBICIDE MIXTURES.
J. Chromatogr., 454, 450-454, (1988).

23. Blessington B., Crabb N., Karkee S., Northage A.
CHROMATOGRAPHIC APPROACHES TO THE QUALITY CONTROL OF CHIRAL PROPRIONATE ANTI-INFLAMMATORY DRUGS AND HERBICIDES.
J. Chromatogr., 469, 183-189, (1989).

24. Boehm R., Martire D.E., Armstrong D.W.
THEORETICAL CONSIDERATIONS CONCERNING THE SEPARATION OF ENANTIOMERIC SOLUTES BY LIQUID CHROMATOGRAPHY.
Anal. Chem. 60, 522-528, (1988).

25. Boomsma F., Van der Hoorn F.A.J., Van in't Veld A.J., Schalekamp M.A.D.H.
DETERMINATION OF D,L-THREO-3,4-DIHYDROXYPHENYLSERINE AND OF THE D- AND L-ENANTIOMERS IN HUMAN PLASMA AND URINE.
J. Chromatogr., 427, 219-227, (1988).

26. Bopp R.J., Kennedy J.H.
PRACTICAL CONSIDERATIONS FOR CHIRAL SEPARATIONS OF PHARMACEUTICAL COMPOUNDS.
LC-GC, 6, 514-522, (1988).

27. Boyd R.A., Chin S.K., Don-Pedro Q., Williams R.L., Giacomini K.M.
THE PHARMACOKINETICS OF THE ENANTIOMERS OF ATENOLOL
Clin. Pharmacol. Ther., 45, 403-419, (1989).

28. Briggs D.A., Homer R.B., Godfrey R.
COMPLEXATION OF DANSYL AMINO ACID ENANTIOMERS BY β-CYCLODEXTRIN STUDIED BY HIGH-PERFORMANCE LIQUID CHROMATOGRAPHY AND FLUORESCENCE MEASUREMENTS.
In "Chiral Separations". Eds: D. Stevenson and I.D. Wilson, Plenum, 61-64, (1988).

29. Bruegger R.R., Marti A.R., Meyer V.R., Arm H.
OPTICAL RESOLUTION OF SAMPLES WITH WEAK INTERACTIONS ON CHIRAL "BRUSH TYPE" CHROMATOGRAPHIC STATIONARY PHASES.
J. Chromatogr., 440, 197-207, (1988).

30. Caccamese S.
RETENTION BEHAVIOR OF DIASTEREOMERIC TRUXILLIC AND TRUXINIC DIAMIDES AND SEPARATION OF AN ENANTIOMERIC PAIR IN HIGH-PERFORMANCE LIQUID CHROMATOGRAPHY.
J. Chromatogr., 457, 366-371, (1988).

31. Caldwell J., Darbyshire J.F., Winter S.M., Hutt A.J.
PITFALLS IN THE ENANTIOSELECTIVE ANALYSIS OF CHIRAL DRUGS.
In "Bioanalysis of Drugs and Metabolites, Especially Anti-Inflammatory and Cardiovascular". Eds: E. Reid, J.D. Robinson and I.D. Wilson, Plenum, 257-261, (1988).

32. Camilleri P., Dyke C., Hossner F.
CHIRAL SEPARATION OF THE OPTICAL ISOMERS OF THE ANTIMALARIAL DRUG HALOFANTRINE.
J. Chromatogr., 477, 471-473, (1989).

33. Camilleri P., Gray A., Weaver K., Bowyer J.R., Williams D.J.
HERBICIDAL DIPHENYL ETHERS: STEREOCHEMICAL STUDIES USING ENANTIOMERS OF A NOVEL DIPHENYL ETHER PHTHALIDE.
J. Agric. Food Chem., 37, 519-523, (1989).

34. Carunchio V., Messina A., Sinibaldi M., Fanali S.
HIGH-PERFORMANCE LIGAND-EXCHANGE CHROMATOGRAPHY OF AMINO ACIDS ON CHIRAL STATIONARY PHASES.
J. High Resolut. Chromatogr. 11, 401-404, (1988).

35. Chae K., Levy L.A., Korach K.S.
CHROMATOGRAPHIC SEPARATION AND ISOLATION OF THE ENANTIOMERS OF DIETHYLSTILBESTROL METABOLITES.
J. Chromatogr., 439, 484-487, (1988).

36. Chan K.C., Yeung E.S.
PEAK IDENTIFICATION OF AMINO ACIDS IN LIQUID CHROMATOGRAPHY BY OPTICAL ACTIVITY DETECTION.
J. Chromatogr., 457, 421-426, (1988).

37. Chin S.K., Hui A.C., Giacomini K.M.
HIGH-PERFORMANCE LIQUID CHROMATOGRAPHIC DETERMINATION OF THE ENANTIOMERS OF β-ADRENOCEPTOR BLOCKING AGENTS IN BIOLOGICAL FLUIDS. II. STUDIES WITH ATENOLOL.
J. Chromatogr., 489, 438-445, (1989).

38. Choi K.E., Schilsky R.L.
RESOLUTION OF THE STEREOISOMERS OF LEUCOVORIN AND 5-METHYL-TETRAHYDROFOLATE BY CHIRAL HIGH-PERFORMANCE LIQUID CHROMATOGRAPHY.
Anal. Biochem., 168, 393-404, (1988).

39. Chou T., Gao C.X., Grinberg N., Krull I.S.
CHIRAL POLYMERIC REAGENTS FOR OFF-LINE AND ON-LINE DERIVATIZATIONS OF ENANTIOMERS IN HIGH-PERFORMANCE LIQUID CHROMATOGRAPHY WITH ULTRAVIOLET AND FLUORESCENCE DETECTION: AN ENANTIOMER RECOGNITION APPROACH.
Anal. Chem., 61, 1548-1558, (1989).

40. Christie W.W., Breckenridge G.H.McG.
SEPARATION OF CIS AND TRANS ISOMERS OF UNSATURATED FATTY ACIDS BY HIGH-PERFORMANCE LIQUID CHROMATOGRAPHY IN THE SILVER ION MODE.
J. Chromatogr., 469, 261-270, (1988).

41. Chu Y.Q., Wainer I.W.
THE MEASUREMENT OF WARFARIN ENANTIOMERS IN SERUM USING COUPLED ACHIRAL/CHIRAL, HIGH-PERFORMANCE LIQUID CHROMATOGRAPHY.
Pharm. Res., 5, 680-683, (1988).

42. Clark T., Deas A.H.B., Vogeler K.
ENANTIOMERS OF FUNGICIDES PLANT GROWTH REGULATORS: APPLICATION IN FUNGAL PLANT AND SOIL METABOLISM STUDIES.
In "Chiral Separations". Eds: D. Stevenson and I.D. Wilson, Plenum, 79-90, (1988).

43. Coors C., Matusch R.
SYNTHESIS OF CHIRAL STATIONARY PHASES FROM (S)-CAMPHOR-10-SULPHONYL DERIVATIVES FOR LIQUID CHROMATOGRAPHY OF DRUGS.
J. High Resolut. Chromatogr., 11, 422-423, (1988).

44. Daeppen R., Karfunkel H.R., Leusen F.J.J.
π-ACCEPTOR AMIDE GROUP FOR LIQUID CHROMATOGRAPHIC CHIRAL SEPARATIONS SPECIAL EMPHASIS ON THE 3,5-DINITROBENZOYL AMIDE.
J. Chromatogr., 469, 101-110, (1989).

45. Daeppen R., Meyer V.R., Arm H.
NEW INSIGHTS INTO THE RECOGNITION MECHANISMS OF CHIRAL "BRUSH-TYPE" LIQUID CHROMATOGRAPHIC STATIONARY PHASES.
J. Chromatogr., 464, 39-47, (1989).

46. Dalgaard L., Hansen J.J., Pederson J.L.
RESOLUTION AND BINDING SITE DETERMINATION OF DL-THYRONINE BY HIGH-PERFORMANCE LIQUID CHROMATOGRAPHY USING IMMOBILIZED ALBUMIN AS CHIRAL STATIONARY PHASES. DETERMINATION OF THE OPTICAL PURITY OF THYROXINE IN TABLETS.
J. Pharm. Biomed. Anal., 7, 361-368, (1989).

47. Davankov V.A.
SEPARATION OF ENANTIOMERIC
COMPOUNDS USING CHIRAL HPLC
SYSTEMS. A BRIEF REVIEW OF
GENERAL PRINCIPLES, ADVANCES AND
DEVELOPMENT TRENDS.
Chromatographia, 27, 475-482, (1989).

48. Davankov V.A., Kurganov A.A.,
Ponomareva T.M.
ENANTIOSELECTIVITY OF COMPLEX
FORMATION IN LIGAND-EXCHANGE
CHROMATOGRAPHIC SYSTEMS WITH
CHIRAL STATIONARY AND/OR CHIRAL
MOBILE PHASES.
J. Chromatogr., 452, 309-316, (1988).

49. Davankov V.A., Navratil J.D., Walton H.F.
LIGAND EXCHANGE CHROMATOGRAPHY.
CRC Press (1988).

50. Delee E., Jullien I., Le Garrec L.
DIRECT HIGH-PERFORMANCE LIQUID
CHROMATOGRAPHIC RESOLUTION OF
DIHYDROPYRIDINE ENANTIOMERS.
J. Chromatogr., 450, 191-197, (1988).

51. Delee E., Jullien I., Le Garrec L., Loupy A.,
Sansoulet A., Zaparucha A.
CHIRAL MICHAEL ADDITIONS OF
ACETAMIDOMALONATE TO α-ENONES.
NUCLEAR MAGNETIC RESONANCE AND
HIGH-PERFORMANCE LIQUID
CHROMATOGRAPHIC RESOLUTION OF
ENANTIOMERS.
J. Chromatogr., 450, 183-189, (1988).

52. Demian I., Gripshover D.F.
ENANTIOMERIC PURITY DETERMINATION
OF 3-AMINOQUINUCLIDINE BY
DIASTEREOMERIC DERIVATIZATION AND
HIGH-PERFORMANCE LIQUID
CHROMATOGRAPHY.
J. Chromatogr., 406, 415-420, (1989).

53. Denissen J.F.
PREPARATIVE SEPARATION OF THE
ENANTIOMERS OF THE
CHOLECYSTOKININ ANTAGONIST (3S)-(+)-
N-(2,3-DIHYDRO-1-([$^3H_3$]METHYL)-2-OXO-
5-PHENYL-1H-1,4-BENZODIAZEPINE-3-
YL)-1H-INDOLE-2-CARBOXAMIDE BY
HIGH-PERFORMANCE LIQUID
CHROMATOGRAPHY.
J. Chromatogr., 462, 454-457, (1989).

54. De Vries J.X., Voelker U.
SEPARATION OF THE ENANTIOMERS OF
PHENPROCOUMON AND WARFARIN BY
HIGH-PERFORMANCE LIQUID
CHROMATOGRAPHY USING A CHIRAL
STATIONARY PHASE. DETERMINATION
OF THE ENANTIOMERIC RATIO OF
PHENPROCOUMON IN HUMAN PLASMA
AND URINE.
J. Chromatogr., 493, 149-156, (1989).

55. Dhanesar S.C., Gisch D.J.
MOBILE PHASE OPTIMIZATION OF A
UREA-LINKED CHIRAL STATIONARY
PHASE FOR THE HIGH-PERFORMANCE
LIQUID CHROMATOGRAPHIC
SEPARATION OF OPTICAL ISOMERS.
J. Chromatogr., 461, 407-418, (1989).

56. Dobashi A., Dobashi Y., Kinoshita K.,
Hara S.
EXTENDED SCOPE OF ENANTIOMER
RESOLUTION WITH CHIRAL DIAMIDE
PHASES IN LIQUID CHROMATOGRAPHY.
Anal. Chem., 60, 1985-1987, (1988).

57. Dolphin J.
ENANTIOMER SEPARATIONS.
Lab. Pract., 38, 71-73, (1989).

58. Doner L.W., Cavender P.J.
CHIRAL CHROMATOGRAPHY FOR
RESOLVING MALIC ACID ENANTIOMERS
IN ADULTERATED APPLE JUICE.
J. Food Sci., 53, 1898-1899, (1988).

59. Drenth B.F.H., Bosman J., Feitsma K.G.,
Van Nijhuis A.
DIRECT DETERMINATION OF THE
ENANTIOMERIC PURITY OF
OXYPHENONIUM USING CHIRAL HPLC
WITH POST-COLUMN EXTRACTION
DETECTION.
Chromatographia, 26, 281-284, (1988).

60. Drummon L., Caldwell J., Wilson H.K.
THE METABOLISM OF ETHYLBENZENE
AND STYRENE TO MANDELIC ACID:
STEREOCHEMICAL CONSIDERATIONS.
Xenobiotica, 19, 199-207, (1989).

61. Duchateau A., Crombach M., Aussems M.,
Bongers J.
DETERMINATION OF THE ENANTIOMERS
OF α-AMINO ACIDS AMIDES BY
HIGH-PERFORMANCE LIQUID
CHROMATOGRAPHY WITH A CHIRAL
MOBILE PHASE.
J. Chromatogr., 461, 419-428, (1989).

62. Duke C.C., Holder G.M.
ENDO-1,4,5,6,7,7-HEXACHLORO-
BICYCLO[2.2.1]HEPT-5-ENE-2-
CARBOXYLIC ACID, A SUPERIOR
RESOLVING AGENT FOR THE HIGH-
PERFORMANCE LIQUID
CHROMATOGRAPHIC SEPARATION OF
ENANTIOMERS OF HYDROXYLATED
DERIVATIVES OF TWO AZAAROMATIC
HYDROCARBONS.
J. Chromatogr., 430, 53-64, (1988).

63. Dyas A.M., Robinson M.L., Fell A.F.
NEW APPROACHES TO CHIRAL
RESOLUTION OF DRUG ENANTIOMERS
AND RELATED COMPOUNDS BY HPLC.
Anal. Proc., 26, 346-348, (1989).

64. Edholm L-E., Lindberg C., Paulson J.,
Walhagen A.
DETERMINATION OF DRUG
ENANTIOMERS IN BIOLOGICAL SAMPLES
BY COUPLED COLUMN LIQUID
CHROMATOGRAPHY AND LIQUID
CHROMATOGRAPHY-MASS
SPECTROMETRY.
J. Chromatogr., 424, 61-72, (1988).

65. Eisenberg E.J., Patterson W.R., Kahn G.C.
HIGH-PERFORMANCE LIQUID
CHROMATOGRAPHIC METHOD FOR THE
SIMULTANEOUS DETERMINATION OF
THE ENANTIOMERS OF CARVEDILOL
AND ITS O-DESMETHYL METABOLITE IN
HUMAN PLASMA AFTER CHIRAL
DERIVATIZATION.
J. Chromatogr., 493, 105-115, (1989).

66. Enquist M., Hermansson J.
COMPARISON BETWEEN TWO METHODS
FOR THE DETERMINATION OF THE
TOTAL AND FREE (R)- AND (S)-
DISOPYRAMIDE IN PLASMA USING AN
α_1-ACID GLYCOPROTEIN COLUMN.
J. Chromatogr., 494, 143-156, (1989).

67. Erlandsson P., Isaksson R., Nilsson I.,
Wold S.
CHEMOMETRIC APPROACH TO EXPLAIN
THE LIQUID CHROMATOGRAPHIC
RETENTION OF SOME CHIRAL INDOLES
ON SWOLLEN MICROCYSTALLINE
TRIACETYLCELLULOSE.
J. Chromatogr., 466, 364-370, (1989).

68. Euerby M.R., Nunn P.B., Partridge L.Z
RESOLUTION OF NEUROEXCITATORY
NON-PROTEIN AMINO ACID
ENANTIOMERS BY HIGH-PERFORMANCE
LIQUID CHROMATOGRAPHY UTILISING
PRE-COLUMN DERIVATISATION WITH O-
PHTHALDIALDEHYDE CHIRAL THIOLS.
APPLICATION TO ω-N-OXALYL DIAMINO
ACIDS.
J. Chromatogr., 466, 407-414, (1989).

69. Euerby M.R., Partridge L.Z., Nunn P.B.
RESOLUTION OF NEUROACTIVE NON-
PROTEIN AMINO ACID ENANTIOMERS BY
HIGH-PERFORMANCE LIQUID
CHROMATOGRAPHY UTILISING PRE-
COLUMN DERIVATISATION WITH O-
PHTHALDIALDEHYDE-CHIRAL THIOLS.
APPLICATION TO 2-AMINO-ω-
PHOSPHONOALKANOIC ACID HOMOLOGS
AND α-AMINO-β-N-
METHYLAMINOPROPANOIC ACID (β-
METHYLAMINO-ALANINE).
J. Chromatogr., 469, 412-419, (1989).

70. Euerby M.R., Partridge L.Z., Rajani P.
RESOLUTION OF LOMBRICINE
ENANTIOMERS BY HIGH-PERFORMANCE
LIQUID CHROMATOGRAPHY UTILIZING
PRE-COLUMN DERIVATIZATION WITH 0-
PHTHALDIALDEHYDE-CHIRAL THIOLS.
J. Chromatogr., 447, 392-397, (1988).

71. Falgueyret J.P., Leblanc Y., Rokach J.,
Riendeau D.
NAD(P)H-DEPENDENT REDUCTION OF
12-KETOEICOSATETRAENOIC ACID TO
12(R)- AND 12(S)-HYDROXYEICOSA-
TETRAENOIC ACID BY RAT LIVER
MICROSOMES.
Biochem. Biophys. Res. Comm., 156,
1083-1089, (1989).

72. Feibush B.
CHIRAL SEPARATION THROUGH A
SINGLE SUBSTITUENT ASSOCIATION.
J. Chromatogr., 436, 517-519, (1988).

73. Feitsma K.G., Drenth B.F.H., De Zeeuw
R.A.
SEPARATION OF ENANTIOMERS OF
OXYPHENONIUM BROMIDE BY HIGH-
PERFORMANCE LIQUID
CHROMATOGRAPHY.
In "Chiral Separations". Eds: D. Stevenson
and I.D. Wilson, Plenum, 37-42, (1988).

74. Feitsma K.G., Drenth B.F.H., De Zeeuw
R.A., Meijer D.K.F.
A NOTE ON CHIRAL DIFFERENCES IN
THE DISPOSITION OF THE QUATERNARY
ANTICHOLINERGIC DRUG
OXYPHENONIUM BROMIDE.
In "Bioanalysis of Drugs and Metabolites,
Especially Anti-Inflammatory and
Cardiovascular". Eds: E. Reid, J.D.
Robinson and I.D. Wilson, Plenum,
265-268, (1988).

75. Fell A.F., Noctor T.A.G.
STRATEGIES FOR OPTIMIZING CHIRAL
SEPARATIONS IN DRUG ANALYSIS.
In "Chiral Separations". Eds: D. Stevenson
and I.D. Wilson, Plenum, 121-126, (1988).

76. Fell A.F., Noctor T.A.G., Mama J.E.,
Clark B.J.
COMPUTER-AIDED OPTIMIZATION OF
DRUG ENANTIOMER SEPARATION IN
CHIRAL HIGH-PERFORMANCE LIQUID
CHROMATOGRAPHY.
J. Chromatogr., 434, 377-384, (1988).

77. Finn J.M.
RATIONAL DESIGN OF PIRKLE-TYPE
CHIRAL STATIONARY PHASES.
Chromatogr. Sci., 40, 53-90, (1988).

78. Fitos I., Simonyi M.
SELECTIVE EFFECT OF CLONAZEPAM
AND (S)-UXEPAM ON THE BINDING OF
WARFARIN ENANTIOMERS TO HUMAN
SERUM ALBUMIN.
J. Chromatogr., 450, 217-220, (1988).

79. Francotte E., Rihs G.
ENANTIOSELECTIVE INCLUSION
COMPLEXATION OF 2,3,4,6-TETRA-O-
ACETYL-D-GLUCOPYRANOSE AS A
CHIRAL HOST WITH (+)-(R)-
PHENYLETHYLAMINE: X-RAY
CHARACTERIZATION OF THE COMPLEX.
Chirality 1, 80-85, (1989).

80. Gaffney M.H., Stiffin R.M., Wainer I.W.
THE EFFECT OF ALCOHOLIC MOBILE
PHASE MODIFIERS ON RETENTION AND
STEREOSELECTIVITY ON A
COMMERCIALLY AVAILABLE
CELLULOSE-BASED HPLC CHIRAL
STATIONARY PHASE: AN UNEXPECTED
REVERSAL IN ENANTIOMERIC ELUTION
ORDER.
Chromatographia., 27, 15-18, (1989).

81. Gal J., Meyer-Lehnert S.
REVERSED-PHASE LIQUID
CHROMATOGRAPHIC SEPARATION OF
ENANTIOMERIC AND DIASTEREOMERIC
BASES RELATED TO CHLORAMPHENICOL
AND THIAMPHENICOL.
J. Pharm. Sci., 77, 1062-1065, (1988).

82. Gaskell R.M., Crooks B.
ANALYTICAL AND PREPARATIVE CHIRAL
RESOLUTION OF SOME AMINO
ALCOHOLS BY ION-PAIR HIGH-
PERFORMANCE LIQUID
CHROMATOGRAPHY.
In "Chiral Separations". Eds: D. Stevenson
and I.D. Wilson, Plenum, 65-70, (1988).

83. Gasparrini F., Misiti D., Villani D., La Torre
F., Sinibaldi M.
HIGH-PERFORMANCE LIQUID
CHROMATOGRAPHY ON CHIRAL PACKED
MICROBORE COLUMNS WITH THE 3,5-
DINITROBENZOYL DERIVATIVE OF
TRANS-1,2-DIAMINOCYCLOHEXANE AS
SELECTOR.
J. Chromatogr., 457, 235-245, (1988).

84. Gazdag M., Szepesi G., Huszar L.
THE α-, β- AND γ-CYCLODEXTRINS AS
MOBILE PHASE ADDITIVES IN THE HIGH
PERFORMANCE LIQUID
CHROMATOGRAPHIC SEPARATION OF
ENANTIOMERIC COMPOUNDS. II.
OPTIMIZATION OF THE SEPARATION
METHOD BY USING α-, β-, AND γ-
CYCLODEXTRINS IN MIXTURE.
J. Chromatogr., 436, 31-38, (1988).

85. Gazdag M., Szepesi G., Mihalyfi K.
SOME ASPECTS OF THE SELECTION OF
HIGH-PERFORMANCE LIQUID
CHROMATOGRAPHIC METHODS FOR THE
SEPARATION OF CHIRAL COMPOUNDS IN
PHARMACEUTICAL ANALYSIS.
J. Chromatogr., 450, 145-156, (1988).

86. Geisslinger G., Dietzel K., Loew D.,
Schuster O., Rau G., Lachmann G.,
Brune K.
HIGH-PERFORMANCE LIQUID
CHROMATOGRAPHIC DETERMINATION
OF IBUPROFEN, ITS METABOLITES AND
ENANTIOMERS IN BIOLOGICAL FLUIDS.
J. Chromatogr., 491, 139-149, (1989).

87. Gerding T.K., Drenth B.F.H., Van de
Grampel V.J.M., Neimeijer N.R., De Zeeuw
R.A, Tepper P.G., Horn A.S.
DETERMINATION OF ENANTIOMERIC
PURITY OF THE NEW D-2 DOPAMINE
AGONIST 2-(N-PROPYL-N-2-
THIENYLETHYLAMINO)-5-
HYDROXYTETRALIN (N-0437) BY
REVERSED-PHASE HIGH-PERFORMANCE
LIQUID CHROMATOGRAPHY AFTER PRE-
COLUMN DERIVATIZATION WITH D(+)-
GLUCURONIC ACID.
J. Chromatogr., 487, 125-134, (1989).

88. Gietl Y., Spahn H., Mutschler E.
SIMULTANEOUS DETERMINATION OF R-
AND S-PRENYLAMINE IN PLASMA AND
URINE BY REVERSED-PHASE HIGH-
PERFORMANCE LIQUID
CHROMATOGRAPHY.
J. Chromatogr., 462, 305-314, (1988).

89. Gill T.S., Hopkins K.J., Rowland M.
STEREOSPECIFIC ASSAY OF
NICOUMALONE: APPLICATION TO
PHARMACOKINETIC STUDIES IN MAN.
Br. J. Clin. Pharmacol., 25, 591-598,
(1988).

90. Goodall D.M., Lloyd D.K.
AN OPTICAL ROTATION DETECTOR FOR
HIGH-PERFORMANCE LIQUID
CHROMATOGRAPHY.
In "Chiral Separations". Eds: D. Stevenson
and I.D. Wilson, Plenum, 131-134, (1988).

91. Gyllenhaal O., Vessman J.
PHOSGENE AS A DERIVATIZING
REAGENT PRIOR TO GAS AND LIQUID
CHROMATOGRAPHY.
J. Chromatogr., 435, 259-269, (1988).

92. Hall M., Grover P.L.
THE USE OF PIRKLE HIGH-
PERFORMANCE LIQUID
CHROMATOGRAPHY PHASES IN THE
RESOLUTION OF ENANTIOMERS OF
POLYCYCLIC AROMATIC HYDROCARBON
METABOLITES.
In "Chiral Separations". Eds: D. Stevenson
and I.D. Wilson, Plenum, 43-54, (1988).

93. Hammonds T.D., Blair I.A., Falck J.R., Capdevila J.H.
RESOLUTION OF EPOXYEICO-SATRIENOATE ENANTIOMERS BY CHIRAL PHASE CHROMATOGRAPHY.
Anal. Biochem., 182, 300-303, (1989).

94. Han S.M., Han Y.I., Armstrong D.W.
STRUCTURAL FACTORS AFFECTING CHIRAL RECOGNITION AND SEPARATION ON β-CYCLODEXTRIN BONDED PHASES.
J. Chromatogr., 441, 376-481, (1988).

95. Hare P.E.
CHIRAL MOBILE PHASES FOR THE ENANTIOMERIC RESOLUTION OF AMINO ACIDS.
Chromatogr. Sci., 40, 165-177, (1988).

96. Hasegawa R., Murai-Kushiya M., Komuro T., Kimura T.
STEREOSELECTIVE DETERMINATION OF PLASMA PINDOLOL IN ENDOTOXIN-PRETREATED RATS BY HIGH-PERFORMANCE LIQUID CHROMATOGRAPHY.
J. Chromatogr., 494, 381-388, (1989).

97. Hawkins, D.J., Kuhn H., Petty E.H., Brash A.R.
RESOLUTION OF ENANTIOMERS OF HYDROXYEICOSATETRAENOATE DERIVATIVES BY CHIRAL PHASE HIGH-PERFORMANCE LIQUID CHROMATOGRAPHY.
Anal. Biochem., 173, 456-462, (1988).

98. Hermansson J.
ENANTIOMERIC SEPARATION OF DRUGS AND RELATED COMPOUNDS BASED ON THEIR INTERACTION WITH α_1-ACID GLYCOPROTEIN.
Trends Anal. Chem., 8, 251-259, (1989).

99. Hermansson J., Schill G.
RESOLUTION OF ENANTIOMERIC COMPOUNDS BY SILICA-BONDED α_1-ACID GLYCOPROTEIN.
J. Chromatogr. Sci., 40, 245-281, (1988).

100. Hirayama C., Ihara H., Tanaka K.
CHROMATOGRAPHIC RESOLUTION OF DIPEPTIDE ENANTIOMERS AND DIASTEREOMERS ON CHIRAL STATIONARY PHASES FROM POLY(L-LEUCINE) OR POLY(L-PHENYLALANINE).
J. Chromatogr., 450, 271-276, (1988).

101. Huffer M., Schreier P.
ANALYTICAL RESOLUTION OF 4(S)-ALKYLATED γ-(D)-LACTONES BY HIGH-PERFORMANCE LIQUID CHROMATOGRAPHY ON A SILICA-BONDED CHIRAL POLYACRYLAMIDE SORBENT. CHROMATOGRAPHIC CHARACTERIZATION OF A STATIONARY PHASE.
J. Chromatogr., 469, 137-141, (1989).

102. Hug E., Rohde B., Tsai W-L., Dreiding A.S.
CHROMATOGRAPHIC FRACTIONATION OF NONRACEMIC MIXTURES OF ENANTIOMERS ON ACHIRAL PHASES.
Chromatographia, 25, 244, (1988).

103. Hutt A.J., Caldwell J.
ENANTIOMERIC ANALYSIS OF 2-PHENYLPROPIONIC ACID NSAID'S IN BIOLOGICAL FLUIDS BY HPLC.
In "Bioanalyis of Drugs and Metabolites, Especially Anti-Inflammatory and Cardiovascular". Eds: E. Reid, J.D. Robinson and I.D. Wilson, Plenum., 115-125, (1988).

104. Hyun M.H., Baik I.K., Pirkle W.H.
LIQUID CHROMATOGRAPHIC RESOLUTION OF ENANTIOMERIC DIPEPTIDES ON THE CHIRAL STATIONARY PHASE DERIVED FROM (S)-1-(6,7-DIMETHYL-1-NAPHTHYL)ISOBUTYLAMINE.
J. Liq. Chromatogr., 11, 1249-1259, (1988).

105. Hyun M.H., Park Y.W., Baik I.K.
LIQUID CHROMATOGRAPHIC RESOLUTION OF RACEMIC KETONES AS THEIR OXIME (3,5-DINITROPHENYL) CARBAMATES ON A CHIRAL STATIONARY PHASE.
Tetrahedron Lett., 29, 4735-4738, (1988).

106. Ichida A., Shibata T.
CELLULOSE DERIVATIVES AS STATIONARY CHIRAL PHASES.
J. Chromatogr. Sci., 40, 219-243, (1988).

107. Ikeda K., Hamasaki T., Kohno H., Ogawa T., Matsumoto T., Sakai J.
DIRECT SEPARATION OF ENANTIOMERS BY REVERSED-PHASE HIGH-PERFORMANCE LIQUID CHROMATOGRAPHY ON CELLULOSE TRIS(3,5-DIMETHYLPHENYLCARBAMATE).
Chem. Lett., 1089-1090, (1989).

108. Inotsume N., Fujii J., Honda M., Nakano M., Higashi A., Matsuda I.
STEREOSELECTIVE ANALYSIS OF THE ENANTIOMERS OF ETHOTOIN IN HUMAN SERUM USING CHIRAL STATIONARY PHASE LIQUID CHROMATOGRAPHY AND GAS CHROMATOGRAPHY-MASS SPECTROMETRY.
J. Chromatogr., 428, 402-407, (1988).

109 Isaksson R., Sandstroem J., Eliaz M., Israely Z., Agranat I.
ENANTIOMER RESOLUTION OF NEFOPAM HYDROCHLORIDE, A NOVEL ANALGESIC: A STUDY BY LIQUID CHROMATOGRAPHY AND CIRCULAR DICHROISM SPECTROSCOPY.
J. Pharm. Pharmacol., 40, 48-50, (1988).

110. Isaksson R., Wennerstroem H., Wennerstroem O.
SYMMETRY AND CHIRAL RECOGNITION: SEPARATION OF ENANTIOMERS ON TRIACETYLCELLULOSE COLUMNS.
Tetrahedron, 44, 1697-1705, (1988).

111. Itabashi Y., Takagi T., Tsuda T.
HIGH PERFORMANCE LIQUID CHROMATOGRAPHIC SEPARATION OF ALKANE-1,2-DIOL ENANTIOMERS ON A CHIRAL SLURRY-PACKED CAPILLARY COLUMN.
J. Chromatogr., 472, 271-276, (1989).

112. Iwakawa S., Suganuma T., Lee S.F., Spahn H., Benet L.Z., Lin E.T.
DIRECT DETERMINATION OF DIASTEREOMETIC CARPROFEN GLUCURONIDES IN HUMAN PLASMA AND URINE AND PRELIMINARY MEASUREMENTS OF STEREOSELECTIVE METABOLIC AND RENAL ELIMINATION AFTER ORAL ADMINISTRATION OF CARPROFEN IN MAN.
Drug. Metab. Dispos., 17, 414-419, (1989)

113. Jadaud P., Thelohan S., Schonbaum G.R., Wainer I.W.
THE STEREOCHEMICAL RESOLUTION OF ENANTIOMERIC FREE AND DERIVATIZED AMINO ACIDS USING AN HPLC CHIRAL STATIONARY PHASE BASED ON IMMOBILIZED α-CHYMOTRYPSIN: CHIRAL SEPARATION DUE TO SOLUTE STRUCTURE OR ENZYME ACTIVITY.
Chirality 1, 38-44, (1989).

114. Jadaud P., Wainer I.W.
STEREOCHEMICAL RECOGNITION OF ENANTIOMERIC AND DIASTEREOMERIC DIPEPTIDES BY HIGH-PERFORMANCE LIQUID CHROMATOGRAPHY OF A CHIRAL STATIONARY PHASE BASED UPON IMMOBILIZED α-CHYMOTRYPSIN.
J. Chromatogr., 476, 165-174, (1988).

115. Jalonen H.G.
SIMULTANEOUS DETERMINATION OF TAMOXIFEN CITRATE AND ITS E ISOMER IMPURITY IN BULK AND TABLETS BY HIGH-PERFORMANCE LIQUID CHROMATOGRAPHY.
J. Pharm. Sci., 77, 810-813, (1988).

116. Jamali F., Mehvar R., Lemko C., Eradiri O.
APPLICATION OF STEREOSPECIFIC HIGH-PERFORMANCE LIQUID CHROMATOGRAPHY ASSAY TO A PHARMACOKINETIC STUDY OF ETODOLAC ENANTIOMERS IN HUMANS.
J. Pharm. Sci., 77, 963-966, (1988).

117. Jegorov A., Triska J., Trnka T., Cerny M.
SEPARATION OF α-AMINO ACID ENANTIOMERS BY REVERSED-PHASE HIGH-PERFORMANCE LIQUID CHROMATOGRAPHY AFTER DERIVATIZATION WITH O-PHTHALDIALDEHYDE AND A SODIUM SALT OF 1-THIO-β-D-GLUCOSE.
J. Chromatogr., 434, 417-422, (1988).

118. Kano S., Hongoh Y., Motoi M., Suda H.
CONVENIENT OPTICAL RESOLUTION OF AXIALLY CHIRAL 1,1'-BINAPHTHYL-2,2'-DICARBOXYLIC ACID.
Bull. Chem. Soc. Jpn., 61, 1032-1034, (1988).

119. Karlsson A., Pettersson C., Bjoerkman S.
DETERMINATION OF (R)- AND (S)-PROPRANOLOL IN PLASMA BY HIGH-PERFORMANCE LIQUID CHROMATOGRAPHY USING N-BENZOXYCARBONYLGLYCYL-L-PROLINE AS A CHIRAL SELECTOR IN THE MOBILE PHASE.
J. Chromatogr., 494, 157-171, (1989).

120. Karlsson A., Pettersson C., Sundell S., Arvidsson L.E.
Hacksell U.
IMPROVED PREPARATION, CHROMATOGRAPHIC SEPARATION AND X-RAYCRYSTALLOGRAPHIC DETERMINATION OF THE ABSOLUTE CONFIGURATION OF THE ENANTIOMERS OF 8-HYDROXY-2 (DIPROPYLAMINO) TETRALIN (8-OHDPAT).
Acta Chem. Scand., Ser. B, B42, 231-236, (1988).

121. Keller J.W., Niwa K.
MILLIGRAM-SCALE SEPARATION OF OPTICAL ISOMERS OF 2-PENTAFLUOROETHYLALANINE AND 2-TRIFLUOROMETHYLALANINE BY MEDIUM-PERFORMANCE REVERSED-PHASE CHROMATOGRAPHY.
J. Chromatogr., 469, 434-439, (1989).

122. Kern J.R., Lokensgard D.M., Manes L.V., Matsuo M., Nakamura T.
SEPARATION OF THE STEREOISOMERS OF AN ALLENIC E-TYPE PROSTAGLANDIN.
J. Chromatogr., 450, 233-240, (1988).

123. Kiniwa H., Baba Y., Ishida T., Katoh H.
GENERAL EVALUATION AND APPLICATION TO TRACE ANALYSIS OF A CHIRAL COLUMN FOR LIGAND-EXCHANGE CHROMATOGRAPHY
J. Chromatogr., 461, 397-405, (1989).

124. Knadler M.P., Hall S.D.
HIGH-PERFORMANCE LIQUID CHROMATOGRAPIC ANALYSIS OF THE ENANTIOMERS OF FLURBIPROFEN AND ITS METABOLITES IN PLASMA AND URINE.
J. Chromatogr., 494, 173-182, (1989).

125. Koine N.
THE FIRST COMPLETE OPTICAL RESOLUTION BASED ON CHIRAL DISCRIMINATION BETWEEN IONS OF THE SAME SIGN, U-FAC-[CO(IDA)$_2$]- AND [SB$_2$(D-TART)$_2$]$^{2-}$.
Chem. Lett., 1547-1550, (1988).

126. Krause M., Galensa R.
DIRECT ENANTIOMERIC SEPARATION OF RACEMIC FLAVANONES BY HIGH-PERFORMANCE LIQUID CHROMATOGRAPHY USING CELLULOSE TRIACETATE AS A CHIRAL STATIONARY PHASE.
J. Chromatogr., 441, 417-422, (1988).

127. Krstulovic A.M.
RACEMATES VERSUS ENANTIOMERICALLY PURE DRUGS: PUTTING HIGH-PERFORMANCE LIQUID CHROMATOGRAPHY TO WORK IN THE SELECTION PROCESS.
J. Chromatogr., 488, 53-72, (1989).

128. Krstulovic A.M., Gianviti J.M., Burke J.T., Mompon B.
ENANTIOMERIC ANALYSIS OF A NEW ANTI-INFLAMMATORY AGENT IN RAT PLASMA USING A CHIRAL β-CYCLODEXTRIN STATIONARY PHASE.
J. Chromatogr., 426, 417-424, (1988).

129. Krstulovic A.M., Fouchet M.H., Burke J.T., Gillet G., Durand A.
DIRECT ENANTIOMERIC SEPARATION OF BETAXOLOL WITH APPLICATIONS TO ANALYSIS OF BULK DRUG AND BIOLOGICAL SAMPLES.
J. Chromatogr., 452, 477-483, (1988).

130. Krstulovic A.M., Bende J.L.
IMPROVED PERFORMANCE OF THE SECOND GENERATION α$_1$-AGP COLUMNS: APPLICATIONS TO THE ROUTINE ASSAY OF PLASMA LEVELS OF ALFUZOSIN HYDROCHLORIDE.
Chirality, 1, 243-245, (1989).

131. Kuhn A.O., Lederer M., Sinibaldi M.
ADSORPTION CHROMATOGRAPHY ON CELLULOSE. IV. SEPARATION OF D- AND L-TRYPTOPHAN AND D- AND L-METHYLTRYPTOPHAN ON CELLULOSE WITH AQUEOUS SOLVENTS.
J. Chromatogr., 469, 253-260, (1988).

132. Laganiere S., Kwong E., Shen D.D.
STEREOSELECTIVE HIGH-PERFORMANCE LIQUID CHROMATOGRAPHIC ASSAY FOR PROPRANOLOL ENANTIOMERS IN SERUM.
J. Chromatogr., 488, 407-416, (1989).

133. Lai J.S., Hung S.S., Unruh L.E., Jung H., Fu P.P.
SEPARATION OF AMINO- AND ACETYLAMINO-POLYCYCLIC AROMATIC HYDROCARBONS BY REVERSED- AND NORMAL-PHASE HIGH-PERFORMANCE LIQUID CHROMATOGRAPHY.
J. Chromatogr., 461, 327-336, (1989).

134. Lalonde R.L., Bottorff M.B., Wainer I.W.
THE STUDY OF CHIRAL CARDIOVASCULAR DRUGS: ANALYTICAL APPROACHES AND SOME PHARMACOLOGICAL CONSEQUENCES.
In "Bioanalysis of Drugs and Metabolites, Especially Anti-Inflammatory and Cardiovascular". Eds: E. Reid, J.D. Robinson and I.D. Wilson, Plenum., 169-177, (1988).

135. Lam S., Malikin G., Murphy M., Freundlich L., Karmen A.
CONFIRMING CHIRAL HIGH-PERFORMANCE LIQUID CHROMATOGRAPHIC SEPARATIONS WITH STEREOSPECIFIC ENZYMES.
J. Chromatogr., 468, 359-364, (1988).

136. Leaver J., Foster G.
THE ENANTIOSELECTIVITY OF ENZYMES MONITORED USING A CHIRAL DETECTOR.
Biotechnol. Tech., 3, 179-184, (1989).

137. Le Corre P., Gibassier D., Sado P.,
Le Verge R.
DIRECT ENANTIOMERIC RESOLUTION OF
DISOPYRAMIDE AND ITS METABOLITE BY
CHIRAL HIGH-PERFORMANCE LIQUID
CHROMATOGRAPHY. APPLICATION TO
STEREOSELECTIVE METABOLISM AND
PHARMACOKINETICS OF RACEMIC
DISOPYRAMIDE IN MAN.
J. Chromatogr., 450, 211-216, (1988).

138. Leeman T., Dayer P.
SIMULTANEOUS ENANTIOSELECTIVE
DETERMINATION OF UNDERIVATIZED
β-BLOCKING AGENTS AND THEIR
METABOLITES IN BIOLOGICAL SAMPLES
BY CHIRAL ION-PAIRING HIGH-
PERFORMANCE LIQUID
CHROMATOGRAPHY.
In "Chiral Separations". Eds: D. Stevenson
and I.D. Wilson, Plenum, 71-78, (1988).

139. Lehr K.H., Damm P.
QUANTIFICATION OF THE ENANTIOMERS
OF OFLOXACIN IN BIOLOGICAL FLUIDS
BY HIGH-PERFORMANCE LIQUID
CHROMATOGRAPHY.
J. Chromatogr., 425, 153-161, (1988).

140. Lenne M., Caude M.,, Rosset R., Tambute
A., Delatour P.
DIRECT SEPARATION OF ALBENDAZOLE
SULFOXIDE ENANTIOMERS BY LIQUID
CHROMATOGRAPHY ON A CHIRAL
COLUMN DERIVING FROM (S)-N-(3,5-
DINITROBENZOYL)TYROSINE:
APPLICATION TO ENANTIOMERIC ASSAYS
ON PLASMA SAMPLES.
Chirality, 1, 142-153, (1989).

141. Liang J.C.
ENANTIOMERIC RESOLUTION OF
FERROELECTRIC LIQUID CRYSTAL
INTERMEDIATES VIA LIQUID
CHROMATOGRAPHY I. ALKYL OR ALKOXY
BUTYRIC ACIDS.
J. Chromatogr. Sci., 27, 38-41, (1989).

142. Lienne M., Caude M., Rosset R.,
Tambute A.
DIRECT ENANTIOMERIC SEPARATION OF
ANTICHOLINERGIC DRUGS DERIVED
FROM (+)-CYCLOHEXYL(2-THIENYL)
GLYCOLIC ACID ON A NOVEL α_1-ACID
GLYCOPROTEIN-BONDED CHIRAL
STATIONARY PHASE (CHIRAL-AGP).
J. Chromatogr., 467, 406-413, (1989).

143. Lienne M., Caude M., Rosset R.,
Tambute A.
DIRECT RESOLUTION OF ANTHELMINTIC
DRUG ENANTIOMERS ON CHIRAL-AGP
PROTEIN-BONDED CHIRAL STATIONARY
PHASE.
J. Chromatogr., 472, 265-270, (1989).

144. Lienne M., Caude M., Rosset R.,
Tambute A.
OPTIMIZATION OF THE DIRECT CHIRAL
SEPARATION OF POTENTIAL CYTOTOXIC
α-METHYLENE-γ-BUTYROLACTONES AND
α-METHYLENE-γ-BUTYROLACTAMS BY
LIQUID CHROMATOGRAPHY.
J. Chromatogr., 448, 55-72, (1988).

145. Lienne M., Macaudiere P., Caude M.,
Rosset R., Tambute A.
EVALUATION OF π-ACID CHIRAL
STATIONARY PHASES DERIVING FROM
TYROSINE AND RELATED AMINO ACIDS
FOR THE CHROMATOGRAPHIC
RESOLUTION OF RACEMATES: SPECIFIC
REQUIREMENTS FOR
ENANTIORECOGNITION ABILITY.
Chirality, 1, 45-56, (1989).

146. Lim W.H., Hooper W.D.
STEREOSELECTIVE METABOLISM AND
PHARMACOKINETICS OF RACEMIC
METHYLPHENOBARBITAL IN HUMANS.
Drug Metab. Dispos., 17, 212-217, (1989).

147. Lindner W.F.
INDIRECT SEPARATION OF
ENANTIOMERS BY LIQUID
CHROMATOGRAPHY.
In "Chromatographic Chiral Separations".
Eds. M. Zief and L.J. Crane, Marcel
Dekker, 11-30, (1988).

148. Lindner W.F, Rath M., Stoschitzky K.,
Semmelrock H.J.
PHARMACOKINETIC DATA OF
PROPRANOLOL ENANTIOMERS IN A
COMPARATIVE HUMAN STUDY WITH (S)-
AND (R,S)-PROPRANOLOL.
Chirality, 1, 10-13, (1989).

149. Lindner W.F, Rath M., Stoschitzky K.,
Uray G.
ENANTIOSELECTIVE DRUG MONITORING
OF (R)- AND (S)-PROPRANOLOL IN HUMAN
PLASMA VIA DERIVATIZATION WITH
OPTICALLY ACTIVE (R,R)-O,O-DIACETYL
TARTARIC ACID ANHYDRIDE.
J. Chromatogr., 487, 375-383, (1989).

150. Lipkowitz K.B., Antell S., Baker B.
ENANTIODIFFERENTIATION IN ROGERS'
BOC-D-VAL CHIRAL STATIONARY PHASE.
J. Org. Chem., 54, 5449-5453, (1989).

151. Lloyd D.K., Goodall D.M. Scrivener H.
DIODE-LASER-BASED OPTICAL
ROTATION DETECTOR FOR HIGH-
PERFORMANCE LIQUID
CHROMATOGRAPHY AND ON-LINE
POLARIMETRIC ANALYSIS.
Anal. Chem. 61, 1238-1243, (1989).

152. Lo L-C., Chen S-T., Wu S-H., Wang K-T.
SEPARATION OF DIASTEREOMERS OF
PROTECTED DIPEPTIDES BY NORMAL-
PHASE HIGH-PERFORMANCE LIQUID
CHROMATOGRAPHY.
J. Chromatogr., 472, 336-339, (1988).

153. Lough W. J. (Ed)
CHIRAL LIQUID CHROMATOGRAPHY.
Blackie, (1989).

154. Macaudiere P., Caude M., Rosset R.,
Tambute A.
CHIRAL RESOLUTION OF A SERIES OF 3-
THIENYLCYCLOHEXYLGLYCOLIC ACIDS
BY LIQUID OR SUBCRITICAL FLUID
CHROMATOGRAPHY. A MECHANISTIC
STUDY.
J. Chromatogr., 450, 255-269, (1988).

155. Macaudiere P., Lienne M., Caude M.,
Rosset R., Tambute A.
RESOLUTION OF πI-ACID RACEMATES
ON πI-ACID CHIRAL STATIONARY
PHASES IN NORMAL-PHASE LIQUID AND
SUBCRITICAL FLUID
CHROMATOGRAPHIC MODES. A UNIQUE
REVERSAL OF ELUTION ORDER ON
CHANGING THE NATURE OF THE
ACHIRAL MODIFIER.
J. Chromatogr., 467, 357-372, (1989).

156. Maibaum J.
INDIRECT HIGH-PERFORMANCE LIQUID
CHROMATOGRAPHIC RESOLUTION OF
RACEMIC TERTIARY AMINES AS THEIR
DIASTEREOMERIC UREA DERIVATIVES
AFTER N-DEALKYLATION.
J. Chromatogr., 436, 269-278, (1988).

157. Mama J.E., Fell A.F., Clark B.J.
PHARMACEUTICAL APPLICATIONS OF
POROUS GRAPHITIC CARBON IN HPLC.
Anal. Proc., 26, 71-73, (1989).

158. Mannschreck A., Andert D., Eiglsperger A.,
Gmahl E.
CHIROPTICAL DETECTION. NOVEL
POSSIBILITIES OF ITS APPLICATION TO
ENANTIOMERS.
Chromatographia, 25, 182-188, (1988).

159. Mannschreck A., Zinner H., Pustet N.
THE SIGNIFICANCE OF THE HPLC TIME
SCALE: AN EXAMPLE OF
INTERCONVERTIBLE ENANTIOMERS.
Chimia, 43, 165-166, (1989).

160. Marle I., Petterson C., Arvidsson T.
DETERMINATION OF BINDING AFFINITY
OF ENANTIOMERS TO ALBUMIN BY
LIQUID CHROMATOGRAPHY.
J. Chromatogr., 456, 323-336, (1988).

161. Masurel D., Wainer I.W.
ANALYTICAL AND PREPARATIVE HIGH-
PERFORMANCE LIQUID
CHROMATOGRAPHIC SEPARATION OF
THE ENANTIOMERS OF IFOSFAMIDE,
CYCLOPHOSPHAMIDE AND
TROFOSFAMIDE AND THEIR
DETERMINATION IN PLASMA.
J. Chromatogr., 490, 133-143, (1989).

162. Matlin S.A., Stacey V.E., Lough W.J.
HEXAHELICENE CHIRAL STATIONARY
PHASE. I. PHASE SYNTHESIS AND USE IN
HIGH-PERFORMANCE LIQUID
CHROMATOGRAPHIC RESOLUTION OF
ENANTIOMERS.
J. Chromatogr., 450, 157-162, (1988).

163. Maurs M., Trigalo F., Azerad R.
RESOLUTION OF α-SUBSTITUTED AMINO
ACID ENANTIOMERS BY HIGH-
PERFORMANCE LIQUID
CHROMATOGRAPHY AFTER
DERIVATIZATION WITH A CHIRAL
ADDUCT OF O-PHTHALALDEHYDE.
APPLICATION TO GLUTAMIC ACID
ANALOGS.
J. Chromatogr., 440, 209-215, (1988).

164. Mehta A.C.
DIRECT SEPARATION OF DRUG
ENANTIOMERS BY HIGH PERFORMANCE
LIQUID CHROMATOGRAPHY WITH
CHIRAL STATIONARY PHASES.
J. Chromatogr., 426, 1-13, (1988).

165. Mehvar R., Jamali F.
STEREOSPECIFIC HIGH-PERFORMANCE
LIQUID CHROMATOGRAPHIC (HPLC)
ASSAY OF FENOPROFEN ENANTIOMERS
IN PLASMA AND URINE.
Pharm. Res., 5, 53-56, (1988).

166. Meinard C., Bruneau P.
SEPARATION AND IDENTIFICATION OF
ENANTIOMERS BY HIGH-PERFORMANCE
LIQUID CHROMATOGRAPHY WITH A
CHIRAL COLUMN AND A POLARIMETRIC
DETECTOR AS APPLIED TO
DELTAMETHRIN.
J. Chromatogr., 450, 169-174, (1988).

167. Michelsen P., Aronsson E., Odham G.,
Akesson B.
DIASTEREOMERIC SEPARATIONS OF
NATURAL GLYCERO DERIVATIVES AS
THEIR 1-(1-NAPHTHYL)ETHYL
CARBAMATES BY HIGH-PERFORMANCE
LIQUID CHROMATOGRAPHY.
J. Chromatogr., 350, 417-426, (1988).

168. Miwa T., Miyakawa T., Miyake Y.
CHARACTERISTICS OF AN AVIDIN-
CONJUGATED COLUMN IN DIRECT
LIQUID CHROMATOGRAPHIC
RESOLUTION OF RACEMIC COMPOUNDS.
J. Chromatogr., 457, 227-233, (1988).

169. Miyazaki A., Nakamura T., Kawaradani M., Marumo S.
RESOLUTION AND BIOLOGICAL ACTIVITY OF BOTH ENANTIOMERS OF METHAMIDOPHOS AND ACEPHATE.
J. Agric. Food Chem., 36, 835-837, (1988).

170. Miyazaki A., Nakamura T., Marumo S.
STEREOSELECTIVITY IN METABOLIC SULFOXIDATION OF PROPAPHOS AND BIOLOGICAL ACTIVITY OF CHIRAL PROPAPHOS SULFOXIDE.
Pestic. Biochem. Physiol., 33, 11-15, (1989).

171. Mueller M.D., Bosshardt H.P.
ENANTIOMER RESOLUTION AND ASSAY OF PROPIONIC ACID-DERIVED HERBICIDES IN FORMULATIONS BY USING CHIRAL LIQUID CHROMATOGRAPHY AND ACHIRAL GAS CHROMATOGRAPHY.
J. Assoc. Off. Anal. Chem., 71, 614-617, (1988).

172. Mularz E.A., Cline-Love L.J., Petersheim M.
STRUCTURAL BASIS FOR ENANTIOMERIC RESOLUTION OF PSEUDOEPHEDRINE AND THE FAILURE TO RESOLVE EPHEDRINE BY USING β-CYCLODEXTRIN MOBILE PHASES.
Anal. Chem., 60, 2751-2755, (1988).

173. Mushtaq M., Weems H.B., Shen K.
STEREOSELECTIVE FORMATIONS OF ENANTIOMERIC K-REGION EPOXIDE AND TRANS-DIHYDRODIOLS IN DIBENZ[A,H]ANTHRACENE METABOLISM.
Chem. Res. Toxicol., 2. 84-93, (1989).

174. Nakamura Y., Yamagishi A., Twamoto T., Koga M.
ADSORPTION PROPERTIES OF MONTOMORILLONITE AND SYNTHETIC SAPONITE AS PACKING MATERIALS IN LIQUID-COLUMN CHROMATOGRAPHY.
Clays Clay Miner., 36, 530-536, (1988).

175. Nation R.L.
ENANTIOSELECTIVE DRUG ANALYSIS: PROBLEMS AND RESOLUTIONS.
Clin. Exp. Pharmacol. Physiol., 16, 471-477, (1989).

176. Nicoll-Griffith D.A., Inaba T., Tang B.K., Kalow W.
METHOD TO DETERMINE THE ENANTIOMERS OF IBUPROFEN FROM HUMAN URINE BY HIGH-PERFORMANCE LIQUID CHROMATOGRAPHY.
J. Chromatogr., 428, 103-112, (1988).

177. Nimura N., Kinoshita T.
O-PHTHALALDEHYDE-N-ACETYL-L-CYSTEINE AS A CHIRAL DERIVATIZATION REAGENT FOR LIQUID CHROMATOGRAPHIC OPTICAL RESOLUTION OF AMINO ACID ENANTIOMERS AND ITS APPLICATION TO CONVENTIONAL AMINO ACID ANALYSIS.
J. Chromatogr., 352 169-177, (1988).

178. Nishi H., Ishii K., Taku K., Shimizu R., Tsumagari N.
NEW CHIRAL DERIVATIZATION REAGENT FOR THE RESOLUTION OF AMINO ACIDS AS DIASTEREOMERS BY TLC AND HPLC.
Chromatographia, 27, 301-305, (1989).

179. Norinder U., Sundholm E.G.
FURTHER USE OF COMPUTER AIDED CHEMISTRY TO PREDICT CHIRAL SEPARATIONS IN LIQUID CHROMATOGRAPHIC: SELECTING THE MOST APPROPRIATE DERIVATIVES.
In "Chiral Separations". Eds: D. Stevenson and I.D. Wilson, Plenum, 127-130, (1988).

180. Nusser E., Nill K., Breyer-Pfaff U.
ENANTIOSELECTIVE FORMATION AND DISPOSITION OF (E)- AND (Z)-10-HYDROXYNORTRIPTYLINE.
Drug. Metab. Dispos., 16, 509-511, (1988).

181. Oi N., Kitahara H., Matsumoto Y., Nakajima H., Horikawa Y.
(R)-N-(3,5-DINITROBENZOYL)-1-NAPHTHYLGLYCINE AS A CHIRAL STATIONARY PHASE FOR THE SEPARATION OF ENANTIOMERS BY HIGH-PERFORMANCE LIQUID CHROMATOGRAPHY.
J. Chromatogr., 462, 382-386, (1989).

182. Oi S., Shijo M., Yamashita J., Miyano S.
AN EFFICIENT CHIRAL STATIONARY PHASE FOR HIGH-PERFORMANCE LIQUID CHROMATOGRAPHIC SEPARATION OF ENANTIOMERIC ALIPHATIC ALCOHOLS.
Chem. Lett., 1545-1546, (1988).

183. Okamoto Y., Aburatani R., Hatada K.
CHROMATOGRAPHIC RESOLUTION. XXI. DIRECT OPTICAL RESOLUTION OF ABSCISIC ACID BY HIGH-PERFORMANCE LIQUID CHROMATOGRAPHY ON CELLULOSE TRIS(3,5-DIMETHYLPHENYL-CARBAMATE).
J. Chromatogr., 448, 454-455, (1988).

184. Okamoto Y., Aburatani R., Hatano K., Hatada K.
OPTICAL RESOLUTION OF RACEMIC DRUGS BY CHIRAL HPLC ON CELLULOSE AND AMYLOSE TRIS(PHENYL-CARBAMATE) DERIVATIVES.
J. Liq. Chromatogr., 11, 2147-2163. (1988).

185. Okamoto Y., Aburatani R., Kaida Y., Hatada K.
CHROMATOGRAPHIC RESOLUTION. 19. DIRECT OPTICAL RESOLUTION OF CARBOXYLIC ACIDS BY CHIRAL HPLC ON TRIS(3,5-DIMETHYLPHENYL-CARBAMATE)S OF CELLULOSE AND AMYLOSE.
Chem. Lett., 1125-1128, (1988).

186. Okamoto Y., Aburatani R., Kaida Y., Hatada K., Inotsume N., Nakano M.
DIRECT CHROMATOGRAPHIC SEPARATION OF 2-ARYLPROPIONIC ACID ENANTIOMERS USING TRIS(3,5-DIMETHYLPHENYLCARBAMATE)S OF CELLULOSE AND α MYLOSE AS CHIRAL STATIONARY PHASES.
Chirality, 1, 239-242, (1989).

187. Okamoto Y., Hatada K.
OPTICALLY ACTIVE POLY(TRIPHENYL-METHYL METHACRYLATE) AS A CHIRAL STATIONARY PHASE.
J. Chromatogr. Sci., 40, 199-218, (1988).

188. Okamoto Y., Hatano K., Aburatani R., Hatada K.
TRIS(4-TERT-BUTYLPHENYL-CARBAMATE)S OF CELLULOSE AND AMYLOSE AS USEFUL CHIRAL STATIONARY PHASES FOR CHROMATOGRAPHIC OPTICAL RESOLUTION.
Chem. Lett., 715-718, (1989).

189. Okamoto Y., Mohri H., Hatada K.
CHROMATOGRAPHIC OPTICAL RESOLUTION BY OPTICALLY ACTIVE POLY(DIPHENYL-2-PYRIDYLMETHYL METHACRYLATE) WITH A HIGHLY ONE-HANDED HELICAL STRUCTURE.
Polym. J., 21, 439-445, (1989).

190. Okamoto Y., Yashima E., Ishikura M., Hatada K.
CHROMATOGRAPHIC RESOLUTION. XVIII. THE CHIRAL RECOGNITION OF OPTICALLY ACTIVE POLY (TRIPHENYLMETHYL METHACRYLATE) DERIVATIVES AS STATIONARY PHASES FOR HPLC.
Bull. Chem. Soc. Jpn., 61, 255-259, (1988).

191. Olieman C., De Vries E.S.
DETERMINATION OF D- AND L-LACTIC ACID IN FERMENTED DAIRY PRODUCTS WITH HPLC.
Neth. Milk Dairy J., 42, 111-120, (1988).

192. Pankonin G., Mahmood S.A., Kuhn H., Schewe T., Pilgrim H., Teuscher E.
INHIBITION OF CELL MIGRATION BY THE LINOLEIC ACID OXYGENATION PRODUCT 9S-HYDROXY-10E,12Z-OCTADECADIENOIC ACID (9-HODE).
Biomed. Biochim. Acta, 47, K17-K21, (1988).

193. Papadopoulou-Mourkidou E.
EFFECT OF COLUMN TEMPERATURE ON THE DIRECT DETERMINATION OF (RS)-α-CYANO(3-PHENOXYPHENYL)METHYL (RS)-2(4-CHLOROPHENYL)-3-METHYL-BUTYRATE OPTICAL ISOMERS BY HIGH-PERFORMANCE LIQUID CHROMATOGRAPHY/DIODE ARRAY SYSTEM.
Anal. Chem., 61, 1149-1151, (1989).

194. Papadopoulou-Mourkidou E.
TEMPERATURE-DEPENDENT SOLUTE RETENTION TIME VARIATIONS, BASE-LINE DRIFTS, AND SOLUTE PEAK SPLITS DURING THE ANALYSIS OF FENVALERATE BY A HIGH-PERFORMANCE LIQUID CHROMATOGRAPIC/DIODE ARRAY SYSTEM.
Anal. Chem., 61, 1152-1158, (1989).

195. Pettersson C.
LIQUID CHROMATOGRAPHIC SEPARATION OF ENANTIOMERS USING CHIRAL ADDITIVES IN THE MOBILE PHASE.
Trends Anal. Chem., 7, 209-217, (1988).

196. Pettersson C., Gioeli C.
IMPROVED RESOLUTION OF ENANTIOMERS OF NAPROXEN BY THE SIMULTANEOUS USE OF A CHIRAL STATIONARY PHASE AND A CHIRAL ADDITIVE IN THE MOBILE PHASE.
J. Chromatogr., 435, 225-228, (1988).

197. Pettersson C., Schill G.
ENANTIOMER SEPARATION IN ION-PAIRING SYSTEMS.
J. Chromatogr. Sci., 40, 283-313, (1988).

198. Pianezzola E., Bellotti V., Fontana E., Moro E., Gal J., Desai D.M.
DETERMINATION OF THE ENANTIOMERIC COMPOSITION OF SALSOLINOL IN BIOLOGICAL SAMPLES BY HIGH-PERFORMANCE LIQUID CHROMATOGRAPHY WITH ELECTROCHEMICAL DETECTION.
J. Chromatogr., 495, 205-214, (1989).

199. Pirkle W.H., Burke J.A.,III.
PREPARATION OF A CHIRAL STATIONARY PHASE FROM AN α-AMINO PHOSPHONATE.
J. Chirality 1, 57-62, (1989).

200. Pirkle W.H., Burke J.A.,III., Wilson R.
X-RAY CRYSTALLOGRAPHIC SUPPORT OF
A CHIRAL RECOGNITION MODEL.
J. Am. Chem., 111, 9222-9223, (1989).

201. Pirkle W.H, Hamper B.C.
CHROMATOGRAPHIC SEPARATION OF
THE ENANTIOMERS OF 1,3-DITHIOLANE-
1-OXIDES.
J. Chromatogr., 450, 199-210, (1988).

202. Pirkle W.H., McCune J.E.
DISCUSSION OF A CONTROVERSIAL
CHIRAL RECOGNITION MODEL.
J. Chromatogr., 469, 67-75, (1989).

203. Pirkle W.H., McCune J.E.
IMPROVED CHIRAL STATIONARY PHASE
FOR THE SEPARATION OF THE
ENANTIOMERS OF CHIRAL ACIDS AS
THEIR ANILIDE DERIVATIVES.
J. Chromatogr., 471, 271-281, (1989).

204. Pirkle W.H., McCune J.E.
LIQUID CHROMATOGRAPHIC
SEPARATION OF THE ENANTIOMERS OF
CHIRAL SECONDARY ALCOHOLS AS
THEIR α-NAPHTHYLURETHANE
DERIVATIVES.
J. Liq. Chromatogr., 11, 2165-2173,
(1988).

205. Pirkle W.H., Pochapsky T.C.
CONSIDERATION OF CHIRAL
RECOGNITION RELEVANT TO THE LIQUID
CHROMATOGRAPHY SEPARATION OF
ENANTIOMERS.
Chem. Rev., 89, 347-362, (1989).

206. Pirkle W.H., Pochapsky T.C.
SEPARATION OF THE STEREOISOMERS
OF A HOMOLOGOUS SERIES OF BIS-
AMIDES ON CHIRAL STATIONARY
PHASES.
Chromatographia, 25, 652-654, (1988).

207. Pirkle W.H., Pochapsky T.C., Burke J.A.,
Deming K.C.
SYSTEMATIC STUDIES OF CHIRAL
RECOGNITION MECHANISMS.
In "Chiral Separations". Eds: D. Stevenson
and I.D. Wilson, Plenum, 23-36, (1988).

208. Porziemsky J.P., Krstulovic A.M., Wick A.,
Barton D.H.R.,
Tachdjian C., Gateau-Olesker A., Gero S.D.
COMPLEMENTARY USE OF REVERSED-
PHASE, NORMAL-PHASE AND CHIRAL
HIGH-PERFORMANCE LIQUID
CHROMATOGRAPHY FOR THE STUDY OF
THE STEREOISOMERS PRODUCED IN
DECARBOXYLATIVE ALKYLATION OF
TARTARIC ACID.
J. Chromatogr., 440, 183-195, (1988).

209. Prakash C., Jajoo H.K., Blair I.A., Mayol
R.F.
RESOLUTION OF ENANTIOMERS OF THE
ANTIARRHYTHMIC DRUG ENCAINIDE
AND ITS MAJOR METABOLITES BY
CHIRAL DERIVATIZATION AND HIGH-
PERFORMANCE LIQUID
CHROMATOGRAPHY.
J. Chromatogr., 493, 325-335, (1989).

210. Prakash C., Koshakji R.P., Wood A.J.J.,
Blair I.A.
SIMULTANEOUS DETERMINATION OF
PROPRANOLOL ENANTIOMERS IN
PLASMA BY HIGH-PERFORMANCE LIQUID
CHROMATOGRAPHY WITH
FLUORESCENCE DETECTION.
J. Pharm. Sci., 78, 771-775, (1989).

211. Rao N.K.R., Towill R.C., Todd B.
THE ROLE OF SOLVENTS AND STERIC
FACTORS IN THE RESOLUTION OF β-
BLOCKER DRUGS ON CHIRAL UREA
PHASES.
In "Chiral Separations". Eds: D. Stevenson
and I.D. Wilson, Plenum, 55-60, (1988).

212. Rathbone E.B., Butters R.W., Cookson D.,
Robinson J.L.
CHIRALITY OF 2-(4-METHOXYPHENOXY)
PROPANOIC ACID IN ROASTED COFFEE
BEANS: ANALYSIS OF THE METHYL
ESTERS BY CHIRAL HIGH-
PERFORMANCE LIQUID
CHROMATOGRAPHY.
J. Agric. Food Chem., 37, 58-60 (1989).

213. Reid J.M., Stobaugh J.F., Sternson L.A.
LIQUID CHROMATOGRAPHIC
DETERMINATION OF
CYCLOPHOSPHAMIDE ENANTIOMERS IN
PLASMA BY PRECOLUMN CHIRAL
DERIVATIZATION.
Anal. Chem. 61, 441-446, (1989).

214. Rizzi A.M.
BAND BROADENING IN HIGH-
PERFORMANCE LIQUID
CHROMATOGRAPHIC SEPARATIONS OF
ENANTIOMERS WITH SWOLLEN
MICROCRYSTALLINE CELLULOSE
TRIACETATE PACKINGS. I. INFLUENCE
OF CAPACITY FACTOR, ANALYTE
STRUCTURE, FLOW VELOCITY AND
COLUMN LOADING.
J. Chromatogr., 478, 71-86, (1989).

215. Rizzi A.M.
BAND BROADENING IN HIGH-
PERFORMANCE LIQUID
CHROMATOGRAPHIC SEPARATIONS OF
ENANTIOMERS WITH SWOLLEN
MICROCRYSTALLINE CELLULOSE
TRIACETATE PACKINGS. II. INFLUENCE
OF ELUENT COMPOSITION,
TEMPERATURE AND PRESSURE.
J. Chromatogr., 478, 87-89, (1989).

216. Rizzi A.M.
EVALUATION OF THE OPTIMIZATION
POTENTIAL IN HIGH-PERFORMANCE
LIQUID CHROMATOGRAPHIC
SEPARATIONS OF OPTICAL ISOMERS
WITH SWOLLEN MICROCRYSTALLINE
CELLULOSE TRIACETATE.
J. Chromatogr., 478, 101-119, (1989).

217. Robinson C.
THE USE OF HPLC IN STUDIES OF THE
STEREOSELECTIVE METABOLISM OF
PROSTAGLANDIN D_2
In "Bioanalysis of Drugs and Metabolites,
Especially Anti-Inflammatory and
Cardiovascular". Eds: E. Reid, J.D.
Robinson and I.D. Wilson, Plenum., 3-13,
(1988).

218. Roussel C., Chemlal A.
RESOLUTION OF N-ARYL-4-THIAZOLINE-
2-THIONE ATROPISOMERS ON
MICROCRYSTALLINE CELLULOSE
TRIACETATE: DETERMINATION OF THE
ABSOLUTE CONFIGURATION OF 3-(2-
TOLYL)-4-TERT-BUTYL-4-THIAZOLINE-2-
THIONE ATROPISOMERS AND THEIR
OXYGEN ANALOGS.
New J. Chem., 12, 947-952, (1988).

219. Roussel C., Stein J.L., Beauvais F.,
Chemlal A.
EXAMPLE OF THE CONCENTRATION
DEPENDENCE OF ELUTION ORDER IN
THE RESOLUTION OF ENANTIOMERS ON
MICROCRYSTALLINE TRIACETYL-
CELLULOSE CHIRAL STATIONARY PHASE.
J. Chromatogr., 462, 95-103, (1989).

220. Ruffing F.J., Lux J.A., Roeder W.,
Schomberg G.
CHIRAL STATIONARY PHASES FOR LC
AND SFC OBTAINED BY "POLYMER
COATING".
Chromatographia, 26, 19-28, (1988).

221. Saigo K., Yuki Y., Kimoto H., Nishida T.,
Hasegawa M.
A NOVEL CHIRAL STATIONARY PHASE
FOR OPTICAL RESOLUTION OF AMINO
ACIDS AND THEIR DERIVATIVES BY
LIGAND EXCHANGE HIGH-
PERFORMANCE LIQUID
CHROMATOGRAPHY.
Bull. Chem. Soc. Jpn., 61, 322-324,
(1988).

222. Salvadori P., Pini D., Rosini C., Uccello-
Barretta G., Bertucci C.
CHIRAL STATIONARY PHASE DERIVED
FROM L-LACTIC ACID FOR THE OPTICAL
RESOLUTION OF N-(3,5-
DINITROBENZOYL)-AMINO ACID METHYL
ESTERS.
J. Chromatogr., 450, 163-168, (1988).

223. Sander L.C., Wise S.A.
RECENT ADVANCES IN BONDED PHASES
FOR LIQUID CHROMATOGRAPHY.
CRC Crit. Rev. Anal. Chem., 18, 299-415,
(1988).

224. Sato Y., Nishikawa M., Shinkai H.
ANALYSIS OF ENANTIOMERS OF A NEW
ANTI-DIABETIC AGENT IN PLASMA BY
HIGH-PERFORMANCE LIQUID
CHROMATOGRAPHY.
J. Liq. Chromatogr. 12, 445-455, (1989).

225. Schmitthenner H.F., Fedorchuk M., Walter
D.J.
RESOLUTION OF ANTIHYPERTENSIVE
ARYLOXYPROPANOLAMINE
ENANTIOMERS BY REVERSED-PHASE
CHROMATOGRAPHY OF (-)-MENTHYL
CHLOROFORMATE DERIVATIVES.
J. Chromatogr., 487, 197-203, (1989).

226. Schomburg G.
STATIONARY PHASES IN HIGH-
PERFORMANCE LIQUID
CHROMATOGRAPHY. CHEMICAL
MODIFICATION BY POLYMER COATING.
LC-GC 6, 37-40, (1988).

227. Schuster D., Modi M.W., Lalka D., Gengo
F.M.
REVERSED-PHASE HIGH-PERFORMANCE
LIQUID CHROMATOGRAPHIC ASSAY TO
QUANTITATE DIASTEREOMERIC
DERIVATIVES OF METOPROLOL
ENANTIOMERS IN PLASMA.
J. Chromatogr., 433, 318-325, (1988).

228. Seeman J.I., Secor H.V., Armstrong D.W.,
Timmons K.D., Ward T.J.
ENANTIOMERIC RESOLUTION AND
CHIRAL RECOGNITION OF RACEMIC
NICOTINE AND NICOTINE ANALOGUES
BY β-CYCLODEXTRIN COMPLEXATION.
STRUCTURE-ENANTIOMERIC
RESOLUTION RELATIONSHIPS IN HOST-
QUEST INTERACTIONS.
Anal. Chem., 60, 2120-2127, (1988).

229. Sellergren B.
MOLECULAR IMPRINTING BY
NONCOVALENT INTERACTIONS: TAILOR-
MADE CHIRAL STATIONARY PHASES OF
HIGH SELECTIVITY AND SAMPLE LOAD
CAPACITY.
Chirality, 1, 63-68, (1989).

230. Shiao M.S., Lin L.J., Chen C.S.
DETERMINATION OF STEREO- AND
POSITIONAL ISOMERS OF OXYGENATED
TRITERPENOIDS BY REVERSED-PHASE
HIGH-PERFORMANCE LIQUID
CHROMATOGRAPHY.
J. Lipid Res., 30, 287-291, (1989).

231. Shinkai H., Nashikawa M., Sato Y.
SEPARATION OF A NEW ANTI-DIABETIC
AGENT N-(TRANS-4-ISOPROPYLCYCLO-
HEXYLCARBONYL)-D-PHENYLALANINE,
AND ITS ISOMERS BY CHIRAL HIGH-
PERFORMANCE LIQUID
CHROMATOGRAPHY.
J. Liq. Chromatogr., 12, 457-464, (1989).

232. Simec Z., Vespalec R.
BOVINE SERUM ALBUMIN BONDED TO
HYDROXYETHYLMETHACRYLATE
POLYMER FOR CHIRAL SEPARATIONS.
J. High Resolut. Chromatogr., 12, 61-62,
(1989).

233. Spahn H.
S-(+)-FLUNOXAPROFEN CHLORIDE AS
CHIRAL FLUORESCENT REAGENT.
J. Chromatogr., 427, 131-137, (1988).

234. Spahn H.
S-(+)-NAPROXEN CHLORIDE AS
ACYLATING AGENT FOR SEPARATING
THE ENANTIOMERS OF CHIRAL AMINES
AND ALCOHOLS.
Arch. Pharm., 321, 847-850, (1988).

235. Spahn H., Iwakawa S., Lin E.T., Benet L.Z.
PROCEDURES TO CHARACTERIZE IN
VIVO AND IN VITRO ENANTIOSELECTIVE
GLUCURONIDATION PROPERLY: STUDIES
WITH BENOXAPROFEN GLUCURONIDES.
Pharm. Res., 6, 125-132, (1989).

236. Spahn H., Krauss D., Mutschler E.
ENANTIOSPECIFIC HIGH-PERFORMANCE
LIQUID CHROMATOGRAPHIC (HPLC)
DETERMINATION OF BACLOFEN AND ITS
FLUORO ANALOG IN BIOLOGICAL FLUID.
Pharm. Res., 5, 107-112, (1988).

237. Spahn H., Spahn I., Pflugmann G.,
Mutschler E., Benet L.Z.
MEASUREMENT OF CARPROFEN
ENANTIOMER CONCENTRATIONS IN
PLASMA AND URINE USING L-
LEUCINAMIDE AS THE CHIRAL
COUPLING COMPONENT.
J. Chromatogr., 433, 331-338, (1988).

238. Spahn H., Wellstein A., Pflugmann G.,
Mutschler E., Palm D.
RADIORECEPTOR ASSAY OF
METOPROLOL IN HUMAN PLASMA:
COMPARISON WITH AN ENANTIO-
SPECIFIC HIGH-PERFORMANCE LIQUID
CHROMATOGRAPHIC (HPLC)
PROCEDURE.
Pharm. Res., 6, 152-155, (1989).

239. Still M.G., Rogers L.B.
MOLECULAR MODELING OF
STRUCTURAL CHANGES WHICH AFFECT
CHROMATOGRAPHIC SELECTIVITY IN
CHIRAL SEPARATIONS.
Talanta, 36, 35-48, (1989).

240. Straka R.J., Lalonde R.L., Wainer I.W.
MEASUREMENT OF UNDERIVATIZED
PROPRANOLOL ENANTIOMERS IN SERUM
USING A CELLULOSE-TRIS(3,5-
DIMETHYLPHENYLCARBAMATE) HIGH
PERFORMANCE LIQUID
CHROMATOGRAPHY (HPLC) CHIRAL
STATIONARY PHASE.
Pharm. Res., 5, 187-189, (1988).

241. Strasak M., Bystricky S.
RAPID OPTICAL RESOLUTION OF
ANIONIC METAL COMPLEXES BY GEL
PERMEATION CHROMATOGRAPHY.
J. Chromatogr., 450, 284-290, (1988).

242. Stuurman H.W., Koehler J., Schomburg G.
HPLC SEPARATION OF ENANTIOMERS
USING QUININE, COVALENTLY BONDED
TO SILICA AS STATIONARY PHASE.
Chromatographia, 25, 265-271, (1988).

243. Szepesi G., Gazdag M.
ENANTIOMERIC SEPARATIONS AND
THEIR APPLICATION IN
PHARMACEUTICAL ANALYSIS USING
CHIRAL ELUENTS.
J. Pharm. Biomed. Anal., 6, 623-639,
(1988).

244. Sztruhar I., Ladanyi L., Simonyi I.,
Furdyga E.
DIRECT DETERMINATION OF THE
RACEMATE RATIO OF LABETALOL
HYDROCHLORIDE BY HPLC.
Chromatographia, 27, 364-366, (1989).

245. Takahashi H., Kanoh S., Ogata H.,
Kashiwada K., Someya K.
DETERMINATION OF PROPRANOLOL
ENANTIOMERS IN HUMAN PLASMA AND
URINE AND IN RAT TISSUES USING
CHIRAL STATIONARY-PHASE LIQUID
CHROMATOGRAPHY.
J. Pharm. Sci., 77, 993-995, (1988).

246. Tambute A.G., Canceill J., Collet A.
OPTICAL RESOLUTION OF C3
CYCLOTRIVERAT'TRYLENES AND D3
CRYPTOPHANES BY LIQUID
CHROMATOGRAPHY ON CHIRAL
STATIONARY PHASE CHIRALPAK-OT(+).
Bull. Chem. Soc. Jpn., 62, 1390-1392,
(1989).

247. Tanaka M., Shono T., Zhu D-Q.,
Kawaguchi Y.
LIQUID CHROMATOGRAPHIC
SEPARATION OF RACEMATES ON
ACETYLATED OR CARBAMOYLATED β-
CYCLODEXTRIN-BONDED STATIONARY
PHASES.
J. Chromatogr., 469, 429-433, (1989).

248. Thompson R.A., Andersson S., Allenmark S.
DIRECT LIQUID CHROMATOGRAPHIC SEPARATION OF ENANTIOMERS ON IMMOBILIZED PROTEIN STATIONARY PHASES.
J. Chromatogr., 465, 263-270, (1989).

249. Topiol S., Sabio M.
COMPUTATIONAL CHEMICAL STUDIES OF CHIRAL STATIONARY PHASE MODELS. COMPLEXES OF METHYL N-(2-NAPHTHYL)ALANINATE WITH N-(3,5-DINITROBENZOYL)LEUCINE N-PROPYLAMIDE.
J. Chromatogr., 461, 129-137, (1989).

250. Topiol S., Sabio M., Moroz J., Caldwell W.B.
COMPUTATIONAL STUDIES OF THE INTERACTIONS OF CHIRAL MOLECULES: COMPLEXES OF METHYL N-(2-NAPHTHYL)ALANINATE WITH N-(3,5-DINITROBENZOYL)LEUCINE N-PROPYLAMIDE AS A MODEL FOR CHIRAL STATIONARY-PHASE INTERACTIONS.
J. Am. Chem. Soc., 110, 8367-8376, (1988).

251. Tsujiyama T., Tsuchiya M., Hamachi Y., Kuriki T., Fukunaga T., Suzuki N.
ION-PAIR CHROMATOGRAPHIC SEPARATION OF NOMIFENSINE MALEATE ENANTIOMERS.
Anal. Sci., 5, 285-288, (1989).

252. Tsukamoto S., Hayashi K., Kaneko K.
IDENTIFICATION OF THE ABSOLUTE STEREOCHEMISTRY OF D- AND L-DIGITOXOSE USING A CHIRAL HIGH-PRESSURE LIQUID CHROMATOGRAPHY COLUMN.
J. Chem. Soc., Perkin Trans, 2621-2624, (1988).

253. Tsukamoto S., Kaneko K., Hayashi K.
A METHOD FOR IDENTIFYING THE ABSOLUTE CONFIGURATION OF RHAMNOSE, LYXOSE, AND 2,6-DIDEOXY SUGARS, CYMAROSE, OLEANDROSE, DIGINOSE AND DIGITOXOSE USING A CHIRAL HIGH-PERFORMANCE LIQUID CHROMATOGRAPHY (HPLC) COLUMN.
Chem. Pharm. Bull., 37, 637-641, (1989).

254. Turk J., Stump W.T., Wolf B.A., Easom R.A., McDaniel M.L.
QUANTITATIVE STEREOCHEMICAL ANALYSIS OF SUBNANOGRAM AMOUNTS OF 12-HYDROXY-(5,8,10,14)-EICOSATETRAENOIC ACID BY SEQUENTIAL CHIRAL PHASE LIQUID CHROMATOGRAPHY AND STABLE ISOTOPE DILUTION MASS SPECTROMETRY.
Anal. Biochem., 174, 580-588, (1988).

255. Van der Haar J., Kip J., Kraak J.C.
EFFECT OF THE MOBILE PHASE COMPOSITION AND LIGAND STRUCTURE ON THE SEPARATION OF D- AND L-DANSYLAMINO ACIDS, AS MIXED METAL COMPLEXES, BY REVERSED PHASE HIGH PERFORMANCE LIQUID CHROMATOGRAPHY.
J. Chromatogr., 445, 219-224, (1988).

256. Vindevogel J., Van Dijck J., Verzele M.
MICRO-LIQUID CHROMATOGRAPHY AND THE CHIRAL RECOGNITION MECHANISM ON ALBUMIN-COATED SILICA GEL. LARGE SELECTIVITY CHANGES WITH SAMPLE SIZE.
J. Chromatogr., 477, 297-303, (1988).

257. Wad N.
SEPARATION OF THE ENANTIOMERS OF PHENETURIDE IN SERUM BY HIGH-PERFORMANCE LIQUID CHROMATOGRAPHY.
J. Liq. Chromatogr., 11, 1107-1116, (1988).

258. Wainer I.W.
A PRACTICAL GUIDE TO THE SELECTION AND USE OF HPLC CHIRAL STATIONARY PHASES.
J.T. Baker (1988).

259. Wainer I.W.
SOME OBSERVATIONS ON CHOOSING AN HPLC CHIRAL STATIONARY PHASE.
LC-GC, 7, 378-382, (1989).

260. Wainer I.W., Alembik M.C.
THE ENANTIOMERIC RESOLUTION OF BIOLOGICALLY ACTIVE MOLECULES ON COMMERCIALLY AVAILABLE LIQUID CHROMATOGRAPHIC CHIRAL STATIONARY PHASES.
J. Chromatogr. Sci., 40, 355-384, (1988).

261. Wainer I.W., Chu Y.Q.
USE OF MOBILE PHASE MODIFIERS TO ALTER RETENTION AND STEREO-SELECTIVITY ON A BOVINE SERUM ALBUMIN HIGH-PERFORMANCE LIQUID CHROMATOGRAPHIC CHIRAL STATIONARY PHASE.
J. Chromatogr., 455, 316-322, (1988).

262. Wainer I.W., Jadaud P., Schombaum G.R., Kadodkar, S.V., Henry M.P.
ENZYMES AS HPLC STATIONARY PHASES FOR CHIRAL RESOLUTIONS: INITIAL INVESTIGATIONS WITH α-CHYMOTRYPSIN.
Chromatographia, 25, 903-907, (1988).

263. Wainer I.W., Stiffin R.M.
DIRECT RESOLUTION OF THE
STEREOISOMERS OF LEUCOVORIN AND
5-METHYLTETRAHYDROFOLATE BY
USING A BOVINE SERUM ALBUMIN HIGH-
PERFORMANCE LIQUID
CHROMATOGRAPHIC CHIRAL
STATIONARY PHASE COUPLED TO AN
ACHIRAL PHENYL COLUMN.
J. Chromatogr., 424, 158-162, (1988).

264. Wainer I.W., Stiffin R.M., Chu Y-Q.
DRUG ANALYSIS USING HIGH-
PERFORMANCE LIQUID
CHROMATOGRAPHIC (HPLC) CHIRAL
STATIONARY PHASES: WHERE TO BEGIN
AND WHICH ONE TO USE.
In "Chiral Separations". Eds: D. Stevenson
and I.D. Wilson, Plenum, 11-22, (1988).

265. Walhagen A., Edholm L-E.,
COUPLED-COLUMN CHROMATOGRAPHY
ON IMMOBILIZED PROTEIN PHASES FOR
DIRECT SEPARATION AND
DETERMINATION OF DRUG
ENANTIOMERS IN PLASMA.
J. Chromatogr., 473, 371-379, (1988).

266. Walhagen A., Edholm L-E., Heeremans
C.E.M., Van der Hoeven R.A.M., Niessen
W.M.A., Tjaden U.R., Van der Greef J.
COUPLED COLUMN CHROMATOGRAPHY-
MASS SPECTROSCOPY. THERMOSPRAY
LIQUID CHROMATOGRAPHIC-MASS
SPECTROMETRIC AND LIQUID
CHROMATOGRAPHIC-TANDEM MASS
SPECTROMETRIC ANALYSIS OF
METOPROLOL ENANTIOMERS IN PLASMA
USING PHASE-SYSTEM SWITCHING.
J. Chromatogr., 474, 257-263, (1988).

267. Walhagen A., Edholm L-E., Kennedy B-M.,
Xiao L.C.
DETERMINATION OF TERBUTALINE
ENANTIOMERS IN BIOLOGICAL SAMPLES
USING LIQUID CHROMATOGRAPHY WITH
COUPLED COLUMNS.
Chirality, 1, 20-26, (1989).

268. Weaner L.E., Hoerr D.C.
SEPARATION OF FATTY ACID ESTER AND
AMIDE ENANTIOMERS BY HIGH
PERFORMANCE LIQUID
CHROMATOGRAPHY ON CHIRAL
STATIONARY PHASES.
J. Chromatogr., 437, 109-119, (1988).

269. Weil A., Caldwell J., Guichard J.P., Picot G.
SPECIES DIFFERENCES IN THE
CHIRALITY OF THE CARBONYL
REDUCTION OF [^{14}C] FENOFIBRATE IN
LABORATORY ANIMALS AND HUMANS.
Chirality, 1, 197-201, (1989).

270. Whelpton R., Jonas G., Buckley D.G.
HIGH PERFORMANCE LIQUID
CHROMATOGRAPHIC RESOLUTION OF
THE ENANTIOMERS OF THIORIDAZINE,
ITS METABOLITES AND RELATED
COMPOUNDS.
J. Chromatogr., 426, 223-228, (1988).

271. Wilson M.J., Ballard K.D., Walle T.
PREPARATIVE RESOLUTION OF THE
ENANTIOMERS OF THE β-BLOCKING
DRUG ATENOLOL BY CHIRAL
DERIVATIZATION AND HIGH-
PERFORMANCE LIQUID
CHROMATOGRAPHY.
J. Chromatogr., 431, 222-227, (1988).

272. Wilson T.D.
SAMPLE SOLVENT EFFECTS IN AN
APPARENT CHIRAL HIGH-PERFORMANCE
LIQUID CHROMATOGRAPHIC
SEPARATION ON β-CYCLODEXTRIN.
J. Chromatogr., 448, 31-39, (1988).

273. Wolf R.M., Francotte E., Lohmann D.
QUANTITATIVE CORRELATION BETWEEN
CALCULATED MOLECULAR PROPERTIES
AND RETENTION OF A SERIES OF
STRUCTURALLY RELATED RACEMATES
ON CELLULOSE TRIACETATE.
J. Chem. Soc., Perkin Trans., 2, 893-901,
(1988).

274. Wu D.F., Reidenberg M.M., Drayer D.E.
DETERMINATION OF GOSSYPOL
ENANTIOMERS IN PLASMA AFTER
ADMINISTRATION OF RACEMATE USING
HIGH-PERFORMANCE LIQUID
CHROMATOGRAPHY WITH PRECOLUMN
CHEMICAL DERIVATIZATION.
J. Chromatogr., 433, 141-148, (1988).

275. Wulff G., Minarik M.
TAILOR-MADE SORBENTS. A MODULAR
APPROACH TO CHIRAL SEPARATIONS.
J. Chromatogr. Sci., 40, 15-52, (1988).

276. Yamashita J., Satoh H., Oi S., Suzuki T.,
Miyano S., Takai N.
HIGH-PERFORMANCE LIQUID
CHROMATOGRAPHIC SEPARATION OF
ENANTIOMERS ON AXIALLY CHIRAL
BINAPHTHALENEDICARBOXYLIC ACID-
CHIRAL PHENYLETHYLAMINE BONDED
TO SILICA GEL.
J. Chromatogr., 464, 411-415, (1989).

277. Yang Q., Sun Z., Ling D.
RESOLUTION OF ENANTIOMERIC DRUGS
OF SOME β-AMINO ALCOHOLS AS THEIR
UREA DERIVATIVES BY HIGH
PERFORMANCE LIQUID
CHROMATOGRAPHY ON A CHIRAL
STATIONARY PHASE.
J. Chromatogr., 447, 208-211, (1988).

278. Yang S.K., Mushtaq M., Bao Z., Weems H.B., Shou M., Lu X.L.
IMPROVED ENANTIOMERIC SEPARATION OF DIHYDRODIOLS OF POLYCYCLIC AROMATIC HYDROCARBONS ON CHIRAL STATIONARY PHASES BY DERIVATIZATION TO O-METHYL ETHERS.
J. Chromatogr., 461, 377-395, (1989).

279. Yang Z., Xu R.
INVESTIGATION ON THE ENANTIOMERIC IMPURITY OF EPINEPHRINE HYDROCHLORIDE INJECTIONS.
Chirality, 1, 92-93, (1989).

280. Yao T., Wasa T.
HIGH-PERFORMANCE LIQUID CHROMATOGRAPHIC DETECTION OF L- AND D-AMINO ACIDS BY USE OF IMMOBILIZED ENZYME ELECTRODES AS DETECTORS.
Anal. Chim. Acta., 209, 259-264, (1988).

281. Zhong-Yuan Y., Ru-Zheng X.
INVESTIGATION ON THE ENANTIOMERIC IMPURITY OF EPINEPHRINE HYDROCHLORIDE INJECTIONS.
Chirality, 1, 92-93, (1989).

282. Zief M.
INFLUENCE OF THE MOBILE PHASE ON CHIRAL LIQUID CHROMATOGRAPHY SEPARATIONS.
J. Chromatogr. Sci., 40, 315-335, (1988).

283. Zolotarev Y.A., Zaitzev D.A., Penkina V.I., Dostavalov I.N., Myasoedov N.G.
LIGAND EXCHANGE CHROMATOGRAPHY FOR ANALYSIS AND PREPARATIVE SEPARATION OF TRITIUM-LABELLED AMINO ACIDS.
J. Radioanal. Nucl. Chem., 121, 469-478, (1988).

284. Zukowski J., Nowakowski R.
DYNAMICALLY GENERATED CHIRAL STATIONARY PHASE SYSTEMS WITH β-CYCLODEXTRIN DERIVATIVES.
J. Liq. Chromatogr., 12, 1545-1569, (1989).

285. Zukowski J., Sybilska D., Bojarski J., Szejtli J.
RESOLUTION OF CHIRAL BARBITURATES INTO ENANTIOMERS BY REVERSED-PHASE HIGH PERFORMANCE LIQUID CHROMATOGRAPHY USING METHYLATED β-CYCLODEXTRINS.
J. Chromatogr., 436, 381-390, (1988).

GAS CHROMATOGRAPHY (GC)

1. Abe I., Tsujioka H., Wasa T.
SOLVENT EXTRACTION CLEAN-UP FOR PRE-TREATMENT IN AMINO ACID ANALYSIS BY GAS CHROMATOGRAPHY APPLICATION TO AGE ESTIMATION FROM THE D/L RATIO OF ASPARTIC ACID IN HUMAN DENTINE.
J. Chromatogr., 449, 165-174, (1988).

2. Alexander G., Juvancz Z., Szejtli J.
CYCLODEXTRINS AND THEIR DERIVATIVES AS STATIONARY PHASES IN GC CAPILLARY COLUMNS.
J. High Resolut. Chromatogr., 11, 110-113, (1988).

3. Aoyama T., Kotaki H., Saitoh Y.
GAS CHROMATOGRAPHIC-MASS SPECTROMETRIC ANALYSIS OF THREO-METHYLPHENIDATE ENANTIOMERS IN PLASMA.
J. Chromatogr., 494, 420-430, (1989).

4. Blessington B., Crabb B., Karkee S., Northage A.
CHROMATOGRAPHIC APPROACHES TO THE QUALITY CONTROL OF CHIRAL PROPIONATE ANTI-INFLAMMATORY DRUGS AND HERBICIDES.
J. Chromatogr., 469, 183-190, (1989).

5. Blink A., Suijerbuijk M.L., Ishiwata T., Geringa B.L.
GAS CHROMATOGRAPHIC METHOD FOR ENANTIOMERIC EXCESS DETERMINATION OF ALCOHOLS NOT REQUIRING CHIRAL AUXILIARY COMPOUNDS OR CHIRAL STATIONARY PHASES.
J. Chromatogr., 467, 285-291, (1989).

6. Brooks C.J.W., Brindle P.A., Cole W.J., Watson D.G.
ANALYTICAL SEPARATIONS OF ENANTIOMERIC PRIMARY AMINES AS PYRROLE DERIVATIVES FORMED BY REACTION WITH POLYGODIAL.
J. Chromatogr., 438, 108-110, (1988).

7. Brueckner H., Hausch M.
GAS CHROMATOGRAPHIC DETECTION OF D-AMINO ACIDS AS COMMON CONSTITUENTS OF FERMENTED FOODS.
Chromatographia, 28, 487-492, (1989).

8. Buyuktimkin N., Buyuktimkin S., Grunow D., Elz S.
THE USE OF ACTIVATED (S)-(+)-NAPROXEN FOR THE GAS CHROMATOGRAPHIC RESOLUTION OF SOME AMINO ACID METHYL ESTERS.
Chromatographia, 25, 925-927, (1988).

9. Buyuktimkin N., Keller F., Schunack W.
GAS-LIQUID CHROMATOGRAPHIC
RESOLUTION OF SOME RACEMIC
SYNTHONS FOR LAMTIDINE ANALOGOUS
HISTAMINE H2-RECEPTOR ANTAGONISTS
VIA DIASTEREOMERIC AMIDES OF (1S)-
(-)-CAMPHANIC ACID.
J. Chromatogr., 467, 402-405, (1989).

10. Clark T., Deas A.H.B., Vogeler K.
CHIRAL GAS AND HIGH-PERFORMANCE
LIQUID CHROMATOGRAPHIC ANALYSIS
OF ENANTIOMERS OF FUNGICIDES AND
PLANT GROWTH REGULATORS:
APPLICATION IN FUNGAL PLANT AND
SOIL METABOLISM STUDIES.
In "Chiral Separations". Eds: D. Stevenson
and I.D. Wilson, Plenum, 79-90, (1988).

11. Davies B.E.
DEVELOPMENT OF A CHIRAL CAPILLARY
GC METHOD FOR THE QUANTITATION OF
THE ENANTIOMERS OF CROMAKALIM IN
BIOLOGICAL FLUIDS.
In "Bioanalysis Of Drugs And Metabolites,
Especially Anti-Inflammatory and
Cardiovascular". Eds: E. Reid, J.D.
Robinson and I.D. Wilson, Plenum,
179-183, (1988).

12. De Bondt M., Couder J., Van der Auwera
L., Van Marsenille M. Elseviers M., Delaet
N., Laus G., Tourwe D., Van Binst G.
DETERMINATION OF THE CHIRAL PURITY
OF DIPEPTIDE ISOSTERES CONTAINING
A REDUCED PEPTIDE BOND BY GAS
CHROMATOGRAPHIC ANALYSIS.
J. Chromatogr., 442, 165-173, (1988).

13. Deger W., Gessner M., Gunther C., Singer
G., Mosandl A.
STEREOISOMERIC FLAVOR
COMPOUNDS. 18. ENANTIO-
DISCRIMINATION OF CHIRAL FLAVOR
COMPOUNDS BY DIASTEREOMERIC
DERIVATIZATION.
J. Agric. Food Chem. 36, 1260-1264,
(1988).

14. Ernst-Cabrera K., Koenig W.A.
ENANTIOMERIC ANALYSIS BY GAS
CHROMATOGRAPHY ON CHIRAL
POLYSILOXANES.
React. Polym., Ion Exch., Sorbents, 6,
267-274, (1988).

15. Feuerbach M., Froehlich O., Schreier P.
CHIRALITY EVALUATION OF 1,4-
DECANOLIDE IN PEACH.
J. Agric. Food Chem., 36, 1236-1237,
(1988).

16. Fitzgerald R.L., Blanke R.V., Glennon R.A.,
Yousif M.Y., Rosecrans J.A., Poklis A.
DETERMINATION OF 3,4-
METHYLENEDIOXYAMPHETAMINE AND
3,4-METHYLENEDIOXY-
METHAMPHETAMINE ENANTIOMERS IN
WHOLE BLOOD.
J. Chromatogr., 490, 59-69, (1989).

17. Frank H.
CHIRAL STATIONARY PHASES FOR
CAPILLARY GAS CHROMATOGRAPHY:
TOWARDS HIGHER SELECTIVTY AND
STABILITY.
J. High Resolut. Chromatogr., 11, 787-792,
(1988).

18. Gyllenhaal O., Vessman J.
PHOSGENE AS A DERIVATIZING
REAGENT PRIOR TO GAS AND LIQUID
CHROMATOGRAPHY.
J. Chromatogr., 435, 259-269, (1988).

19. Hosten N., Anteunis M.J.O.
ENANTIOMERIC QUANTIFICATIONS OF
AMINO ACIDS THROUGH THEIR N^a-ACYL
AMIDES BY GAS CHROMATOGRAPHY.
Bull. Soc. Chim. Belg., 97, 45-50, (1988).

20. Inotsume N., Fujii J., Honda M., Nakano
M., Higashi A., Matsuda I.
STEREOSELECTIVE ANALYSIS OF THE
ENANTIOMERS OF ETHOTOIN IN HUMAN
SERUM USING CHIRAL STATIONARY
PHASE LIQUID CHROMATOGRAPHY AND
GAS CHROMATOGRAPHY-MASS
SPECTROMETRY.
J. Chromatogr., 428, 402-407, (1988).

21. Jacob P., Benowitz N.L., Copeland J.R.,
Risner M.E., Cone E.J.
DISPOSITION KINETICS OF NICOTINE
AND COTININE ENANTIOMERS IN
RABBITS AND BEAGLE DOGS.
J. Pharm. Sci., 77, 396-400, (1988).

22. Joshi N.N., Srebnik M.
RESOLUTION OF RAC-1,2-HALOHYDRINS
BY CHIRAL COMPLEXATION GAS
CHROMATOGRAPHY.
J. Chromatogr., 462, 458-460, (1989).

23. Kitching W., Lewis J.A., Perkins M.V., Drew
R., Moore C.J., Schurig V., Koenig W.A.,
Francke W.
CHEMISTRY OF FRUIT FLIES.
COMPOSITION OF THE RECTAL GLAND
SECRETION OF (MALE) DACUS CUCUMIS
(CUCUMBER FLY) AND DACUS
HALFORDIAE. CHARACTERIZATION OF
(Z,Z)-2,8-DIMETHYL-1,7-
DIOXASPIRO[5.5]UNDECANE.
J. Org. Chem., 54, 3893-3902, (1989).

24. Koenig W.A., Lutz S., Colberg C., Schmidt N., Wenz G., Von der Bey E., Mosandl A., Gunther C., Kustermann A.
CYCLODEXTRINS AS CHIRAL STATIONARY PHASES IN CAPILLARY GAS CHROMATOGRAPHY. PART III: HEXAKIS(3-O-ACETYL-2,6-DI-O-PENTYL)-α-CYCLODEXTRIN.
J. High Resolut. Chromatogr., 621-625, (1988).

25. Koenig W.A., Lutz S., Mischnick-Luebbecke P., Brassat B.,
IMPROVED GAS CHROMATOGRAPHIC SEPARATION OF ENANTIOMERIC CARBOHYDRATE DERIVATIVES USING A NEW CHIRAL STATIONARY PHASE.
Carbohydr. Res., 183, 11-17. (1988).

26. Koenig W.A., Lutz S., Mischnick-Luebbecke P., Brassat B., Wenz G.
CYCLODEXTRINS AS CHIRAL STATIONARY PHASES IN CAPILLARY GAS CHROMATOGRAPHY. I. PENTYLATED α-CYCLODEXTRIN.
J. Chromatogr., 447, 193-197, (1988).

27. Koenig W.A., Lutz S., Wenz G., Von der Bey E.
CYCLODEXTRINS AS CHIRAL STATIONARY PHASES IN CAPILLARY GAS CHROMATOGRAPHY. II: HEPTAKIS(3-O-ACETYL-2,6-DI-O-PENTYL)-β-CYCLODEXTRIN.
J. High Resolut. Chromatogr., 506-509, (1988).

28. Koppenhoefer B., Allmendinger H., Chang L.P., Cheng L.B.
RESOLUTION OF STEREOISOMERS OF DIPEPTIDES BY GAS CHROMATOGRAPHY ON CHIRASIL-VAL.
J. Chromatogr., 441, 89-98, (1988).

29. Lai G., Nicholson G., Bayer E.
IMMOBILIZATION OF CHIRASIL-VAL ON GLASS CAPILLARIES.
Chromatographia, 26, 229-233, (1988).

30. Liu D.
GAS CHROMATOGRAPHIC RESOLUTION OF ENANTIOMERIC AND DIASTEREO-ISOMERIC AMINO ACID ESTERS ON CHIRASIL-VAL CAPILLARY COLUMN.
Chromatographia, 25, 393-396, (1988).

31. Miwa B.J., Choma N., Brown S.Y., Keigher N., Fukuda E.K.
QUANTIATION OF THE ENANTIOMERS OF RIMANTADINE IN HUMAN PLASMA AND URINE BY GAS CHROMATOGRAPHY-MASS SPECTOMETRY.
J. Chromatogr., 431, 343-352, (1988).

32. Mosandl A., Hagenauer-Hener U.
STEREOISOMERIC FLAVOR COMPOUNDS. PART XXVI. HRGC-ANALYSIS OF CHIRAL 1,3-DIOXOLANES.
J. High Resolut. Chromatogr., 744-749, (1988).

33. Mueller M.D., Bosshardt H.P.
ENANTIOMER RESOLUTION AND ASSAY OF PROPIONIC ACID-DERIVED HERBICIDES IN FORMULATIONS BY USING CHIRAL LIQUID CHROMATOGRAPHY AND ACHIRAL GAS CHROMATOGRAPHY. J. Assoc. Off. Anal. Chem., 71, 614-617, (1988).

34. Nakamura K., Hara S., Dobashi Y.
CHIRAL POLYSILOXANES DERIVED FROM (R,R)-TARTRAMIDE FOR THE GAS CHROMATOGRAPHIC SEPARATION OF ENANTIOMERS.
Anal. Chem., 61, 2121-2124, (1989).

35. Prestwich G.D., Graham S.M.G., Koenig W.A.
ENANTIOSELECTIVE OPENING OF (+) AND (-)-DISPARLURE BY EPOXIDE IN GYPSY MOTH ANTENNAE.
Chem. Commun., 575-577, (1989).

36. Roy S.D., Lim H.K.
SEPARATION OF OPTICAL ISOMERS OF METHOXYPHENAMINE AND ITS METABOLITES AS N-HEPTA-FLUOROBUTYRYL-L-PROLYL DERIVATIVES BY FUSED-SILICA CAPILLARY GAS CHROMATOGRAPHY.
J. Chromatogr., 431, 210-215, (1988).

37. Rieck M., Hagen M., Lutz S., Koenig W.A.
ENANTIOMER SEPARATION OF ACYLOINS AND CYANOHYDRINS BY ENANTIO-SELECTIVE CAPILLARY GAS CHROMATOGRAPHY.
J. Chromatogr., 439, 301-306, (1988).

38. Schurig V.
ENANTIOMER ANALYSIS BY COMPLEXATION GAS CHROMATOGRAPHY. SCOPE, MERITS AND LIMITATIONS.
J. Chromatogr., 441, 135-153, (1988).

39. Schurig V.
ENANTIOMER SEPARATION BY COMPLEXATION GAS CHROMATOGRAPHY - APPLICATIONS IN CHIRAL ANALYSIS OF PHEROMONES AND FLAVORS.
"Bioflavour '87", Eds: P. Schreier, de Gruyter, 35-54, (1988).

40. Schurig V., Link R.
RECENT DEVELOPMENTS IN
ENANTIOMER SEPARATION BY
COMPLEXATION GAS
CHROMATOGRAPHY.
In "Chiral Separations". Eds: D. Stevenson
and I.D. Wilson, Plenum, 91-114, (1988).

41. Schurig V., Nowotny H.P.
SEPARATION OF ENANTIOMERS ON
DILUTED PERMETHYLATED β-
CYCLODEXTRIN BY HIGH-RESOLUTION
GAS CHROMATOGRAPHY.
J. Chromatogr., 441, 155-163, (1988).

42. Schurig V., Ossig A., Link R.
ENANTIOMER SEPARATION OF 2-
HALOCARBOXYLIC ACID ESTERS BY
CHIRAL COMPLEXATION GAS
CHROMATOGRAPHY.
J. High Resolut. Chromatogr., 11, 89-93,
(1988).

43. Shiner C.S., Berks A.H.
PREPARATION OF (+)- AND (-)-N,S-
DIMETHYL-S-PHENYLSULFOXIMINE VIA
AN IMPROVED RESOLUTION. ACCURATE
DETERMINATION OF VERY HIGH
ENANTIOMERIC PURITIES BY ON-
COLUMN GC ANALYSIS OF
DIASTEREOMERIC DERIVATIVES.
J. Org. Chem., 53, 5542-5545, (1988).

44. Sioufi A., Kaiser G., Leroux F., Dubois J.P.
DETERMINATION OF THE S(+)- AND R(-)-
ENANTIOMERS OF BACLOFEN IN PLASMA
AND URINE BY GAS CHROMATOGRAPHY
USING A CHIRAL FUSED-SILICA
CAPILLARY COLUMN AND AN ELECTRON-
CAPTURE DETECTOR.
J. Chromatogr., 450, 221-232, (1988).

45. Spisni A., Corradini R., Marchelli R.,
Dossena A.
CHIRAL RECOGNITION OF AMINO ACID
DERIVATIVES: AN NMR INVESTIGATION
OF THE SELECTOR AND THE
DIASTEREOMERIC COMPLEXES.
J. Org. Chem., 54, 684-688, (1989).

46. Srinivas N.R., Cooper J.K., Hubbard J.W.,
Midha K.K.
ISOTHERMAL CAPILLARY GAS
CHROMATOGRAPHY WITH ELECTRON-
CAPTURE DETECTION OF
HEPTAFLUOROBUTYRYL-L-PROLYL
DERIVATIVES OF CHIRAL
AMPHETAMINES.
J. Chromatogr., 491, 262-264, (1989).

47. Srinivas N.R., Hubbard J.W., Cooper J.K.,
Midha K.K.
ENANTIOSELECTIVE GAS
CHROMATOGRAPHIC ASSAY WITH
ELECTRON-CAPTURE DETECTION FOR
DL-FENFLURAMINE AND DL-
NORFENFLURAMINE IN PLASMA.
J. Chromatogr., 433, 105-117, (1988).

48. Srinivas N.R., Hubbard J.W., Hawes E.M.,
McKay G., Midha K.K.
ENANTIOSELECTIVE GAS-
CHROMATOGRAPHIC ASSAYS WITH
ELECTRON- CAPTURE DETECTION FOR
METHOXYPHENAMINE AND ITS THREE
PRIMARY METABOLITES IN HUMAN
URINE.
J. Chromatogr., 487, 61-72, (1989).

49. Turk J., Stump W.T., Wolf B.A., Easom
R.A., McDaniel M.L.
QUANTITATIVE STEREOCHEMICAL
ANALYSIS OF SUBNANOGRAM AMOUNTS
OF 12-HYDROXY-(5,8,10,14)-
EICOSATERAENOIC ACID BY
SEQUENTIAL CHIRAL PHASE LIQUID
CHROMATOGRAPHY AND STABLE
ISOTOPE DILUTION MASS
SPECTROMETRY.
Anal. Biochem., 174, 580-588, (1988).

50. Wagner J., Gaget C., Heintzelmann B.,
Wolf E.
RESOLUTION OF THE ENANTIOMERS OF
VARIOUS A-SUBSTITUTED ORNITHINE
AND LYSINE ANALOGS BY HIGH
PERFORMANCE LIQUID
CHROMATOGRAPHY WITH CHIRAL
ELUANT AND BY GAS
CHROMATOGRAPHY ON CHIRASIL-VAL.
Anal. Biochem., 164, 102-116, (1987).

51. Watabe K., Gil-Av E., Hobo T., Suzuki S.
EFFECT OF APOLAR DILUENTS ON THE
BEHAVIOR OF CHIRAL STATIONARY
PHASES IN GAS CHROMATOGRAPHY.
BINARY MIXTURES OF N-LAUROYL-L-
VALINE-TERT-BUTYLAMIDE WITH
SQUALANE AND N-TERTACOSANE.
Anal. Chem., 61, 126-132, (1989).

52. Wong B., Castellanos M.
ENANTIOSELECTIVE MEASUREMENT OF
THE CANDIDA METABOLITE D-
ARABINITOL IN HUMAN SERUM USING
MULTIDIMENSIONAL GAS
CHROMATOGRAPHY AND A NEW CHIRAL
PHASE.
J. Chromatogr., 495, 21-30, (1989).

53. Young C.L., Frank H., Stewart C.R., Wainer I.W.
THE DETERMINATION OF (-)-(S)- and (+)-(R)-IFOSFAMIDE IN PLASMA USING ENANTIOSELECTIVE GAS CHROMATOGRAPHY: A VALIDATED ASSAY FOR PHARMACOKINETIC AND CLINICAL STUDIES.
Chirality, 1, 235-238, (1989).

SUPER- AND SUBCRITICAL FLUID CHROMATOGRAPHY (SFC)

1. Dobashi A., Dobashi Y., Ono T., Hara S., Saito M., Higashidate S., Yamauchi Y.
ENANTIOMER RESOLUTION OF D-AND L-A-AMINO ACIDS DERIVATIVES BY SUPERCRITICAL FLUID CHROMATOGRAPHY ON NOVEL CHIRAL DIAMIDE PHASES WITH CARBON DIOXIDE.
J. Chromatogr., 461, 121-127, (1989).

2. Lienne M., Macaudiere P., Caude M., Rosset R., Tambute A.
EVALUATION OF P1-ACID CHIRAL STATIONARY PHASES DERIVING FROM TYROSINE AND RELATED AMINO ACIDS FOR THE CHROMATOGRAPHIC RESOLUTION OF RACEMATES: SPECIFIC REQUIREMENTS FOR ENANTIO-RECOGNITION ABILITY.
Chirality, 1, 45-56, (1989).

3. Macaudiere P., Caude M., Rosset R., Tambute A.
CHIRAL RESOLUTION OF A SERIES OF 3-THIENYLCYCLOHEXYLGLYCOLIC ACIDS BY LIQUID OR SUBCRITICAL FLUID CHROMATOGRAPHY. A MECHANISTIC STUDY.
J. Chromatogr., 450, 255-269, (1988).

4. Macaudiere P., Caude M., Rosset R., Tambute A.
USE OF VARIOUS COMMERCIALLY AVAILABLE CHIRAL STATIONARY PHASES IN SUPERCRITICAL FLUID CHROMATOGRAPHY.
In "Chiral Separations". Eds: D. Stevenson and I.D. Wilson, Plenum, 115-120, (1988).

5. Macaudiere P., Lienne M., Caude M., Rosset R., Tambute A.
RESOLUTION OF P1-ACID RACEMATES ON P1-ACID CHIRAL STATIONARY PHASES IN NORMAL-PHASE LIQUID AND SUBCRITICAL FLUID CHROMATOGRAPHIC MODES.
J. Chromatogr., 467, 357-372, (1989).

6. Ruffing F.J., Lux J.A., Roeder W., Schomburg G.
CHIRAL STATIONARY PHASES FOR LC AND SFC OBTAINED BY "POLYMER COATING".
Chromatographia, 26, 19-28, (1988).

7. Steuer W., Schindler M., Schill G., Erni F.
SUPERCRITICAL FLUID CHROMATOGRAPHY WITH ION-PAIRING MODIFERS. SEPARATION OF ENANTIOMERIC OF 1,2-AMINOALCOHOLS AS DIASTEREOMERIC ION PAIRS.
J. Chromatogr., 447, 287-296, (1988).

8. Veuthey J.L., Caude M., Rosset R.
SEPARATION OF SOME AMINO ACIDS BY SUPERCRITICAL FLUID CHROMATOGRAPHY AFTER A PREDERIVATIZATION STEP WITH CLASSICAL REAGENTS.
Chromatographia, 27, 105-108, (1989).

CAPILLARY ZONE ELECTROPHORESIS (AND RELATED TOPICS) (CZE)

1. Dobashi A., Ono T., Hara S., Yamaguchi J.
OPTICAL RESOLUTION OF ENANTIOMERS WITH CHIRAL MIXED MICELLES BY ELECTROKINETIC CHROMATOGRAPHY.
Anal. Chem., 61, 1984-1986, (1989).

2. Guttmann A., Paulus A., Cohen A., Grinberg N., Karger B.L.
USE OF COMPLEXING AGENTS FOR SELECTIVE SEPARATION IN HIGH-PERFORMANCE CAPILLARY ELECTROPHORESIS. CHIRAL RESOLUTION VIA CYCLODEXTRINS INCORPORATED WITHIN POLY ACRYLAMIDE GEL COLUMNS.
J. Chromatogr., 448, 41-53, (1988).

3. Terabe S., Shibata M., Miyashita Y.
CHIRAL SEPARATION BY ELECTROKINETIC CHROMATOGRAPHY WITH BILE SALT MICELLES.
J. Chromatogr., 480, 403-411, (1989).

ISOTACOPHORESIS (IS)

1. Jelinek I., Dohnal J., Snopek J., Smolkova-Keulemansova E.
USE OF CYCLODEXTRINS IN ISOTACOPHORESIS. VII. RESOLUTION OF STRUCTURALLY RELATED AND CHIRAL PHENOTHIAZINES.
J. Chromatogr., 464, 139-147, (1989).

2. Snopek J., Jelinek I., Smolkova-Keulemansova E.
USE OF CYCLODEXTRINS IN ISOTACOPHORESIS. IV. THE INFLUENCE OF CYCLODEXTRINS ON THE CHIRAL RESOLUTION OF EPHEDRINE ALKALOID ENANTIOMERS.
J. Chromatogr., 438, 211-218, (1988).

3. Snopek J., Jelinek I., Smolkova-Keulemansova E.
USE OF CYCLODEXTRINS IN ISOTACOPHORESIS. VIII. TWO-DIMENSIONAL CHIRAL SEPARATION IN ISOTACHOPHORESIS.
J. Chromatogr., 472, 308-313, (1989).

PREPARATIVE CHROMATOGRAPHY (PREP)

1. Akanya J.N., Taylor D.R.
ATTEMPTS ON THE SEMIPREPARATIVE RESOLUTION OF RACEMIC ESTERS OF AMINO ACIDS ON A CHIRAL CHROMATOGRAPHIC COLUMN DERIVED FROM N-FORMYLISOLEUCINE.
Chromatographia, 25, 923-924, (1988).

2. Denissen J.F.
PREPARATIVE SEPARATION OF THE ENANTIOMERS OF THE CHOLECYSTOKININ ANTAGONIST (3S)-(±)-N-(2,3-DIHYDRO-1-(($^3H_3$)METHYL)-2-OXO-5-PHENYL-1H-1,4-BENZODIAZEPINE-3-YL)-1H-INDOLE-2-CARBOXAMINE BY HIGH-PERFORMANCE LIQUID CHROMATOGRAPHY.
J. Chromatogr., 462, 454-457, (1989).

3. Isaksson R., Sandstroem J., Eliaz M., Israely Z., Agranat I.
ENANTIOMER RESOLUTION OF NEFOPAM HYDROCHLORIDE, A NOVEL ANALGESIC: A STUDY BY LIQUID CHROMATOGRAPHY AND CIRCULAR DICHROISM SPECTROSCOPY.
J. Pharm. Pharmacol., 40, 48-50, (1988).

4. Masurel D., Wainer I.W.
ANALYTICAL AND PREPARATIVE HIGH-PERFORMANCE LIQUID CHROMATOGRAPHIC SEPARATION OF THE ENANTIOMERS OF IFOSFAMIDE, CYCLOPHOSPHAMIDE AND TROFOSFAMIDE AND THEIR DETERMINATION IN PLASMA.
J. Chromatogr., 490, 133-143, (1989).

5. Wilson M.J., Ballard K.D., Walle T.
PREPARATIVE RESOLUTION OF THE ENANTIOMERS OF THE β-BLOCKING DRUG ATENOLOL BY CHIRAL DERIVATIZATION AND HIGH-PERFORMANCE LIQUID CHROMATOGRAPHY.
J. Chromatogr., 431, 222-237, (1988).

6. Zief M.
PREPARATIVE ENANTIOMERIC SEPARATION.
J. Chromatogr. Sci., 40, 337-353, (1988).

7. Zolotarev Y.A., Zaitzev D.A.M., Penkina V.I., Dostavalov I.N., Myasoedov N.F.
LIQUID EXCHANGE CHROMATOGRAPHY FOR ANALYSIS AND PREPARATIVE SEPARATION OF TRITIUM-LABELLED AMINO ACIDS.
J. Radioanal. Nucl. Chem., 121, 469-478, (1988).

THIN-LAYER CHROMATOGRAPHY (TLC)

1. Armstrong D.W., Faulkner J.R., James R. Jnr., Han S.M.
USE OF HYDROXYPROPYL- AND HYDROXYETHYL-DERIVATIZED β-CYCLODEXTRINS FOR THE THIN LAYER CHROMATOGRAPHIC SEPARATION OF ENANTIOMERS AND DIASTEREOMERS.
J. Chromatogr., 452, 323-330, (1988).

2. Armstrong D.W., He F.Y., Han S.M.
PLANAR CHROMATOGRAPHIC SEPARATION OF ENANTIOMERS AND DIASTEREOMERS WITH CYCLODEXTRIN MOBILE PHASE ADDITIVES.
J. Chromatogr., 448, 345-354, (1988).

3. Bhushan R.
METHODS OF TLC RESOLUTION OF ENANTIOMERIC AMINO ACIDS AND THEIR DERIVATIVES.
J. Liq. Chromatogr., 11, 3049-3065, (1988).

4. Bhushan R.
RESOLUTION OF ENANTIOMERIC MIXTURES OF AMINO ACIDS ON BERBERINE-IMPREGNATED SILICA GEL PLATES.
Fresenius' Z. Anal Chem., 333, 144, (1989).

5. Bhushan R., Ali I.
RESOLUTION OF ENANTIOMERIC AMINO ACIDS ON BERBERINE-IMPREGNATED SILICA PLATES.
Fresenius' Z. Anal. Chem., 329, 793, (1988).

6. Brunner C.A., Wainer I.W.
DIRECT STEREOCHEMICAL RESOLUTION OF ENANTIOMERIC AMIDES VIA THIN-LAYER CHROMATOGRAPHY ON A COVALENTLY BONDED CHIRAL STATIONARY PHASES.
J. Chromatogr., 472, 277-283, (1989).

7. Buyutimkin N., Buschauer A.
SEPARATION AND DETERMINATION OF
SOME AMINO ACID ESTER
ENANTIOMERS BY THIN-LAYER
CHROMATOGRAPHY AFTER
DERIVATIZATION WITH (S)-(+)-
NAPROXEN.
J. Chromatogr., 450, 281-283, (1988).

8. Gunther K.
THIN-LAYER CHROMATOGRAPHIC
ENANTIOMERIC RESOLUTION VIA
LIGAND EXCHANGE.
J. Chromatogr., 448, 11-30, (1988).

9. Mack M., Hauck H.E.
SEPARATION OF ENANTIOMERS BY THIN-
LAYER CHROMATOGRAPHY.
J. Planar Chromatogr., 2, 190-193, (1989).

10. Mack M., Hauck H.E.
SEPARATION OF ENANTIOMERS IN THIN-
LAYER CHROMATOGRAPHY.
Chromatographia, 26, 197-205, (1988).

11. Mack M., Hauck H.E., Herbert H.
ENANTIOMERIC SEPARATION IN TLC
WITH THE NEW HPTLC PRECOATED
PLATE CHIR WITH CONCENTRATING
ZONE.
J. Planar Chromatogr., 1, 304-308, (1988).

12. Nyiredy Sz., Dallenbach-Toelke K.,
Sticher O.
APPLICABILITY OF FORCED-FLOW
PLANAR CHROMATOGRAPHIC METHODS
FOR THE SEPARATION OF ENANTIOMERS
ON CHIRALPLATE.
J. Chromatogr., 450, 241-252, (1988).

13. Palamareva M.
CHROMATOGRAPHIC BEHAVIOUR OF
DIASTEREOISOMERS. IX. APPLICATION
OF A MICROCOMPUTER PROGRAM TO
THE THIN-LAYER CHROMATOGRAPHIC
SEPARATION OF SOME
DIASTEREOISOMERS ON SILICA.
J. Chromatogr., 438, 219-224, (1988).

14. Rausch R.
THE IMPORTANCE OF THIN-LAYER
CHROMATOGRAPHY AS A RAPID
METHOD FOR THE CONTROL OF
OPTICAL PURITY.
In "Recent Advances in Thin-Layer
Chromatography". Eds: F.A.A. Dallas,
H. Read, R.J. Ruane and I.D. Wilson,
Plenum, 151-161 (1988).

15. Sinibaldi M., Messina A., Girelli A-M.
SEPARATION OF DANSYLAMINO ACID
ENANTIOMERS BY THIN-LAYER
CHROMATOGRAPHY.
Analyst, 113, 1245-1247, (1988).

16. Tivert A-M., Backman A.
ENANTIOMERIC SEPARATION OF
AMINOALCOHOLS BY TLC USING A
CHIRAL COUNTER-ION IN THE MOBILE
PHASE.
J. Planar Chromatogr., 2, 472-473, (1989).

17. Wall P.E.
PREPARATION AND APPLICATION OF
HPTLC PLATES FOR ENANTIOMER
SEPARATION.
J. Planar Chromatogr., 2, 228-232, (1989).

18. Wilson I.D., Ruane R.J.
PROSPECTS FOR CHIRAL THIN-LAYER
CHROMATOGRAPHY.
In "Chiral Separations". Eds: D. Stevenson
and I.D. Wilson, Plenum, 135-144, (1988).

REVIATIONS

	CAPILLARY ZONE ELECTROPHORESIS (AND RELATED TOPICS)
	GAS CHROMATOGRAPHY
REV	GENERAL REVIEWS
C/LC	HIGH-PERFORMANCE LIQUID CHROMATOGRAPHY/LIQUID CHROMATOGRAPHY
	ISOTACOPHORESIS
P	PREPARATIVE CHROMATOGRAPHY
	SUPER- AND SUBCRITICAL FLUID CHROMATOGRAPHY
	THIN-LAYER CHROMATOGRAPHY

Brune K.	HPLC/LC 86	Davankov V.A.	GEN REV 9
Bruneau P.	HPLC/LC 166		HPLC/LC 47, 48, 49
Brunner C.A.	TLC 6	Davies B.E.	GC 11
Buckley D.G.	HPLC/LC 270	Dayer P.	HPLC/LC 138
Burke J.A. III	HPLC/LC 199, 200, 207	Deas A.H.B.	GC 10 HPLC/LC 42
Burke J.T.	HPLC/LC 128, 129	De Bondt M.	GC 12
Buschauer A.	TLC 7	Deger W.	GC 13
Butters R.W.	HPLC/LC 212	Delaet N.	GC 12
Buyuktimkin N.	GC 8, 9	Delatour P.	HPLC/LC 140
	TLC 7	Delee E.	HPLC/LC 50, 51
Buyuktimkin S.	GC 8	Demian I.	HPLC/LC 52
Bystricky S.	HPLC/LC 241	Deming K.C	HPLC/LC 207
		Denissen J.F.	HPLC/LC 53
Caccamese S.	HPLC/LC 30		PREP 2
Caldwell J.	GEN REV 7, 16	Desai D.M.	HPLC/LC 198
	HPLC/LC 31, 60, 103, 269	De Vries E.S.	HPLC/LC 191
		De Vries J.X.	HPLC/LC 54
Caldwell W.B.	HPLC/LC 250	De Zeeuw R.A.	HPLC/LC 73, 74, 87
Camilleri P.	HPLC/LC 32, 33	Dhanesar S.C.	HPLC/LC 55
Canceill J.	HPLC/LC 246	Dietzel K.	HPLC/LC 86
Capdevila J.H.	HPLC/LC 93	Dobashi A.	CZE 1
Carunchio V.	HPLC/LC 34		HPLC/LC 56
Castellanos M.	GC 52		SFC 1
Caude M.	HPLC/LC 140, 142, 143, 144, 145, 154, 155	Dobashi Y.	GC 34 HPLC/LC 56 SFC 1
	SFC 2, 3, 4, 5, 8	Dohnal J.	IS 1
Cavender P.J.	HPLC/LC 58	Dolphin J.	GEN REV 10
Cerny M.	HPLC/LC 117		HPLC/LC 57
Chae K.	HPLC/LC 35	Doner L.W.	HPLC/LC 58
Chan K.C.	HPLC/LC 36	Don-Pedro O.	HPLC/LC 27
Chang L.P.	GC 28	Dossena A.	GC 45
Chemlal A.	HPLC/LC 18, 218, 219		HPLC/LC 14, 15
Chen C.S.	HPLC/LC 230	Dostavalov I.N.	HPLC/LC 283
Chen S.-T.	HPLC/LC 152		PREP 7
Cheng L.B.	GC 28	Doyle T.D.	GEN REV 11
Chin S.K.	HPLC/LC 27, 37	Drayer D.E.	HPLC/LC 274
Choi K.E.	HPLC/LC 38	Dreiding A.S.	HPLC/LC 102
Choma N.	GC 31	Drenth B.F.H.	GEN REV 13
Chou T.	HPLC/LC 39		HPLC/LC 59, 73, 74, 87
Christie W.W.	HPLC/LC 40		
Chu Y.Q.	GEN REV 42	Drew R.	GC 23
	HPLC/LC 41, 261, 264	Drummond L.	HPLC/LC 60
Claramunt R.M.	HPLC/LC 18	Dubois J.P.	GC 44
Clark B.J.	HPLC/LC 76, 157	Duchateao A.	HPLC/LC 61
Clark T.	GC 10	Duke C.C.	HPLC/LC 62
	HPLC/LC 42	Durand A.	HPLC/LC 129
Cline-Love L.J.	HPLC/LC 172	Dyas A.M.	GEN REV 12
Cohen A.	CZE 2		HPLC/LC 63
Colberg C.	GC 24	Dyke C.	HPLC/LC 32
Cole W.J.	GC 6		
Collet A.	HPLC/LC 246	Easom R.A.	GC 149
Cone E.J.	GC 21		HPLC/LC 254
Cookson D.	HPLC/LC 212	Edholm L.-E.	HPLC/LC 64, 265, 266, 267
Cooper J.K.	GC 46, 47		
Coors C.	HPLC/LC 43	Eiglsperger A.	HPLC/LC 158
Copeland J.R.	GC 21	Einarsson S.	GEN REV 1
Corradini R.	GC 45	Eisenberg E.J.	HPLC/LC 65
Couder J.	GC 12	Elguero J.	HPLC/LC 18
Coventry L.	GEN REV 8	Eliaz M.	HPLC/LC 109
Crabb B.	GC 4		PREP 3
Crabb M.	HPLC/LC 22	Elseviers M.	GC 12
Crabb N.	HPLC/LC 23	Elz S.	GC 8
Crombach M.	HPLC/LC 61	Enquist M.	HPLC/LC 66
Crooks B.	HPLC/LC 82	Eradiri O.	HPLC/LC 116
		Erlandsson P.	HPLC/LC 67
Daeppen R.	HPLC/LC 44, 45	Erni F.	SFC 7
Dalgaard L.	HPLC/LC 46	Ernst-Cabrera K.	GC 14
Dallenbach-Toelke K.	TLC 12	Euerby M.R.	HPLC/LC 68, 69, 70
Damm P.	HPLC/LC 139		
Darbyshire J.F.	GEN REV 7	Falgueyret J.P.	HPLC/LC 71
	HPLC/LC 31	Falk J.R.	HPLC/LC 93
		Fanali S.	HPLC/LC 34

Mahmood S.A.	HPLC/LC 192
Maibaum J.	HPLC/LC 156
Malikin G.	HPLC/LC 135
Mama J.E.	HPLC/LC 76, 157
Manes L.V.	HPLC/LC 122
Mannschreck A.	HPLC/LC 158, 159
Marchelli R.	GC 45
	HPLC/LC 14, 15
Marle I.	HPLC/LC 160
Marti A.R.	HPLC/LC 29
Martire D.E.	HPLC/LC 24
Marumo S.	HPLC/LC 169, 170
Masurel D.	HPLC/LC 161
	PREP 4
Matlin S.A.	GEN REV 24
	HPLC/LC 162
Matsuda H.	HPLC/LC 13
Matsuda I.	GC 20
	HPLC/LC 108
Matsumoto T.	HPLC/LC 107
Matsumoto Y.	HPLC/LC 181
Matsuo M.	HPLC/LC 122
Matusch R.	HPLC/LC 43
Maurs M.	HPLC/LC 163
Mayol R.F.	HPLC/LC 209
McCune J.E.	HPLC/LC 202, 203,
	204
McDaniel M.L.	GC 49
	HPLC/LC 254
McKay G.	GC 48
Mehta A.C.	GEN REV 27
	HPLC/LC 164
Mehvar R.	HPLC/LC 116, 165
Meijer D.K.F.	HPLC/LC 74
Meinard C.	HPLC/LC 166
Messina A.	HPLC/LC 34
	TLC 15
Meyer V.R.	HPLC/LC 29, 45
Meyer-Lehnert S.	HPLC/LC 81
Michelsen P.	HPLC/LC 167
Midha K.K.	GC 46, 47, 48
Mihalyfi K.	HPLC/LC 85
Minarik M.	HPLC/LC 275
Mischnick-Luebbecke P.	GC 25, 26
Misiti D.	HPLC/LC 83
Miwa B.J.	GC 31
Miwa T.	HPLC/LC 168
Miyakawa T.	HPLC/LC 169
Miyake Y.	HPLC/LC 168
Miyano S.	HPLC/LC 182, 276
Miyashita Y.	CZE 3
Miyazaki A.	HPLC/LC 169, 170
Modi M.W.	HPLC/LC 227
Mohri H.	HPLC/LC 189
Mompon B.	HPLC/LC 128
Moore C.J.	GC 23
Moro E.	HPLC/LC 198
Moroz J.	HPLC/LC 250
Mosandl A.	GC 13, 24, 32
	GEN REV 28
Motoi M.	HPLC/LC 118
Mueller M.D.	GC 33
	HPLC/LC 171
Mularz E.A.	HPLC/LC 172
Murai-Kushiya M.	HPLC/LC 96
Murphy M.	HPLC/LC 135
Mushtaq M.	HPLC/LC 173, 278
Mutschler E.	HPLC/LC 88, 236, 237,
	238
Myasoedov N.F.	HPLC/LC 283
	PREP 7
Nakajima H.	HPLC/LC 181
Nakamura K.	GC 34

Nakamura T.	HPLC/LC 122, 169,
	170
Nakamura Y.	HPLC/LC 174
Nakano M.	GC 20
	HPLC/LC 108, 186
Nashikawa M.	HPLC/LC 231
Nation R.L.	GEN REV 29
	HPLC/LC 175
Navratil J.D.	HPLC/LC 49
Neimeijer N.R.	HPLC/LC 87
Nicoll-Griffith D.A.	HPLC/LC 176
Nicholson G.	GC 29
Niessen W.M.A.	HPLC/LC 266
Nill K.	HPLC/LC 180
Nilsson I.	HPLC/LC 67
Nimura N.	HPLC/LC 177
Nishi H.	HPLC/LC 178
Nishida T.	HPLC/LC 221
Nishikawa M.	HPLC/LC 224
Niwa K.	HPLC/LC 121
Noctor T.A.G.	GEN REV 30
	HPLC/LC 75, 76
Norinder U.	HPLC/LC 179
Northage A.	GC 4
	HPLC/LC 23
Nowakowski R.	HPLC/LC 284
Nowotny H.P.	GC 41
Nunn P.B.	HPLC/LC 68, 69
Nusser E.	HPLC/LC 180
Nyiredy Sz.	TLC 12
Odham G.	HPLC/LC 167
Ogata H.	HPLC/LC 245
Ogawa T.	HPLC/LC 107
Oi N.	HPLC/LC 181
Oi S.	HPLC/LC 182, 276
Oizumi H.	HPLC/LC 12, 13
Okamoto Y.	GEN REV 31
	HPLC/LC 183, 184,
	185, 186, 187, 188,
	189, 190
Olieman C.	HPLC/LC 191
Ono T.	CZE 1
	SFC 1
Ossig A.	GC 42
Palamareva M.	TLC 13
Palm D.	HPLC 238
Pankonin G.	HPLC/LC 192
Papadopoulou-	
Mourkidou E.	HPLC/LC 193, 194
Park Y.W.	HPLC/LC 105
Partridge L.Z.	HPLC/LC 68, 69, 70
Patterson W.R.	HPLC/LC 65
Paulson J.	HPLC/LC 64
Paulus A.	CZE 2
Pederson J.L.	HPLC/LC 46
Penkina V.I.	HPLC/LC 283
	PREP 7
Perkins M.V.	GC 23
Persson B.A.	HPLC/LC 19
Petersheim M.	HPLC 172
Pettersson C.	GEN REV 32, 33
	HPLC/LC 119, 120,
	160, 195, 196, 197
Petty E.H.	HPLC/LC 97
Pflugman G.	HPLC/LC 237, 238
Pianezzola E.	HPLC/LC 198
Picot G.	HPLC/LC 269
Pilgrim H.	HPLC/LC 192
Pini D.	HPLC/LC 222

Pirkle W.H.	GEN REV 34
	HPLC/LC 104, 199, 200, 201, 202, 203, 204, 205, 206, 207
Pochapsky T.C.	GEN REV 34
	HPLC/LC 205, 206, 207
Poklis A.	GC 16
Ponomareva T.M.	HPLC/LC 48
Porziemsky J.P.	HPLC/LC 208
Prakash C.	HPLC/LC 209, 210
Prestwich G.D.	GC 35
Pryde A.	GEN REV 36
Purdie N.	GEN REV 35
Pustet N.	HPLC/LC 159
Rajani P.	HPLC/LC 70
Rao N.K.R	HPLC/LC 211
Rath M.	HPLC/LC 148, 149
Rathbone E.B.	HPLC/LC 212
Rau G.	HPLC/LC 86
Rausch R.	TLC 14
Reid J.M.	HPLC/LC 213
Reidenberg M.M.	HPLC/LC 274
Rieck M.	GC 37
Riendeau D	HPLC/LC 71
Rihs G.	HPLC/LC 79
Risner M.E.	GC 21
Rizzi A.M.	HPLC/LC 214, 215, 216
Robinson C.	HPLC/LC 217
Robinson J.L.	HPLC/LC 212
Robinson M.L.	GEN REV 12
	HPLC/LC 63
Roeder W.	HPLC/LC 220
	SFC 6
Rogers L.B.	HPLC/LC 239
Rohde B.	HPLC/LC 102
Rokach J.	HPLC/LC 71
Rosecrans J.A.	GC 16
Rosini C.	HPLC/LC 222
Rosset R.	HPLC/LC 140, 142, 143, 144, 145, 154,155
	SFC 2, 3, 4, 5, 8
Roussel C.	HPLC/LC 18, 218, 219
Rowland M.	HPLC/LC 89
Roy S.D.	GC 36
Ruane R.J.	GEN REV 43
	TLC 18
Ruffing F.J.	HPLC/LC 220
	SFC 6
Ru-Zheng X.	HPLC/LC 281
Sabio M.	HPLC/LC 249, 250
Sado P.	HPLC/LC 137
Saigo K.	HPLC/LC 221
Saito M.	SFC 1
Saitoh Y.	GC 3
Sakai J.	HPLC/LC 107
Salvadori P.	HPLC/LC 222
Sander L.C.	HPLC/LC 223
Sandstroem J.	HPLC/LC 109
	PREP 3
Sansoulet J.	HPLC/LC 51
Sato Y.	HPLC/LC 224, 231
Satoh H.	HPLC/LC 276
Schalekamp M.A.D.H.	HPLC/LC 25
Schewe T.	HPLC/LC 192
Schill G.	GEN REV 33
	HPLC/LC 19, 99, 197
	SFC 7
Schilsky R.L.	HPLC/LC 38
Schindler M.	SFC 7

Schmidt N.	GC 24
Schmitthenner H.F.	HPLC/LC 225
Schonbaum G.R.	HPLC/LC 113, 262
Schomburg G.	HPLC/LC 220, 226, 242
	SFC 6
Schreier P.	GC 15
	HPLC/LC 101
Schunack W.	GC 9
Schurig V.	GC 23, 38, 39, 40, 41, 42
Schuster D.	HPLC/LC 227
Schuster O.	HPLC/LC 86
Scrivener H.	HPLC/LC 151
Secor H.V.	HPLC/LC 228
Seeman J.I.	HPLC/LC 228
Sellergren B.	HPLC/LC 229
Semmelrock H.J.	HPLC/LC 148
Shen D.D.	HPLC/LC 132
Shen K.	HPLC/LC 173
Shiao M.S.	HPLC/LC 230
Shibata M.	CZE 3
Shibata T.	GEN REV 18
	HPLC/LC 106
Shijo M.	HPLC/LC 182
Shimizu R.	HPLC/LC 178
Shiner C.S.	GC 43
Shinkai H.	HPLC/LC 224, 231
Shono T.	HPLC/LC 247
Shou M.	HPLC/LC 278
Simek Z.	HPLC/LC 232
Simonyi I.	HPLC/LC 244
Simonyi M.	HPLC 78
Singer G.	GC 13
Sinibaldi M.	HPLC/LC 34, 83, 131
	TLC 15
Sioufi A.	GC 44
Smolkova-Keulemansova E.	IS 1, 2, 3
Snopek J.	IS 1, 2, 3
Someya K.	HPLC/LC 245
Spahn H.	HPLC/LC 88, 112, 233, 234, 235, 236, 237, 238
Spahn I.	HPLC/LC 237
Spisni A.	GC 45
Srebnik M.	GC 22
Srinivas N.R.	GC 46, 47, 48
Stacey V.E.	HPLC/LC 162
Stein J.L.	HPLC/LC 219
Sternson L.A.	HPLC/LC 213
Steuer W.	SFC 7
Stevenson D.	GEN REV 37
Stewart C.R.	GC 53
Sticher O.	TLC 12
Stiffin R.M.	GEN REV 42
	HPLC/LC 80, 263, 264
Still M.G.	HPLC/LC 239
Stobaugh J.F.	HPLC/LC 213
Stoschitzky K.	HPLC/LC 148, 149
Straka R.J.	HPLC/LC 240
Strasak M.	HPLC/LC 241
Stump W.T.	GC 49
	HPLC/LC 254
Stuurman H.W.	HPLC/LC 242
Suda H.	HPLC/LC 118
Suganuma T.	HPLC/LC 112
Suijerbuijk M.L.	GC 5
Sun Z.	HPLC/LC 277
Sundell S.	HPLC/LC 120
Sundholm E.G.	HPLC/LC 179
	HPLC/LC 251
Suzuki N.	GC 51
Suzuki S.	
Suzuki T.	HPLC/LC 276

Swallows K.A.	GEN REV 35
Sybilska D.	HPLC/LC 285
Szejtli J.	GC 2
	HPLC/LC 285
Szepesi G.	GEN REV 38
	HPLC/LC 84, 85, 243
Sztruhar I.	HPLC/LC 244
Tachdjian C.	HPLC/LC 208
Takagi T.	HPLC/LC 111
Takahashi H.	HPLC/LC 245
Takai N.	HPLC/LC 276
Taku K.	HPLC/LC 178
Tambute A.	HPLC/LC 140, 142,
	143, 144, 145, 154,
	155, 246
	SFC 2, 3, 4, 5
Tanaka K.	HPLC/LC 100
Tanaka M.	HPLC/LC 247
Tang B.K.	HPLC/LC 176
Taylor D.R.	GEN REV 39,
	HPLC/LC 3, 4, 5
	PREP 1
Tepper P.G.	HPLC/LC 87
Terabe S.	CZE 3
Teuscher E.	HPLC/LC 192
Thelohan S.	HPLC/LC 113
Thompson R.A.	HPLC/LC 248
Timmons K.D.	HPLC/LC 228
Tivert A.-M.	TLC 16
Tjaden U.R.	HPLC/LC 266
Todd B.	HPLC/LC 211
Topiol S.	HPLC/LC 249, 250
Tourwe D.	GC 12
Towill R.C.	HPLC/LC 211
Trigalo F.	HPLC/LC 163
Triska J.	HPLC/LC 117
Trnka T.	HPLC/LC 117
Tsai W.-L.	HPLC/LC 102
Tsuchiya M.	HPLC/LC 251
Tsuda T.	HPLC/LC 111
Tsujioka H.	GC 1
Tsujiyama T.	HPLC/LC 251
Tsukamoto S.	HPLC/LC 252, 253
Tsumagari N.	HPLC/LC 178
Turk J.	GC 49
	HPLC/LC 254
Twamoto T.	HPLC/LC 174
Uccello-Barretta G.	HPLC/LC 222
Unruh L.E.	HPLC/LC 133
Uray G.	HPLC/LC 149
Van Binst G.	GC 12
Van de Grampel V.J.M.	HPLC 87
Van der Auwera L.	GC 12
Van der Greef J.	HPLC/LC 266
Van der Haar J.	HPLC/LC 255
Van der Hoeven R.A.M.	HPLC/LC 266
Van der Hoorn F.A.J.	HPLC/LC 25
Van Dijck J.	HPLC/LC 256
Van in't Veld A.J.	HPLC/LC 25
Van Marsenille M.	GC 12
Van Nijhuis A.	HPLC/LC 59
Verzele M.	HPLC/LC 256
Vespalec R.	HPLC/LC 232
Vessman J.	GC 18
	HPLC/LC 91
Veuthey J.L.	SFC 8
Villani D.	HPLC/LC 83
Vindevogel J.	HPLC/LC 256
Virgili R.	HPLC/LC 15
Voelker U.	HPLC/LC 54

Vogeler K	GC 10
	HPLC/LC 42
Von der Bey E.	GC 24, 27
Wad N.	HPLC/LC 257
Wagner J.	GC 50
Wainer I.W.	GC 53
	GEN REV 25, 40, 41,
	42
	HPLC/LC 6, 41, 80,
	113, 114, 134,
	161, 240, 258, 259,
	260, 261, 262, 263,
	264
	PREP 4
	TLC 6
Walhagen A.	HPLC/LC 64, 265, 266,
	267
Wall P.E.	TLC 17
Walle T.	HPLC/LC 271
	PREP 5
Walter D.J.	HPLC/LC 225
Walton H.F.	HPLC/LC 49
Wang K.-T.	HPLC/LC 152
Ward T.J.	HPLC/LC 228
Wasa T.	GC 1
	HPLC/LC 280
Watabe K.	GC 51
Watson D.G.	GC 6
Weaner L.E.	HPLC/LC 268
Weaver K.	HPLC/LC 33
Weems H.B.	HPLC/LC 173, 278
Weil A.	HPLC/LC 269
Wellstein A.	HPLC 238
Wennerstroem H.	HPLC/LC 110
Wennerstroem O.	HPLC/LC 110
Wenz G.	GC 24, 26, 27
Whelpton R.	HPLC/LC 270
Wick A.	HPLC/LC 208
Williams D.J.	HPLC/LC 33
Williams G.	GEN REV 37
Williams R.L.	HPLC/LC 27
Wilson H.K.	HPLC/LC 60
Wilson I.D.	GEN REV 43
	TLC 18
Wilson M.J.	HPLC/LC 271
	PREP 5
Wilson R.	HPLC/LC 200
Wilson T.D.	HPLC/LC 272
Winter S.M.	GEN REV 7
	HPLC/LC 31
Wise S.A.	HPLC/LC 223
Wold S.	HPLC/LC 67
Wolf B.A.	GC 49
	HPLC/LC 254
Wolf E.	GC 50
Wolf R.M.	HPLC/LC 273
Wong B.	GC 52
Wood A.J.J.	HPLC/LC 210
Wu D.F.	HPLC/LC 274
Wu S.-H.	HPLC/LC 152
Wulff G.	HPLC/LC 275
Xiao L.C.	HPLC/LC 267
Xu R	HPLC/LC 279
Yamagishi A.	HPLC/LC 174
Yamaguchi J.	CZE 1
Yamashita J.	HPLC/LC 182, 276
Yamauchi Y.	SFC 1
Yang Q.	HPLC/LC 277
Yang S.K.	HPLC/LC 278
Yang Z.	HPLC/LC 279
Yao T.	HPLC/LC 280

COMPOUNDS TO PAPERS CROSS-REFERENCE

Abscisic acid	HPLC/LC 183
Acephate	HPLC/LC 169
π-Acid racemates	SFC 5
Acids	HPLC/LC 203
Acyloins	GC 37
β-Adrenoceptor blocking agents	HPLC/LC 37
Albendazole sulfoxide	HPLC/LC 140
Alcohols	GC 5
	HPLC/LC 234
Alfuzosin hydrochloride	HPLC/LC 130
Aliphatic alcohols	HPLC/LC 182
Alkane-1,2-diol enantiomers	HPLC/LC 111
4(5)-Alkylated G-(D)-lactones	HPLC/LC 101
Alkyl or alkoxy butyric acids	HPLC/LC 141
Amides	TLC 6
Amines	GC 6
	HPLC/LC 234
α-Amino acid amides	HPLC/LC 61
Amino acid derivatives	GC 45
α-Amino acid enantiomers	HPLC/LC 117
Amino acid esters	GC 30
	HPLC/LC 3, 5, 10
	PREP 1
	TLC 7
Amino acid methyl esters	GC 8
Amino acids	GEN REV 14
	GC 1, 19
	HPLC/LC 10, 34, 36, 95, 113, 177, 178, 283
	SFC 8
	TLC 4, 5
Amino acids and their derivatives	GEN REV 4
	HPLC/LC 222
	TLC 3
D-Amino acids	GC 7
D- and L-α-amino acid derivatives	SFC 1
L- and D-Amino acids	HPLC/LC 280
Amino acids (tritium labelled)	PREP 7
Amino alcohols	HPLC/LC 82
β-Amino alcohols	HPLC/LC 277
1,2-Aminoalcohols	SFC 7
Amino- and acetylamino-polycyclic aromatic hydrocarbons	HPLC/LC 133
α-Amino-β-N-methylaminopropanoic acid (β-methylaminoalanine)	HPLC/LC 69
Aminoglutethimide and its acetylated metabolite	HPLC/LC 1
3-Aminoquinuclidine	HPLC/LC 52
2-Amino-W-phosphonoalkanoic acid homologs	HPLC/LC 69
Amphetamines	GC 46
	HPLC/LC 6
Anionic metal complexes	HPLC/LC 241
Anthelmintic drug enantiomers	HPLC/LC 143
Anticholinergic drugs	HPLC/LC 142
D-Arabinitol	GC 52
Aryloxypropanolamine enantiomers	HPLC/LC 225
Arylpropionic acid enantiomers	HPLC/LC 186
N-Aryl-4-thiazoline-2-thione atropisomers	HPLC/LC 218
Aspartic acid	GC 1
Atenolol	HPLC/LC 27, 37, 271
	PREP 5

213

DERIVATIZATION REACTION INDEX

Montomorillonite	HPLC/LC 74
Pentylated α-cyclodextrin	GC 26
Permethylated β-cyclodextrin	GC 41
Pirkle high-performance liquid chromatography phases	HPLC/LC 11, 92
Pirkle-type chiral stationary phases	HPLC/LC 77
Polyacrylamide	HPLC/LC 101
Polyacrylamides (substituted)	GEN REV 5
	HPLC/LC 21
Poly(diphenyl-2-pyridylmethyl methacrylate)	HPLC/LC 189
Poly(L-leucine)	HPLC/LC 100
Poly(L-phenylalanine)	HPLC/LC 100
Polymer coating	SFC 6
Polysiloxanes	GC 14
Polysiloxanes derived from (R,R)- tartramide	GC 34
Poly(triphenylmethyl methacrylate)	GEN REV 31
Poly(triphenylmethyl methacrylate) as a chiral stationary phase	HPLC/LC 187
Poly(triphenylmethyl methacrylate) derivatives as a stationary phase for HPLC	HPLC/LC 190
Protein stationary phases	HPLC/LC 9, 248, 265
Proteins	GEN REV 40
Quinine bonded to silica	HPLC/LC 242
Saponite	HPLC/LC 174
BSA-Silica (Resolvosil columns)	HPLC/LC 8
Synthetic multi-interaction chiral bonded phases	GEN REV 11
Synthetic polymers	GEN REV 20
Triacetylcellulose stationary phase	HPLC/LC 18, 67, 110, 219
Tris(3,5-dimethylphenylcarbamate)s of cellulose and amylose	HPLC/LC 185
Tris(3,5-dimethylphenylcarbamate)s of cellulose and amylose as chiral stationary phases	HPLC/LC 186
Tris(4-tert-butylphenylcarbamate)s of cellulose and amylose	HPLC/LC 188
Urea-linked chiral stationary phase	HPLC/LC 55
Urea phases	HPLC/LC 211

AUTHOR INDEX

COMPOUND INDEX

Human serum albumin, 16
α-Hydroxy carboxylic acids, 151, 160
3-Hydroxy-1,2-benzocycloheptene, 49
1-Hydroxy-1,2-dihydronaphthalene, 46, 49
1-Hydroxy-1,4-dihydronaphthalene, 46
[1R]-1-Hydroxy-1,2-dihydronaphthalene, 45
2-Hydroxy-1,2-dihydronaphthalene, 45, 46
(2S, 4R, 2'RS)-4-Hydroxy-1-(2'-hydroxydodecyl)-proline/copper(II) acetate, 160
Hydroxy-N-methyl morphinan, 2
α-Hydroxyacids, 160
2-Hydroxydodecyl-4-hydroxyproline, 151
Hydroxyethyl-β-cyclodextrins, 166
Hydroxyethyl-derivatised β-cyclodextrin, 164
1-Hydroxyindene, 49
Hydroxyisoleucine, 161
Hydroxyleucine (sodium salt), 161
3-Hydroxymandelic acid, 161
4-Hydroxymandelic acid, 161
Hydroxymethionine, 161
Hydroxyphenylalanine, 161
Hydroxypropyl-β-cyclodextrins, 166
Hydroxypropyl-derivatised β-cyclodextrin, 164
Hydroxyvaline, 161
Hyoscyamine, 2

Ibuprofen, 2
    derivatives, 173
    structure, 8
Indene, 46
Inderal, 68
Isocyanate, 36
Isoleucine, 161, 163
Isopropyl oxirane, 126
Isothiocyanate, 36

Ketamine, 2, 138
Ketoprofen, 17

Lactams, 173
Lactones, 173
Lauric acid, 17, 18
Lawesson's reagent, 111
Leucine, 33, 161, 163
Leucovorin, 18, 21
Levomethorphan, 3
Lorazepam, 17, 18, 21, 162, 165

Mandelic acid, 161
    and derivatives, 160
Mefloquine, 63
(1S,2R,5S)-(+)-Menthyl-(R)-p-toluenesulfinate, 166
(1R,2S,5R)-(-)-Menthyl-(S)-p-toluenesulfinate, 166
Mephenytoin, 166, 167
Mephobarbital, 173
3-Mercaptopropylsilane, 79
3-Mercaptopropyltrimethoxysilane, 39, 78
Methaqualone, 173

Methionine, 163
DL-Methionine-B-naphthylamide, 167
Methsuximide, 173
Methyl phenoxypropionate, 11
2-Methyl-1,3-benzodithiole, 52, 53
2-Methyl-1-phenyl-1-propanol, 150
rac-2-Methyl-1-phenyl-1-propanol, 148
α-Methyl-D-glucoside, 19
α-Methyl-D-mannoside, 19
E-5-Methyl-hept-2-en-4-one, hazelnut flavour compound Filbertone, 120, 122
α-Methylamino acids, 160
α-Methylarylacetic acids, 151
4-Methylbenzoate, 173
R(+)-α-Methylbenzylamine, 77, 79
5-(4-Methylphenyl)-5-phenylhydantoin, 167
Metoprolol, 67, 162, 165

Naphthalene, 47, 48
Naphthalene 1,2-oxide, 48
(±),1'-bi-2-Naphthol, 152, 154, 155, 156
1,1'-bi-2-Naphthol, 151, 165
1-Naphthoyl chloride, 68, 111
α-Naphthoyl chloride, 109
α-Naphthyl, 31
R-(-)-1-(1-Naphthyl) ethyl urea, 151
N-Naphthyl-(S)-valine, 12
Naphthylalanine, 7, 8
1-(1-Naphthyl)ethanol, 173
α-Naphthylisocyanate, 31, 32
N'-(2-Naphthylmethyl)nornicotine, 166
Naphthylvaline, 7, 12
(S)-Naphthylurea, 65
Neomenthol, 123
Nickel(II)-bis(3-heptafluorobutanoyl-(1R)-8-isobutylene-camphor, 126
Nickel(II)-bis(3-heptafluorobutanoyl-(1R)-8-methyl-camphor, 126
p-Nitrobenzoyl fenchelyl nitroxide, 149
p-Nitrophenyl-α-D-galactopyranoside, 19
p-Nitrophenyl-α-D-glucopyranoside, 19
p-Nitrophenyl-α-D-mannopyranoside, 19
Non-steroidal antiinflammatory agents, 109
Norleucine, 163

n-Octane, 125, 127
n-Octanol, 15
Omeprazole, 16
rac-2-Oxa-bicyclo[3,3,0]oct-7-en-3-one, 123
Oxaminiquine, 172
Oxazepam, 162, 165, 173
(+)-Oxazepam hemisuccinate, 17
(-)-Oxazepam hemisuccinate, 17
Oxiranes, aromatic, 121

Paclobutrazol, 143, 144, 146
Penicillamine, 144, 151
2,3,4,5,6-Pentafluorobenzoyl chloride, 109, 111
Peptides, 151
Per-n-pentylated amylose, 119
Per-n-pentylated cyclodextrins, 119

225